Fungi in Coastal and Oceanic Marine Ecosystems

Seshagiri Raghukumar

# Fungi in Coastal and Oceanic Marine Ecosystems

## Marine Fungi

Springer

Seshagiri Raghukumar
Formerly of CSIR-National Institute
of Oceanography
Vainguinnim Valley
Dona Paula, Goa
India

ISBN 978-3-319-54303-1 ISBN 978-3-319-54304-8 (eBook)
DOI 10.1007/978-3-319-54304-8

Library of Congress Control Number: 2017937746



Printed on acid-free paper

This Springer imprint is published by Springer Nature
The registered company is Springer International Publishing AG
The registered company address is: Gewerbestrasse 11, 6330 Cham, Switzerland

**This Book is Dedicated to my Gurus**
**Prof. C.V. Subramanian**
**And**
**Dr Alwin Gaertner**
**Who Introduced me to the Wonderful**
**World of Marine Fungi**

# Foreword

I am honored to provide a foreword to this well-conceived and authoritative book on marine fungi written by a world-renowned expert from India, Dr Seshagiri Raghukumar who made the first authentic and exhaustive floristic survey of filamentous marine fungi in the country, the results of which were published in the first volume of KAVAKA, the Transactions of Mycological Society of India, in 1973.

Fungi in general colonize, grow, and multiply on a variety of substrates and habitats. They are the major players in the cycling of organic materials and energy around the world. They play an important role in agriculture, biodegradation, bioremediation, food, medicine, pharmaceutical industries, waste management, and other activities. Marine fungi are special in that they grow and reproduce in seawater and their surroundings. Marine mycologists have been researching on their identity and utility since the discovery of fungi in marine environments. This book is essentially an effort to showcase the development, current status, and excitement on all aspects of marine fungi.

The book opens up with an introduction to the marine environment and the role of fungi. In all, the book contains 15 chapters which deal with the history of marine mycology; allochthonous wood in coastal waters; mangrove, salt marsh, seagrass, macroalgal, coral reef, coastal, benthic, and pelagic ecosystems; extreme marine environments; physiology, biochemistry, and biotechnology; and origin and evolution of marine fungi, besides an exhaustive list of references. The narrations are authenticated by photo-illustrations and data. The last chapter details out the methods used to study the marine fungi. The facts and statements provided in this book are well supported by latest references.

I have known Seshagiri Raghukumar as a scholar of marine fungi since our student days in the early 1970s when we worked for our doctoral degrees in the same lab under the guidance of late Professor C.V. Subramanian, University of Madras, India. Subsequently, Seshagiri Raghukumar went on to study the zoosporic thraustochytrids with Dr Alwin Gaertner at the Institute für Meeresforschung, Bremerhaven, Germany. His wife Dr Chandralata Raghukumar too made outstanding contributions in marine fungi. Together, they established an excellent research lab for marine mycology at the National Institute of Oceanography, Goa, India, where they worked for nearly three decades. I have seen Seshagiri Raghukumar's dedicated work culture and sincere contributions from close proximity.

This book is a masterly account on marine fungi. I commend the author for his diligence, foresight, and painstaking and sincere efforts in the production of this invaluable contribution to marine mycology. This book will be most useful to those interested in marine-related fungi, especially to the new generation of mycologists world over.

I compliment and congratulate Seshagiri Raghukumar for his notable venture.

Formerly, Department of Botany
Goa University
Goa, India

D. Jayarama Bhat

Visiting Professor, Mae Fah Luang University
Chiang Rai, Thailand

Azad Housing Society
Curca, Goa Velha, India

# Preface

Mycologists have been fascinated with marine fungi for more than 70 years now. An enormous volume of information on marine fungi is presently available. We can now say with certainty that there are special fungi inhabiting marine, lignocellulosic substrates and that they attain substantial biomass in detritus of salt marsh grass and mangroves. A number of algal and animal parasites have been described, some of which cause significant mortalities. Corals harbor endobiontic fungi. Recent studies have also shown that fungi may not be an insignificant component of the pelagic ecosystem. Diversity and role of fungi in coral reefs and the deep sea have begun to attract attention of late. Marine fungi are some of the most significant sources of novel metabolites and biotechnological products.

Surprisingly, fungi still do not find much space in textbooks on marine biology and microbiology. This is probably because fungi occupy a peculiar niche in the environment, a niche which can be easily overlooked by aquatic biologists. The vast body of water in the ocean prompts one to look at organisms living directly therein. Thus, we know a great deal about free-living plankton and bacteria, which on average number up to a million per ml. We look at large marine organisms that swim freely or are suspended in the water column. We consider organisms that are attached to other substrates. Seldom does one pause to look at fungi that live concealed within living and dead organisms. That is exactly where one would find mycelia, as well as unicellular and endobiontic fungi, whose niche is to penetrate and live within organic matter. Therefore, the approach to study fungi is quite different from that toward bacteria. Add to it the fact that many unicellular fungi are the same size as other protists and one could easily overlook them. Happily, this scenario is changing fast and we are learning more and more about fungi in the marine ecosystem.

This book cannot replace the detailed scholarly books and reviews published by a number of established and emerging marine mycologists. The book by T.W. Johnson and F.K. Sparrow Jr. titled "Fungi in Oceans and Estuaries" published in 1961 is a masterly treatise of those times that inspired marine mycologists. Many others, such as the scholarly book of Jan and Erika Kohlmeyer on "higher marine fungi" published in 1979, as well as others edited by eminent marine mycologists have appeared from time to time. Those interested in specific topics introduced in this book are strongly advised to supplement reading of this book by referring to such publications.

What this book intends to do is to introduce the fascinating world of marine fungi to marine biologists and microbiologists. At the same time, it aims to encourage aspiring marine mycologists to explore the amazing diversity of marine ecosystems where one could look for fungi.

This book also provides interesting topics for future research. I hope it will also be a useful reference book for established marine mycologists.

Fungi are highly adaptable and may even have originated in the sea. Fungi have not restricted themselves to land. Nor have mycologists. It is time to wade in the pleasant oceans.

Goa, India Seshagiri Raghukumar

# Acknowledgments

I am grateful to a number of colleagues and friends at the former Institut für Meeresforschung, Bremerhaven, Germany (now Alfred-Wegener Institut für Polar- und Meeresforschung) and at the CSIR-National Institute of Oceanography (NIO), Goa, India, who helped me to expand my scope as a marine biologist, a microbiologist, and a biological oceanographer. The concept of this book would have been impossible if I had not been associated with them. My colleagues at CSIR-NIO, N. Ramaiah and A.C. Anil, have been a source of encouragement. A number of former students and young associates of myself and my wife, Chandralata, including Veena Sathe-Pathak, Sumita Sharma, Ruchi Jain, Samir Damare, Varada Damare, J. Ravindran, V.V.S. Sarma, Lucia Bongiorni, Donna D'Souza, Ashutosh Verma, and many others contributed much to the information that I have drawn upon. My friend and fellow mycologist, Prof. D. Jayarama Bhat, has been a source of encouragement and advice. My wife, Chandralata, has been a constant source of companionship and inspiration. To all these, and many more, I am deeply indebted.

# Contents

# About the Author

**Seshagiri Raghukumar** born in 1946, has an experience of over 40 years in studying marine fungi. He obtained his Ph.D. in 1973 from Madras University for a thesis on marine fungi, under the supervision of the famous mycologist, Prof. C.V. Subramanian. He was subsequently groomed by Dr Alwin Gaertner of the Institut für Meeresforsching (presently Alfred-Wegener Institute for Polar and Marine Research), at Bremerhaven, Germany, with whom he worked for 5 years on taxonomy and ecology of thraustochytrids. He later joined the CSIR-National Institute of Oceanography, Goa, India, in 1982 where he worked as a Scientist for 23 years. S. Raghukumar has been associated with Myko Tech Private Limited, India, a microbial biotechnology company since 2005.

Seshagiri Raghukumar's research spans a wide variety of topics. These include marine lignicolous fungi, the role of fungi in marine, macrophyte detrital processes, biotechnology, and the biology of thraustochytrids. He is particularly well known for his pioneering ecological studies on labyrinthulomycetes, particularly thraustochytrids and aplanochytrids. Raghukumar participated in over 30 oceanographic cruises during his career and headed the Marine Biodiversity Project at the CSIR-NIO. He has published over 80 research papers on marine fungi and holds several patents on their applications. He taught several courses in marine microbiology and mycology at the Goa University and has guided a number of M.Sc. and Ph.D. students. Seshagiri Raghukumar was the past Secretary and President of the Mycological Society of India and is presently a Member of the Board of the International Marine Biotechnology Association.

# Chapter 1
# Fungi: Characteristics and Classification

*Fungi are the interface organisms between life and death.*
Paul Stamets

**We study fungi because they are fascinating organisms that play a key role in sustaining life on earth.** Fungi are the second most diverse eukaryotic organisms on earth, next only to insects. Fungi together with bacteria are the major decomposers of dead organic material and therefore are essential for recycling of nutrients on earth. They are also of tremendous relevance to human society.

On the one hand, fungi and bacteria contribute enormously to our welfare in the form of various foods, beverages, drugs, industrial products, and chemicals. We are very familiar with the traditional use of yeasts, a group of fungi in leavening of bread and fermentation to produce alcohol. We also know of the discovery of the fabulous antibiotic penicillin. Many other drugs have subsequently helped mankind. Fungal enzymes are used in many industries to speed up chemical processes.

On the other hand, many members of the two groups cause serious diseases in plants and animals. The potato blight in mid-nineteenth century caused a large-scale famine and devastation in Ireland. Many other plant diseases as well as human ailments are caused by fungi.

We know a great deal about fungi in freshwater and terrestrial environments. Fungi in marine environments are relatively less known. It is becoming increasingly evident in recent years that fungi are present in almost all marine habitats and play several key ecological roles. In order to appreciate the role of fungi in the marine ecosystem, it is necessary to understand their basic characteristics and also consider the different habitats of the marine environment where one would find them.

Fungi differ from bacteria in many fundamental ways that allow them to coexist and yet function in different ecological niches. A holistic understanding of the microbial component of an ecosystem can be achieved only by considering the distinct individual roles of fungi and bacteria and their combined effect.



S. Raghukumar, *Fungi in Coastal and Oceanic Marine Ecosystems*,
DOI 10.1007/978-3-319-54304-8_1

## 1.1 Basic Characteristics of Fungi

**Fungi are eukaryotic microorganisms with an osmoheterotrophic nutrition**. They obtain their food by absorbing dissolved organic substances through their cell surface. Osmoheterotrophy appeared as a convergent evolution in many groups of eukaryotes. Many mycologists consider only those osmoheterotrophic organisms that fall under the Supergroup Opisthokonta as fungi and apply the term "Kingdom Fungi" to them. However, the term "fungi" in this book is used in a broad ecological sense for osmoheterotrophic organisms both in the Opisthokonta and in the Supergroup SAR (Straminipila, Alveolata, and Rhizaria) (Fig. 1.1). This is the same as the approach of Barr (1992) and Webster and Weber (2007).

**Thus, "fungi" are polyphyletic, belonging to two separate groups.** One of these comprises fungi belonging to Opisthokonta and commonly termed the "Kingdom Fungi." This is termed the **Kingdom Mycetae** in this book. The term "Mycetae" was first used by the famous evolutionary biologists R.H. Whittaker

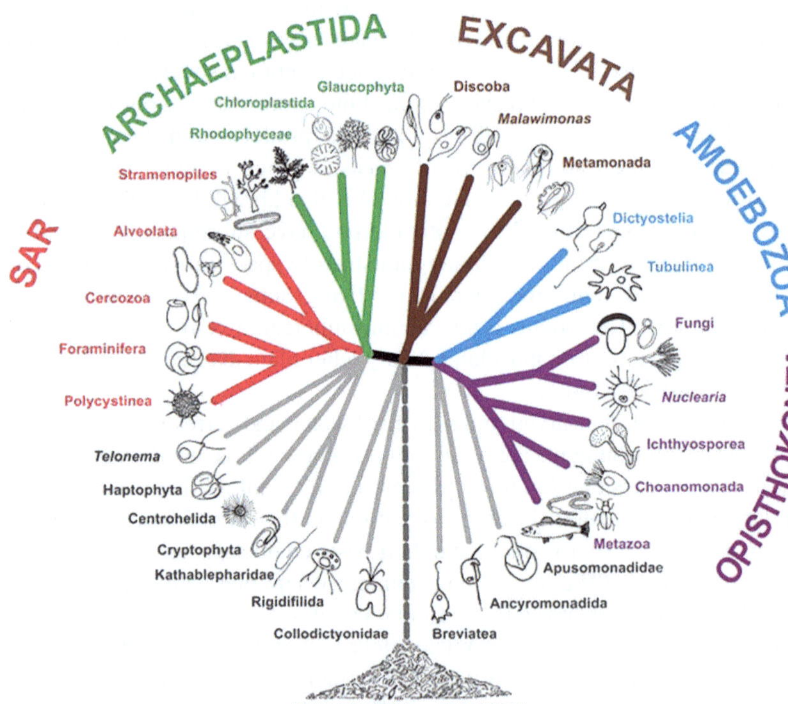


**Fig. 1.1** A general eukaryote phylogenetic tree. The clade 'Fungi' under Opisthokonta represents the Kingdom Mycetae and the clade 'Stramenopiles' under SAR represents the Kingdom Straminipila (Source: Adl, S.M. et al. J. Eukaryot. Microbiol. 59(5), 2012 pp. 429–493. Journal of Eukaryotic Microbiology © 2012 International Society of Protistologists. Reproduced with kind permission from John Wiley and Sons)

and Lynn Margulis in 1978. The second, belonging to the SAR group and frequently addressed as "fungal-like organisms" are considered in this book as fungi belonging to the **Kingdom Straminipila**. Thus, fungi belong to Kingdom Mycetae and to the Kingdom Straminpila.

The phylogenetic positions of The Kingdom Mycetae and the Kingdom Straminipila within the eukaryotes are given in Fig. 1.1 (Adl et al. 2012). The Kingdom Mycetae of the supergroup Opisthokonta is shown as "Fungi" in the figure. Opisthokonta also includes the Kingdom Metazoa or Animalia. All members of the Kingdom Mycetae are fungi, being strictly osmoheterotrophic in nutrition.

The Kingdom Straminipila of the SAR supergroup is shown as "Stramenopiles." These are diverse in their nutritional mode, comprising photosynthetic algae and phagotrophic protists, as well as osmoheterotrophic fungi.

Fungi belonging to the Kingdom Mycetae are referred to as mycetaen fungi and those belonging to the Kingdom Straminipila are referred to as stramninipilan fungi in the rest of this book.

**The vegetative structure (thallus) in fungi follows three basic patterns.**

1. **Most mycetaen fungi and many straminipilan fungi are characterized by microscopic filaments called hyphae.** The hyphae (sing. hypha) form a network called the mycelium (pl.: mycelia) (Fig. 1.2a, b). Hyphae are characterized by apical growth. Apical growth generates high turgor pressure at the hyphal tips adpressed against solid substrata. This helps the hyphae penetrate such substrates with ease, live endobiontically, and derive nutrition from within the particle (Fig. 1.2c). Additionally, apical growth allows fungi to traverse through microscopic aerial spaces and reach out to other organic particles separated distally. They can also spread within particles filled with pores. Mycelial growth is also useful in the growth of fungi in soils, where they can widely ramify in the substratum from a given location and utilize dissolved and solid organic nutrients present therein.
2. **Many fungi are unicellular and colonize the surfaces of particles, but produce extensions that penetrate the solid particle and extract nutrition from within.** In the case of chytrids belonging to the mycetaen fungi and hyphochytriomycetes of the straminipilan fungi, these are extensions of the cell covered by a cell wall and are called "rhizoids" (Fig. 1.2d). The straminipilan fungi belonging to Labyrinthulomycetes produce extensions of the plasma membrane and are called the "ectoplasmic net elements."
3. **Many mycetaen fungi are unicellular and lack penetrative ability. Many of these are "yeasts" belonging to the phyla Ascomycota and Basidiomycota** (see below). Yeasts grow mostly on the surfaces of particles or live suspended in an aqueous milieu (Fig. 1.2e).

Unicellular lifestyle allows the organism to live planktonically in water and to spread easily in the aqueous medium.

It is important to remember that there are often overlaps of these various forms of vegetative structure in fungi.

**The capability to penetrate solid particles is a key characteristic of many fungi that distinguishes them from unicellular bacteria.**

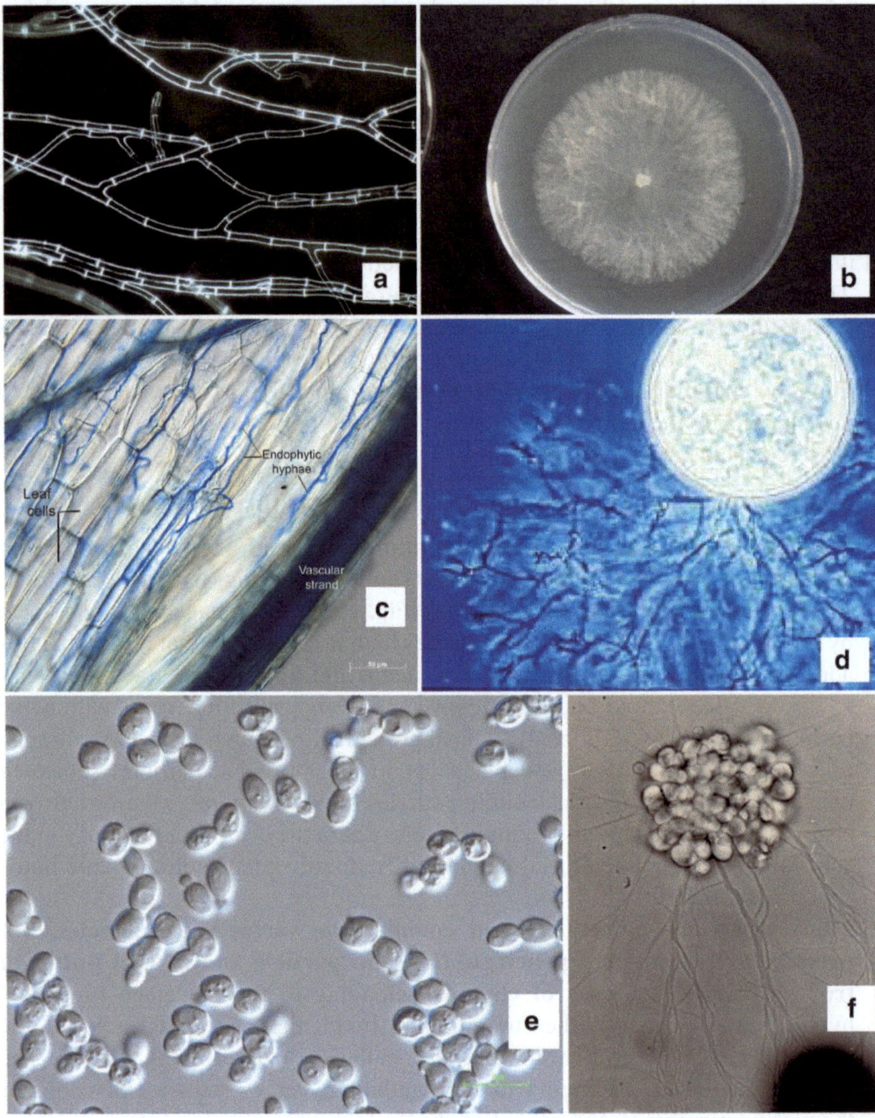


**Fig. 1.2** Vegetative structures of fungi. (**a**) Branched, multicellular, fungal hyphae with septa (cross walls). (Courtesy of Dr George Barron, Atrium Collection, University of Guelph); (**b**) A fungal colony in culture showing filamentous mycelium (S. Raghukumar). (**c**) Hyphae of an endophytic fungus living inside a leaf. (Courtesy: Dr William L'Amoreaux, Advanced Imaging Facility of the College of Staten Island, USA). (**d**) Unicellular vegetative thallus of a chytrid fungus showing branched rhizoids. (Courtesy: Prof. Bryce Kendrick, Mycologue Publications and Consulting; Canada). (**e**) Unicellular fungi represented by the yeast *Zygosporium rouxii*. (Photomicrographs by Petra Wrent and Maria Isabel de Siloniz, Universidad Complutense de Madrid, Spain. With permission from Dra. Susana Serrano Barrero). (**f**) Cluster of unicellular vegetative cells of the straminipilan fungus Schizochytrium aggregatum (thraustochytrid) with branched ectoplasmic net elements (arrow) (S. Raghukumar)

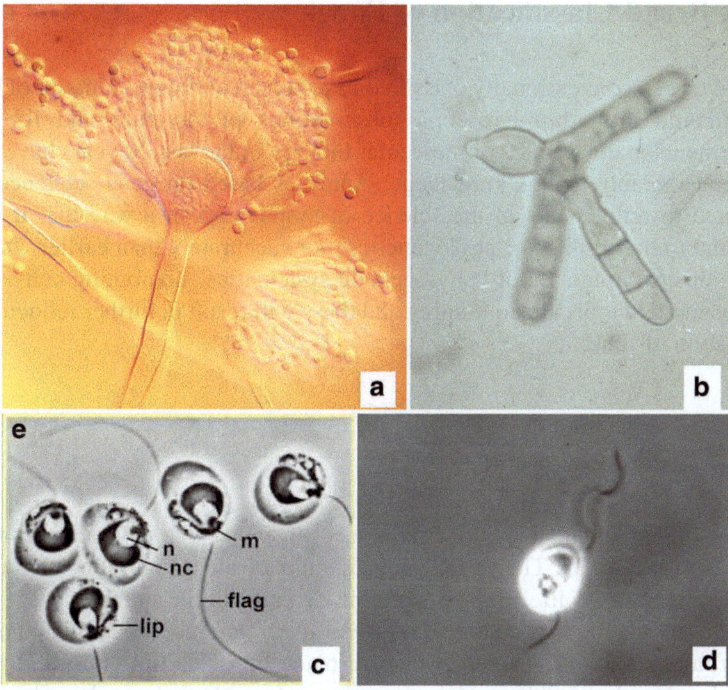


**Fig. 1.3** Reproductive spores of fungi. (**a**) The fungus *Aspergillus terreus* produces aerially dispersed spores (conidia) on spore-bearing structures. (Source: Photograph by J. Scott, at Medmyco at English Wikipedia made available to Creative Commons). (**b**) Tetraradiate spore of the marine fungus *Clavatospora bulbosa* (S. Raghukumar). (**c**): Water dispersed, posteriorly uniflagellate zoospores of the aquatic fungus *Blastocladiella emersonii*. (Source: Reichle, R.E. and M.S. Fuller. 1964. The Fine Structure of *Blastocladiella emersonii* zoospores. American Journal of Botany 54: 81–92. Kind permission of the Botanical Society of America). (**d**) Water-dispersed, biflagellate zoospores of a thraustochytrid, a straminipilan fungus (S. Raghukumar)

**Fungi reproduce and disperse by producing spores** (Fig. 1.3)**.** Spores may be produced through sexual or asexual reproduction. Many fungi have lost the ability to reproduce sexually and produce only asexual spores.

Spores help disperse the fungi in the environment. **Most terrestrial species have spores that are dispersed by wind. Fungi in freshwater and marine habitats produce spores that are adapted to dispersal in aquatic conditions.** Many possess spores that typically have a high surface-to-volume ratio that enables them to float in water. Others have spores with flagella that actively swim in water till they find a suitable substratum. These are called zoospores. Zoospores of Chytridiomycota (Kingdom Mycetae) are posteriorly uniflagellate, while those of Kingdom Straminipila have two flagella.

## 1.2 A Broad Classification of Fungi

Fungi have been conventionally classified based on an enormous amount of knowledge gathered through microscopic observations of structure and life cycle. Advancements in the science of molecular biology in the last 20 years have helped to add considerably to this knowledge. Analyses based on gene sequences have contributed to refinements in the science of systematics in a way that it reflects evolutionary relationships or phylogenetics more accurately than earlier. Developments in this exciting area of "phylogenetic systematics" combining conventional as well as modern tools are continuing and these will result in further refinements in classification of fungi.

### *1.2.1 Kingdom Mycetae*

Fungi belonging to the Kingdom Mycetae (Fig. 1.4), the mycetaen fungi, are mycelial or unicellular. Their cells may be uninucleate or multinucleate (coenocytic). Their mode of life may be saprobic or parasitic in animals and plants or mutualistic with other organisms. Mycetaen fungi reproduce sexually and/or asexually through spores. Their cell walls are made up of chitin, chitosan, or polysaccharides. Nearly 85,000 species of mycetaen fungi have been described so far. The classification given here is based on several sources (James et al. 2006; Hibbett et al. 2007; Ebersberger et al. 2012; https://www.britannica.com/science/fungus/Annotated-classification; http://shigen.nig.ac.jp/algae_tree/FungiE.html). The Kingdom Mycetae can be broadly divided into 9 phyla. An enormous amount of literature on taxonomy and biology of different groups of mycetaen fungi is available (Kendrick 2000; Webster and Weber 2007).

1. **Phylum Chytridiomycota** (Chytrids) (Fig. 1.5a): The thallus is made of single cells (monocentric) or several cells connected by rhizomycelia (polycentric). Each cell transforms into reproductive zoosporangia. Many thalli produce rhizoids. Thalli in some are filamentous, bearing a distinct zoosporangium. Asexual reproduction is by posteriorly uniflagellate zoospores. The flagellate cell possesses a centriole and microbody-lipid globule complex. Sexual reproduction is by isogamy or anisogamy (fusion of two flagellate gametes of equal or unequal size). Chytrids are zoosporic fungi living mostly in freshwater or soil. A few are marine. They may be saprobic or parasitic in algae, plants, animals, and other fungi. Classes: Chytridiomycetes and Monoblephariomycetes.
2. **Phylum Neocallimastigomycota:** The thallus may be monocentric or polycentric (see Chytridiomycota). These are obligate anaerobic fungi living in the rumen and cecum of herbivorous animals such as cows, sheeps, horses, elephants, and kangaroos. They are extremely efficient lignocelluloses degraders. Neocallimastigomycetes lack mitochondria but contain hydrogenosomes of mitochondrial origin which form hydrogen and carbon dioxide from pyruvate

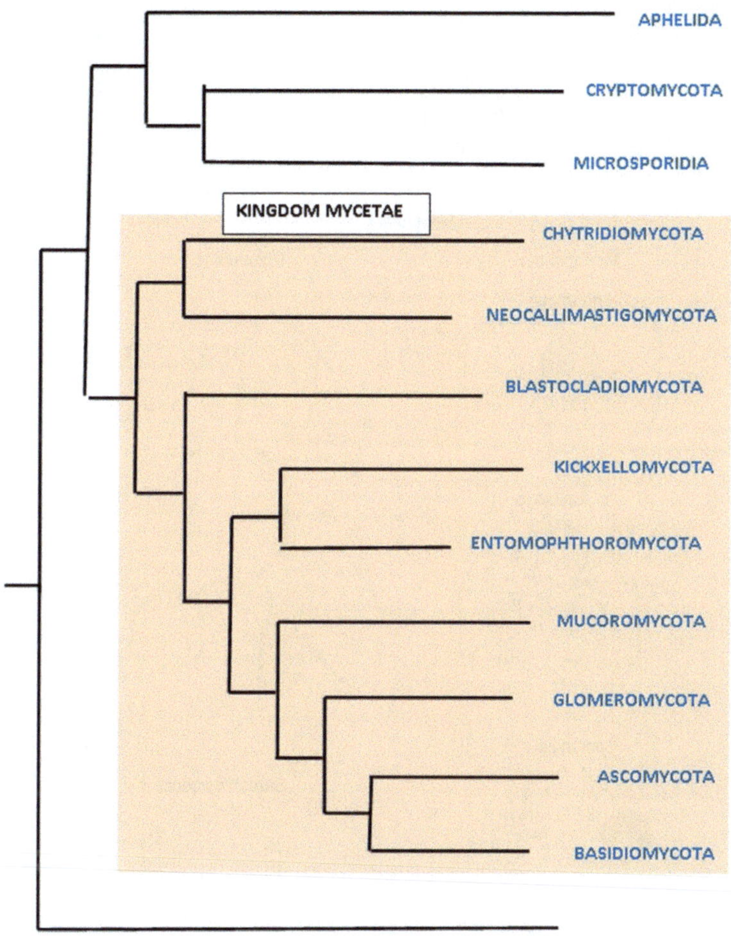


**Fig. 1.4** A broad classification of the Kingdom Mycetae

and malate during fermentation of carbohydrates. Asexual reproduction is by means of zoospores produced in zoosporangia. Zoospores are posteriorly unflagellate or polyflagellate. Class: Neocallimastigomycetes.

3. **Phylum Blastocladiomycota**: The thallus is monocentric or polycentric (see Chytridiomycota). *Coelomomyces* produces a tubular unwalled thallus in its host. Asexual reproduction is by posteriorly uniflagellate zoospores or spores without flagella. Sexual reproduction is by isogamy or anisogamy. Some species show alternation of sporophytic and gametophytic generations. Class: Blastocladiomycetes.
4. **Phylum Kickxellomycota**: The thallus is made of branched or unbranched, septate mycelium. They are often attached by means of a holdfast to other fungi on which they are parasites (mycoparasite). Septa possess median,

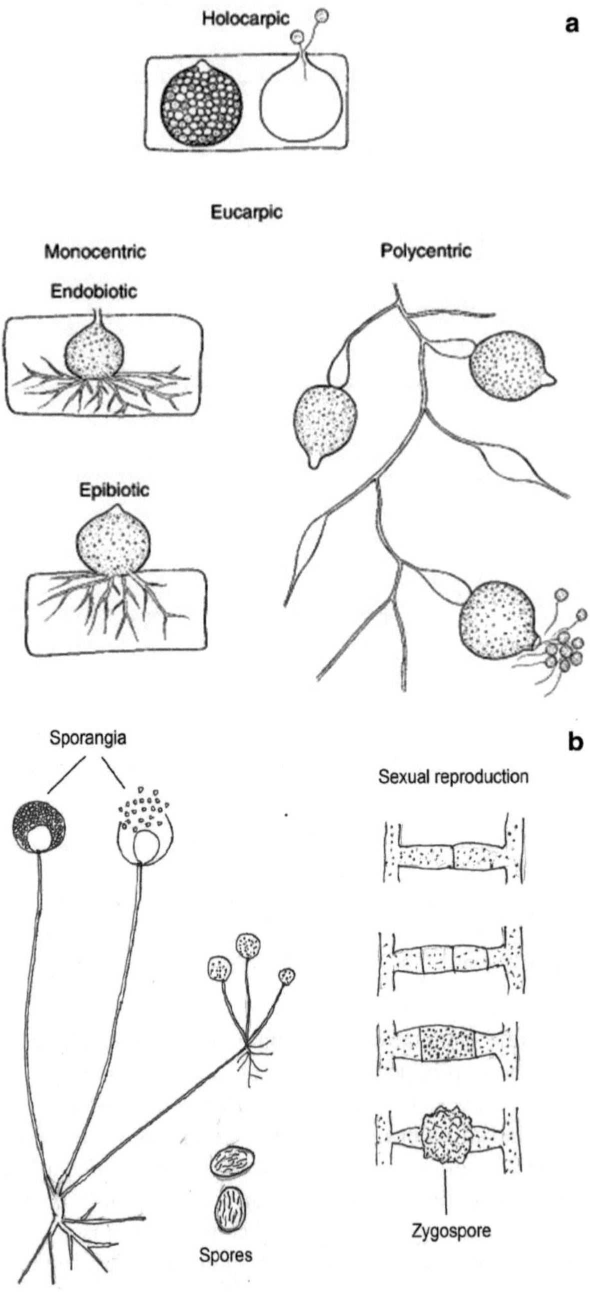


**Fig. 1.5** Important structural features of some mycetaen fungal groups. (**a**) Phylum Chytridiomycota: Different structures of the vegetative thallus. (**b**) Phylum Mucoromycota, *Rhizopus* sp. showing sporangiophores bearing sporangia with spores. (**c**) Phylum Ascomycota: Various ascocarps. (**d**) Structure and life cycle of *Coprinus,* a typical member of the Basidiomycota (Source: **a** and **d** Webster, J. and Weber, R.W.S. 2007. Introduction to Fungi. With Kind permission from Cambridge University Press; **c** Pöggeler, S. 2006. Fig. 16.1. in Mycota. Kind permission of Springer)

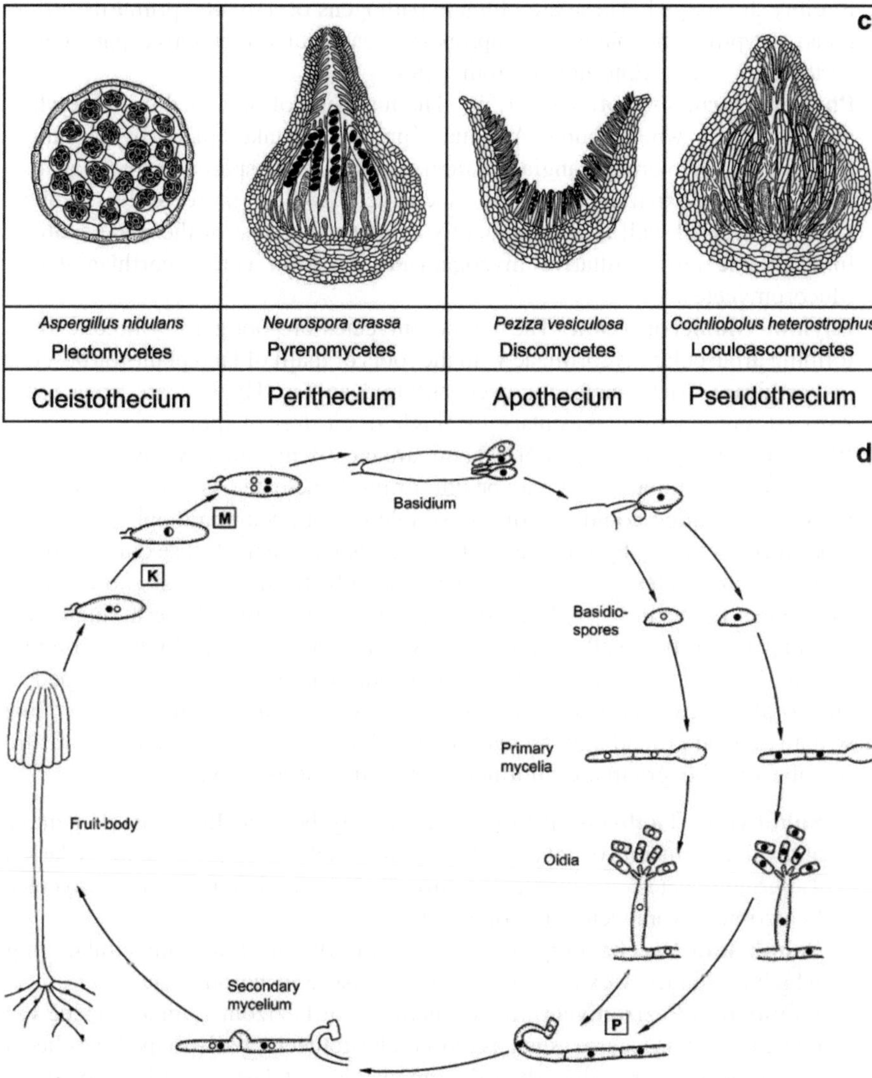


**Fig. 1.5** (continued)

disciform cavities containing plugs. Asexual reproduction by arthrospores. Sexual reproduction is by fusion of two gametangial cells resulting in "zygospores." Fungi that belong to this group are saprobes, mycoparasites, or obligate symbionts. Class: Kickxellomycetes

5. **Phylum Entomophthoromycota**: The thallus is made of a well-defined mycelium, which is coenocytic or septate. Cells may be walled or amoeboid and changeable in shape. Some taxa form cystidia or rhizoids. Branched or unbranched spore (conidia) bearing conidiophores are produced. Conidia are

forcibly discharged. These are obligate pathogens of animals (primarily arthropods), cryptogamic plants, or saprobes; occasionally facultative parasites of vertebrates. Class: Entomophthoromycetes.

6. **Phylum Mucoromycota** (Fig. 1.5b): Thallus made of branched mycelia which are coenocytic when young. Asexual reproduction takes place by sporangia, sporangiola, or merosporangia, or rarely by chlamydospores, arthrospores, or blastospores. Sexual reproduction is characterized by more or less globose zygospores. Cell wall is predominantly of chitosan. Most of these are saprobic fungi. Some are facultative mycoparasites or form ectomycorrhiza. Class: Mucoromycetes.
7. **Phylum Glomeromycota**: The glomeromycetes are obligate symbiotic fungi forming arbuscular mycorrhizae with the roots or thalli of land plants (arbuscular mycorrhizae). They have not been cultured so far. Hyphae are coenocytic. Asexual reproduction takes place by single or clusters of spores.
8. **Phylum Ascomycota** (Fig. 1.5c): These are mostly mycelial. Many, namely the yeasts are unicellular. Cells may be uni- or multinucleate. Asexual reproduction is by spores called conidia borne on simple or complex conidiophores. Yeasts reproduce asexually by budding or fission. Characteristically, sexual reproduction in ascomycetes results in eight meiotically formed sexual spores called ascospores borne in a sac-like ascus (pl. asci); Asci may have a single wall (unitunicate) or two walls (bitunicate). Asci are mostly formed within complex fruiting bodies called ascocarps or ascomata. The asexual state is termed an anamorph. Many members of Ascomycota have lost their ability to reproduce sexually and are found only as anamorphs. These fungi are saprobic or parasitic. Members of this group are informally referred to as ascomycetes.

   a. **Subphylum Taphrinomycotina**: These may be unicellular or filamentous and often parasitic on plants. Asci mostly not in fruiting bodies. Classes: Taphrinomycetes; Neolectomycetes; Pneumocystidiomycetes; Schizosaccharomycetes (fission yeasts).
   b. **Subphylum Saccharomycotina** (true yeasts): These are unicellular fungi which reproduce asexually by budding. Class: Saccharomycetes.
   c. **Subphylum Pezizomycotina**: All members of Pezizomycotina produce various kinds of ascocarps, such as closed cleistothecia, flask-shaped perithecia, or saucer- or disc-shaped apothecia. Class: Arthoniomycetes; Dothidiomycetes; Eurotiomycetes; Laboulbeniomycetes; Lecanaromycets; Leotiomycetes; Lichinomycetes; Orbiliomycetes; Pezizomycetes; Sordariomycetes.

9. **Phylum Basidiomycota** (Fig. 1.5d): These include mostly mycelial forms. Many also exist as unicellular yeasts. Reproduction takes place by sexual and/or asexual spores. Sexual reproduction results from karyogamy in binucleate cells and leads to the formation of 4 meiotically formed basidiospores borne on basidia (sing. basidium). Basidia are formed in a wide variety of basidiocarps or basidiomata (fruiting bodies of basidiomycetes), such as in mushrooms, puffballs, and jelly fungi. Most species are terrestrial, but some are aquatic. They are

saprobic or symbiotic. The asexual state is termed an anamorph. Members of this group are informally referred to as basidiomycetes.

a. **Subphylum Pucciniomycotina**: These are the rust fungi, which are parasitic on land plants. Class: Pucciniomycetes; Cystobasidiomycetes; Agaricostilbomycetes; Microbotryomycetes; Atractiellomycetes; Classiculomycetes; Mixiomycetes; Cryptomycocolacomycetes.
b. **Subphylum Ustilaginomycotina**: These are the smut fungi, which are parasitic on land plants. Class: Ustilaginomycetes; Exobasidiomycetes.
c. **Subphylum Agaricomycotina**: Basidia in these fungi are undivided or divided longitudinally or transversely. This subphylum includes mushrooms, bracket fungi, and puffballs. Class: Tremellomycetes; Dacrymycetes; Agaricomycetes.

Three interesting sister groups of fungi also need to be considered in a discussion of fungi belonging to Mycetae (Fig. 1.4). These are organisms belonging to the **Phylum Microsporidia, the Phylum Cryptomycota or Rozellida, and the Phylum Aphelida, together called the Opisthosporidia** (Neuhauser et al. 2012; Karpov et al. 2014). These organisms are important in undertstanding the evolution of fungi and are discussed in Chap. 10.

### 1.2.2 *Kingdom Straminipila*

Members of the Kingdom Straminipila (Fig. 1.6) may be photosynthetic or heterotrophic. Heterotrophic straminipiles comprise both phagotrophic organisms, as well as osmoheterotrophic fungi.

Organisms belonging to Straminipila, the straminipiles, may be single-celled and microscopic or filamentous or thalloid and macroscopic as in brown seaweeds. Motile zoospores when present are characterized by a pair of two unequal or heterokont flagella. The anterior flagellum is longer and is called the tinsel flagellum because it is covered by lateral hairs. The posterior one is shorter and is called the whiplash flagellum, being smooth and devoid of lateral hairs. The osmoheterotrophs belonging to the Kingdom Straminipila are called the **straminipilan fungi**. These are represented by three groups, the Oomycetes, Hyphochytriomycetes, and Labyrinthulomycetes. A good account of the systematics of these fungi is given by Beakes et al. (2014).

**Straminipilan fungi fall under three groups** (Figs. 1.6 and 1.7).

1. **Hyphochytriomycetes**: These are microscopic, unicellular forms with rhizoids. Cytoplasm of the thallus is converted into zoospores. Although they resemble the chytrids of the Kingdom Mycetae in these characters, the zoospores are anteriorly uniflagellate and the flagellum is of the tinsel type, possessing lateral hairs. Cell walls contain both chitin and cellulose. Hyphochytriomycetes are mostly marine. This group contains a single Order Hyphochytriales.

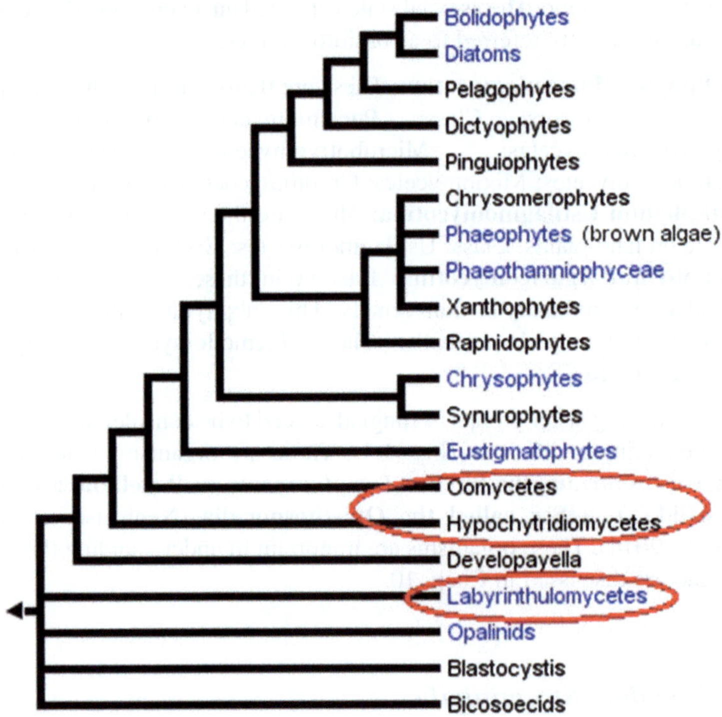


**Fig. 1.6** The Kingdom Straminipila: A broad classification showing the distribution of straminipilan fungi within the Kingdom, shown within red circles. (Courtesy: Dr J. Craig Bailey, Center for Marine Science, University of North Carolina-Wilmington, USA. *Red circles* are superimposed upon the original figure)

2. **Oomycetes**: These fungi possess unicellular or filamentous or unicellular thalli. When unicellular, the entire cell is converted into a zoosporangium (holocarpic thalli). A special zoosporangium is formed in filamentous forms (eucarpic thalli). Zoospores are produced within the zoosporangia and often sexually by gametangia. Zoospores are biflagellate, with a long anterior and a shorter posterior flagellum. The longer flagellum is a tinsel flagellum bearing lateral hairs, while the shorter one is of the whiplash, smooth type. Cell walls are made of cellulose. Contains the Classes Saprolegniomycetes, Peronosporomycetes, and a group of uncertain affinities (Class Incertae sedis), with organisms belonging to an early diverging evolutionary clade of the Oomycota.
3. **Labyrinthulomycetes**: These fungi are single-celled. They are characterized by plasma membrane extensions called ectoplasmic net (EN) elements, produced from a special organelle called sagenogenetosome or bothrosome. The EN resemble the rhizoids of chytrids, help in absorbing nutrients from the surrounding milieu, and are capable of penetrating solid particles. At maturity, the single cell is converted into a sporangium. Cell walls are made up of sulfated

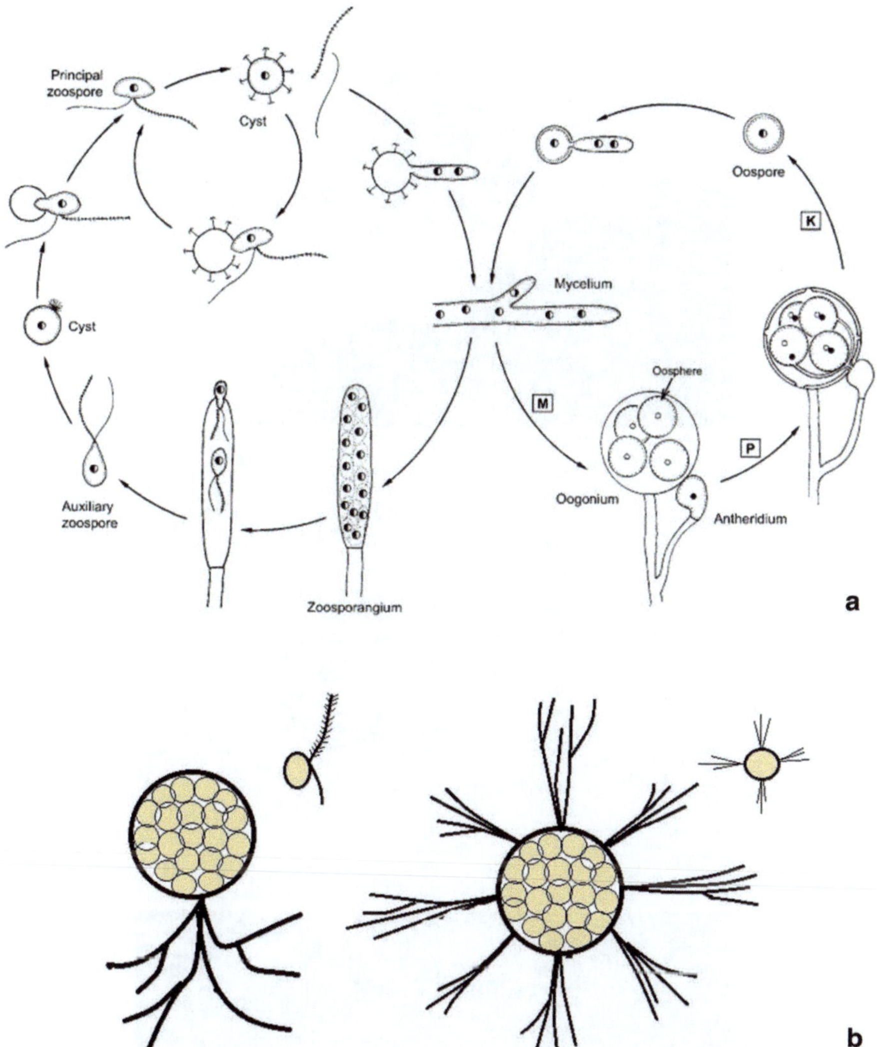


**Fig. 1.7** Morphological and developmental characteristics of some straminipilan fungi. (**a**) Life cycle of the oomycete *Saprolegnia*. (Reproduced from: John Webster and Roland Weber. 2007. Introduction to Fungi. 3rd edition. With permission from Cambridge University Press). (**b–e**) Fungi belonging to Labyrinthulomycetes. (**b**) Schematic diagram of a: Thraustochytrid with ectoplasmic net (EN) and biflagellage zoospore; (**b**) Aplanochytrid with EN and aflagellate, motile spore. (**c**) Zoosporangium of the thraustochytrid *Thraustochytrium motivum* with fully developed zoospores about to be liberated (S. Raghukumar). (**d**): Different stages of *Aplanochytrium kerguelense,* showing sporangia with spores and spores moving out using EN (*arrow*) (Varada Damare and S. Raghukumar) (**e**) Colony of *Labyrinthula* cells moving through EN (Courtesy: Daniel Martin, University of South Alabama, USA)

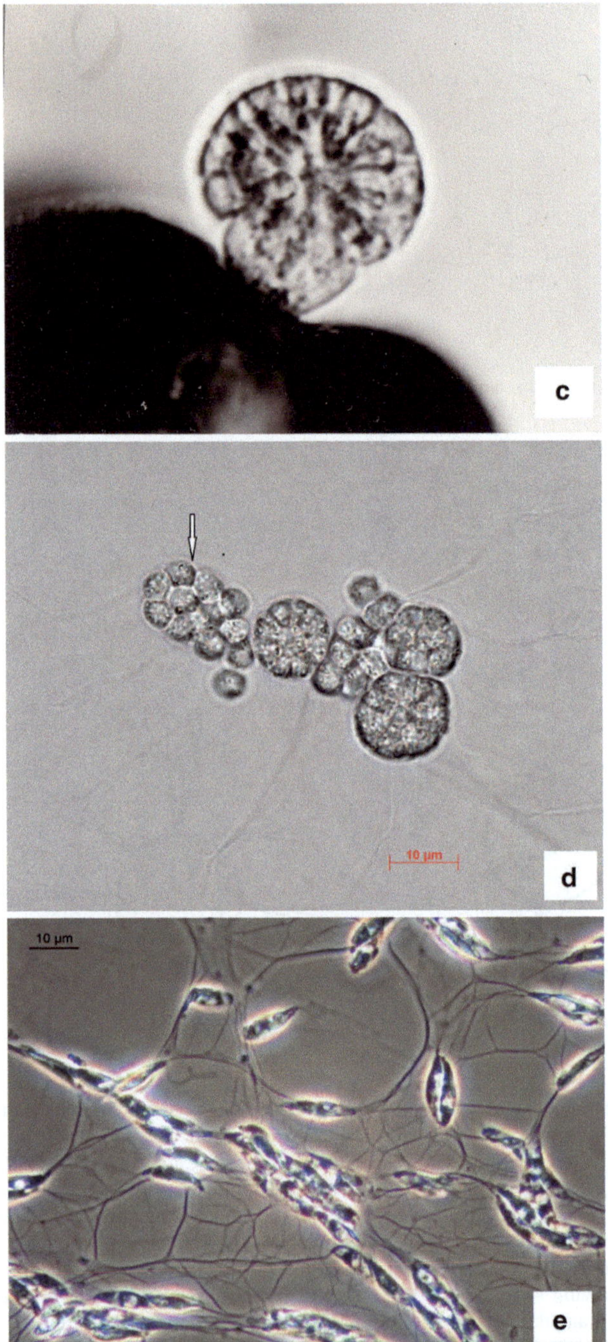


**Fig. 1.7** (continued)

polysaccharides. Contains three Orders, Thraustochytridiales, Labyrinthulales, and Aplanochytriales. Members of Labyrinthulomycetes are typically marine.

a. Thraustochytridiales (Thraustochytrids): These are unicellular. Sporangia are zoosporangia that produce biflagellate zoospores. As with Oomycetes, these are biflagellate, with a long anterior and a shorter posterior flagellum. The longer flagellum is a tinsel flagellum bearing lateral hairs, while the shorter one is of the whiplash, smooth type.
b. Aplanochytriales (aplanochytrids): These are unicellular fungi, where the sporangia produce non-flagellate spores. These spores produce EN and move in a crawling manner using these.
c. Labyrinthulales (Labyrinthulids): These fungi produce single cells that are enclosed within the EN and live as a colony. The cells move rapidly within the EN. Labyrinthulales reproduce by biflagellate zoospores or by cell division to form spores.

- Fungi are fascinating organisms that play a key role in sustaining life on earth.
- Fungi are eukaryotic microorganisms with an osmoheterotrophic nutrition.
- Fungi are polyphyletic and fall under two separate Kingdoms, the Kingdom Mycetae (also called Kingdom Fungi) and the Kingdom Straminipila
- The vegetative structure or thallus in most fungi is filamentous. Many are also unicellular.
- Fungi reproduce sexually or asexually by producing spores.
- Fungi in freshwater and marine habitats produce spores that are adapted to dispersal in aquatic conditions.
- Cell walls of fungi belonging to Mycetae are made up of chitin, chitosan, or polysaccharides.
- All members of the Kingdom Mycetae are osmoheterotrophic. Most are mycelial, while the yeasts are unicellular. Mycetae contains the Phyla Chytridiomycota, Neocallimastigomycota, Blastocladiomycota, Kickexellomycota, Entomophthoromycota, Mucoromycota, Glomeromycota, Ascomycota, and Basidiomycota.
- The phyla Microsporidia, Cryptomycota (Rozellida), and Aphelida are enigmatic sister groups of Mycetae.
- Kingdom Straminipila contains photosynthetic algae, phagotrophic protists, and osmoheterotrophic fungi. Straminipilan fungi are found in the phyla Hyphochytriomycetes, Oomycetes, and Labyrinthulomycetes.
- Every group of fungi is characterized by its own structure, morphology, and life cycle efficiently adapted to its environment.

# Chapter 2
# The Marine Environment and the Role of Fungi

*How inappropriate to call this planet Earth, When it is quite clearly Ocean...*

Arthur C. Clarke

Oceans cover more than 70% of the Earth's surface. The marine ecosystem is driven by an interaction between the physicochemical environment of the oceans and the organisms that live there (Lalli and Parsons 1997). Biological processes of the marine ecosystem have an impact on climate and human society.

**Physical and chemical properties of seawater are crucial in determining marine ecological processes.** The average depth of the oceans is about 4000 m, and the volume covered by seawater is nearly $1.35 \times 10^9$ cubic kilometers. Seawater has an average salt content of about 3.5%, expressed as salinity of 35 ppt (parts per thousand). The pH of seawater ranges from 7.5 to 8.5. Sodium followed by magnesium are the major cations in seawater, while chloride and sulfate are the major anions. Many other macro- and microelements contribute to the salinity of seawater (Table 2.1).

Temperature of seawater varies from about 30 °C in warm tropical surface waters to around −2 °C at the poles. Temperature also varies according to depth, the deep-sea environment being about 2–4 °C. Carbon dioxide and oxygen are the major dissolved gases in seawater. Oxygen levels vary from a maximum of 20 parts per million (ppm) to nearly 0 ppm at some sites. Most life processes in the ocean are supported by the photosynthetic, primary producers, the phytoplankton in the open ocean or macroalgae in coastal waters. The main nutrients for plant growth are nitrogen (as nitrate, nitrite and ammonia), phosphorus (as phosphate), potassium, sulfur, magnesium, and calcium. Iron is an essential trace element. Diatoms, the important primary producers, also require silicon.



S. Raghukumar, *Fungi in Coastal and Oceanic Marine Ecosystems*,
DOI 10.1007/978-3-319-54304-8_2

**Table 2.1** Composition of seawater

| Chemical ion | Concentration in parts per thousand (ppt) | Proportion of total salinity |
|---|---|---|
| Chloride | 19.245 | 55.03 |
| Sodium | 10.752 | 30.59 |
| Sulfate | 2.701 | 7.68 |
| Magnesium | 1.295 | 3.68 |
| Calcium | 0.416 | 1.18 |
| Potassium | 0.390 | 1.11 |
| Bicarbonate | 0.145 | 0.41 |
| Bromide | 0.066 | 0.19 |
| Borate | 0.027 | 0.08 |
| Strontium | 0.013 | 0.04 |
| Fluoride | 0.001 | 0.003 |
| Other | Less than 0.001 | Less than 0.001 |

## 2.1 Divisions of the Marine Environment

**The marine environment is divided into various zones** (Fig. 2.1)**. Each of these is defined by distinct physical and chemical characteristics that determine the organisms that live in it, their interactions, and ecosystem functioning.**

Horizontally, the oceans are divided into neritic or coastal and oceanic or offshore habitats.

- **The neritic or coastal habitat** is strongly influenced by land. This region is generally understood to extend from the high tide level to the farther edge of the continental shelf that lies approximately at 200 m depth. The continental shelf lies beyond the edge of the continental margin that marks the edge of the continental shelf. The region between the high tide and low tide levels is the littoral.
- **The offshore or oceanic habitat** is much less influenced by land than the neritic. It encompasses the region extending beyond the continental margin at about 200 m depth.

On a vertical basis, the oceans are divided into the open water or pelagic and the sea bottom or benthic.

- **The pelagic habitat:** When we say that the oceans cover 70% of the earth, we are referring to the vast amount of water column or the open sea that fills most parts of the earth. The pelagic comprises the open waters of the oceans, which are divided into several vertical zones. The epipelagic region covers approximately the upper 200 m of the oceans, at level with the upper margin of the continental shelf. The well-lit upper zone of the epipelagic from the surface to roughly 80 m depth is called the euphotic. Below the epipelagic lies the

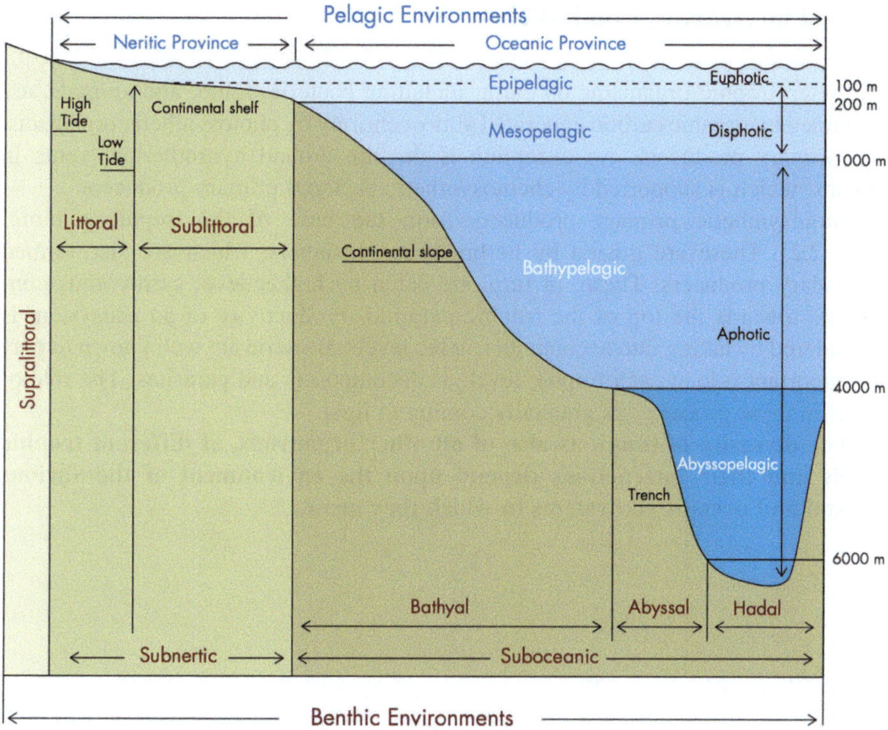


**Fig. 2.1** The various divisions of the marine environment (Illustration by Craig Gladding)

mesopelagic till about 1000 m. This zone is dimly lit, while the deep-water or bathypelagic below 1000 m is almost totally dark or aphotic. The abyssopelagic zone lies below 4000 m, which is the average depth of the oceans.

- **The sea bottom constitutes the benthic zone**. The benthic zone extending from the high tide level to the low tide level is the **littoral benthic** and that below the low tide level to a depth of 200 m depth corresponding to the edge of the continental shelf is the **sublittoral benthic**. The bottom of the sloping continental shelf till about 4000 m is the **bathyal**, beyond which lie the **abyssal** and **hadal** zones. The average depth of the oceans is approximately 4000 m, the deepest part of the oceans being about 10,500 m. The deep sea is an extreme environment with temperatures of about 2–4 °C and high hydrostatic pressures that affect life processes. Hydrothermal vents occur in many parts of the deep-sea. These are characterized by a flow of super-heated seawater carrying a variety of dissolved minerals (see Sect. 12.2).

## 2.2 The Marine Food Web

All heterotrophic organisms on earth, including bacteria, fungi, and animals, are sustained by organic carbon generated almost entirely by photosynthetic organisms, the primary producers. An exception is the life around hydrothermal vents in oceans, which is supported by chemosynthetic bacterial primary producers.

Photosynthetic primary producers form the base of the trophic pyramid (Fig. 2.2). These are grazed by herbivorous consumers, which are also termed secondary producers. These, in turn, are eaten by higher level carnivorous consumers towards the top of the trophic pyramid. Productivity of an ecosystem is considered by taking into account all trophic levels. Bacteria are well known to play an important role at each trophic level, as decomposers and parasites. The role of fungi in these processes is gradually coming to light.

**The diversity of fungi, as also of all other organisms, at different trophic levels and their interactions depend upon the environment of the various coastal and oceanic ecosystems in which they grow.**

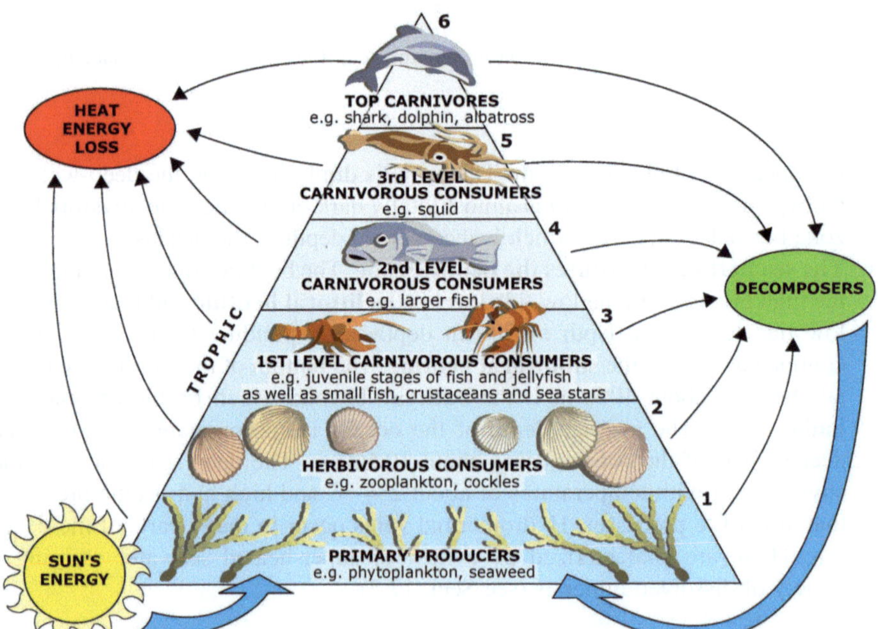


**Fig. 2.2** The marine trophic pyramid (Copyright: University of Waikato. All Rights Reserved. http://sciencelearn.org.nz)

### 2.2.1 Primary Production in Marine Ecosystems

Primary production (PP) is the amount of organic matter fixed by photosynthetic organisms. Primary productivity is the rate of primary production. PP is usually expressed in terms of the amount of carbon fixed. The total amount of carbon fixed is the Gross Primary Production or GPP. Not all of GPP is converted into biomass. Part of the energy is used for respiration. **The amount of photosynthetically fixed energy that is converted into biomass is net primary production or NPP.** The level of primary productivity varies depending on the composition of photosynthetic organisms and varies among ecosystems and indicates how biologically productive an ecosystem is.

Part of the NPP is grazed by herbivores and enters the trophic chain, while the rest undergoes death and enters the microbial decomposition pathway. **The proportion of NPP that is consumed by grazers, or enters the detrital food chain upon their death, or is stored in sediments depends on the nature of the primary producers, which in turn is determined by the ecosystem.**

Coastal and oceanic waters together contribute a total NPP of nearly $52.14 \times 10^9$ metric tons of carbon per year (Fig. 2.3; Duarte and Cebrian 1996; Duarte et al. 2005; Bouillon et al. 2008). This roughly amounts to 50% of the Earth's photosynthetically fixed carbon, the rest coming from land plants.

Single-celled phytoplankton are the major primary producers in coastal as well as pelagic ecosystems. Of the 52.14 trillion metric tons of NPP produced every year in the oceans, phytoplankton are responsible for 89% or a total of $47.5 \times 10^9$ metric tons of carbon. The rest comes from vascular plants and macroalgae. About 40–50% of phytoplankton are consumed by herbivores. Coral reef algae, macroalgae, seagrasses, marsh plants, and mangroves are eaten by herbivores to a much lesser extent. These largely enter the decompositional, heterotrophic pathway rather than the grazer pathway.

## 2.3 The Major Marine Ecosystems

The marine environment may be broadly divided into two distinct categories, the coastal and oceanic environments. Each contains distinct ecosystems that are driven by certain key groups of organisms.

### 2.3.1 Coastal Environments

Coastal fisheries and maritime activities have an enormous impact on human society. Coastal productivity, therefore, is of tremendous importance. **A variety**

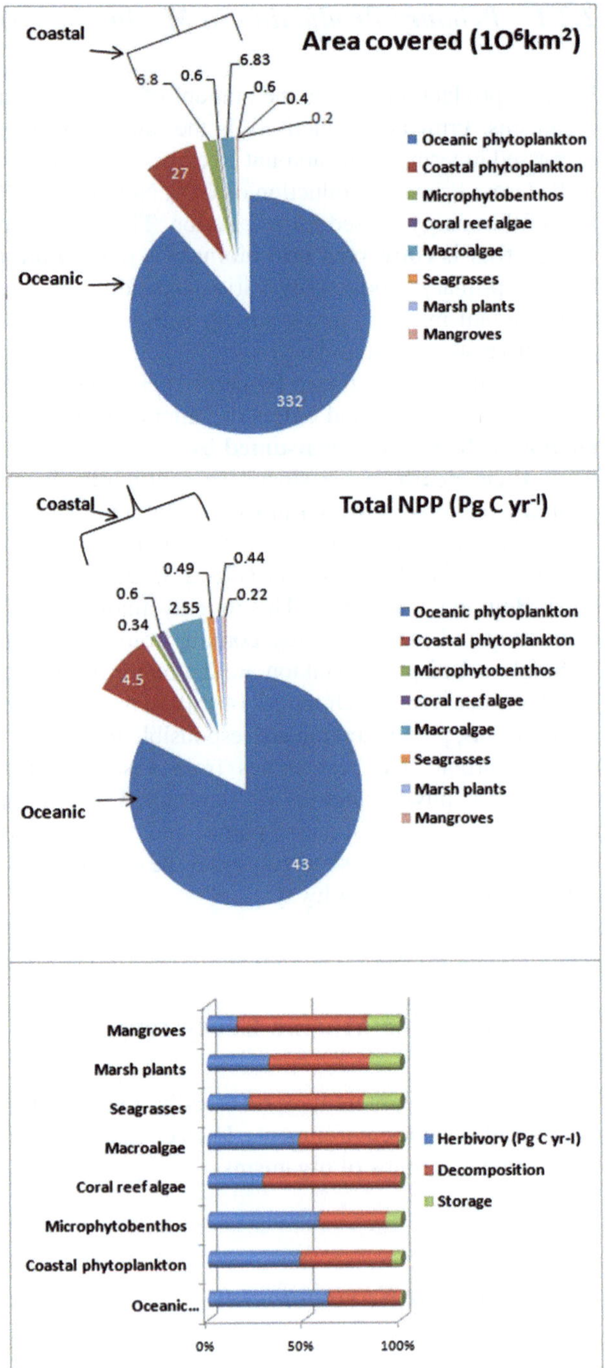


**Fig. 2.3** Major marine environments, their net primary productivity (NPP) and the fate of NPP (Based on information from Duarte and Cebrian 1996)

**of coastal habitats, each with its unique characteristics, is found along the edges of continents.**

**Coastal oceans are biologically much more productive than the open oceans** (Fig. 2.3).

- Coastal environment forms only about 12% of the ocean's area available for primary production. However, they are responsible for 18% of the NPP of the entire oceans. Net primary production in coastal environments equals nearly $9.14 \times 10^9$ metric tons of carbon.
- Of the above NPP, nearly 48% comes from mangroves, salt marsh plants, seagrass beds, coral reef algae, and macroalgae. Phytoplankton contribute around 52% of coastal NPP.
- NPP arising from vascular plants in mangroves, salt marsh plants, and seagrass beds is difficult to digest by herbivores, and only part of the NPP from this source enters the trophic web. Biomass produced by macroalgae and coastal phytoplankton are consumed to a larger extent by herbivores.
- A large portion of biomass produced by all primary producers enters the decompositional pathway.
- A greater portion of the NPP of vascular plants is recalcitrant and becomes stored in the sediments, compared to biomass from macroalgae and phytoplankton.
- Phytoplankton are easily grazed upon. Nearly 40% of the phytoplankton produced is grazed upon by herbivores. This, in turn, is passed on to higher trophic levels.

**The following are the major coastal ecosystems.**

- **The estuarine environment**, where a river meets the sea is characterized by a gradient in salinity, from about 35 ppt at the mouth of the estuary meeting the sea to nearly 0 ppt towards the upper reaches of the estuary. Riverine discharge from the estuary into the sea brings enormous amounts of nutrients into coastal waters. Wood from dead trees in coastal areas is often washed into the sea (see Chap. 4). Lagoons and networks of inland backwaters are other important coastal marine habitats.
- **Mangroves, salt marshes, and seagrasses** are important coastal vegetations (Chaps 5–7). Dense stands of mangrove trees are often found lining the estuarine margins and backwaters. Mangrove trees contribute to enormous organic material in the form of leaves and woody material that transport nutrients to coastal waters. Salt marshes are characteristic of lower salinity, brackish water regions. Seagrass beds are found in many shallow, sublittoral waters.
- **Coral reefs** are found in coastal waters of middle latitudes from 30° N to 30° S of the equator (Chap. 9). Fringing coral reefs along many coasts rival tropical rain forests in productivity and biodiversity. Hermatypic or reef-building corals that secrete calcium carbonate are the keystone species of this ecosystem, which is extremely rich in biodiversity. Coral reefs are found typically in coasts with clear waters and low nutrients.

- **Algal beds**, comprising both micro- and macroalgae are often found in intertidal, rocky beaches (Chap. 8). Dense algal vegetations are often found in the sublittoral zones.
- **Coastal pelagic:** Primary productivity differs in various regions of the oceans, depending on the availability of nutrients and light and the presence of a suitable temperature (Fig. 2.4). Coastal waters are typically richer in nutrients than the oceanic. Therefore, primary production in the euphotic zone of the coastal pelagic is higher than in offshore pelagic. Phytoplankton, dominated by diatoms, drive the coastal pelagic ecosystem. Coastal waters of many continents experience seasonal or perennial upwelling that bring nutrients to the euphotic zone and are sites of intense primary production. Many coasts close to estuaries are enriched by nutrients brought from land by rivers resulting in enhanced productivity (Brink 2004). The coastal marine environment is extremely diverse in terms of its physicochemical characteristics and ecosystems. Primary production varies vastly in the euphotic pelagic of different oceanic regions depending on available light at different latitudes.

### *2.3.2 Offshore Pelagic, Benthic, and Deep-sea Habitats*

Although seawater covers most part of the earth, **the coastal (neritic) pelagic and offshore (oceanic) pelagic are very different from each other in their biological characteristics.** The sheer size of the oceanic waters is responsible for much of the global carbon budget.

- Nearly 81% of ocean's primary production and about 40% of global primary production take place in the offshore pelagic realm (Fig. 2.3).
- About 88% of herbivores depend upon oceanically produced NPP. This amounts to approximately $25.0 \times 10^6$ Pg C consumed by herbivores in the entire marine environment.
- Oceanic waters contribute nearly 26.0% of all oceanically produced NPP to the storage pool, most of these being deposited in the deep sea.
- Because of the size, NPP, and associated processes, oceanic waters are extremely important determinants of global climate.

Many parts of the oceanic pelagic waters, such as the North and South Pacific gyres, the North and South Atlantic gyres, and the South Indian gyre are perennially oligotrophic. As a result, primary production is extremely low in these regions. Microorganisms play a unique role in the food web even in such a harsh environment. No primary production can take place below the photic zone. Yet a high diversity of life and biological processes take place in the deep sea. Life in the aphotic deep sea is sustained because of a process termed the "biological pump" that transports dead organic matter from the euphotic zone to the deep sea.

**The marine, benthic environment offers the largest surface area on earth for living organisms.** Benthic diatoms and other algae form the food for consumers in shallow, well-lit littoral sediments. However, most of the benthic animals as well

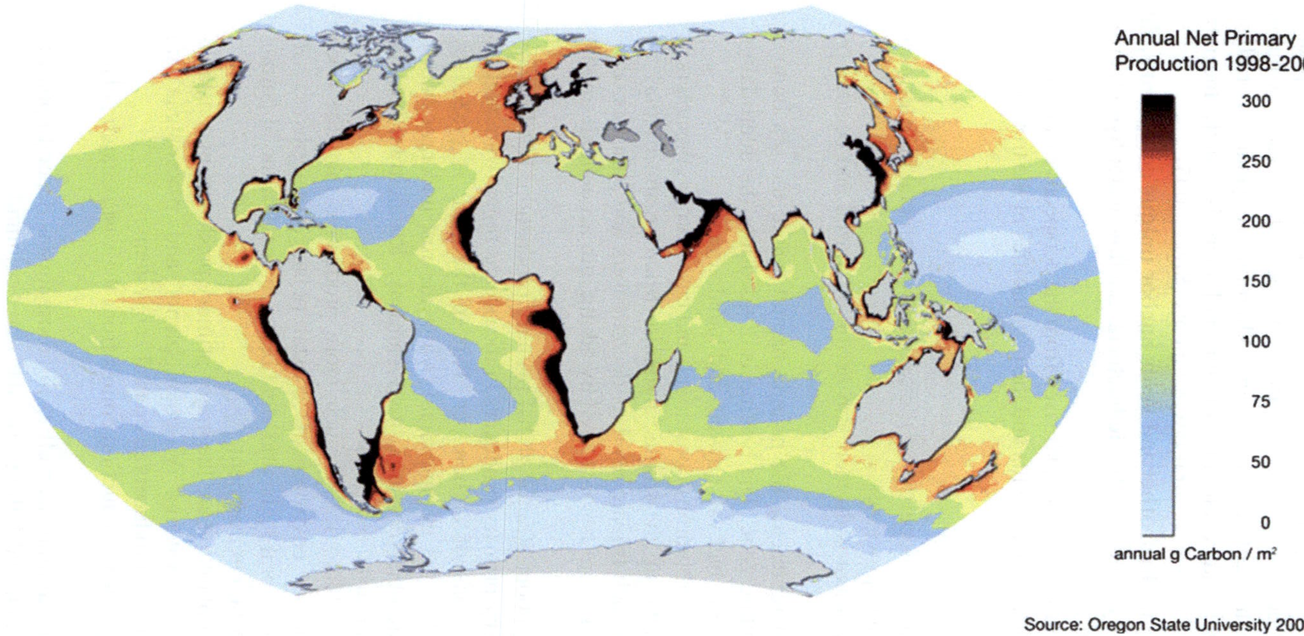


**Fig. 2.4** Primary productivity in the global oceans. Note that the productivity is extremely high in certain coasts. Many parts of the world's oceans have extremely low primary productivity (Courtesy: GRID-Arendal. http://www.grida.no/graphicslib/detail/natural-resources-marine-resources_132a)

as microorganisms living in this zone **depend upon the supply of organic material falling from the water column**. The labile components of the deposited organic material are easily degraded or consumed by detritivores. Compounds which are more slowly degraded, or recalcitrant to degradation, such as humic material accumulates in the benthic environment. Microbial enzymatic hydrolysis of high-molecular weight and recalcitrant organic polymers is usually the rate-limiting step in mineralization in natural environments (Sigee 2005).

**Deep-sea sediments and hydrothermal vents harbor unique ecosystems, the latter being supported by chemosynthetic bacteria.** Shallow water hydrothermal vents are not uncommon.

The study of marine fungi has to be addressed from the perspective of the diverse coastal and oceanic marine environments and the peculiarities of the trophic web in each of these.

- Physical and chemical properties of the seawater are crucial in determining marine ecological processes.
- The marine environment is divided into neritic or coastal and oceanic or offshore habitats.
- The water column represents the pelagic habitat, while the sea bottom constitutes the benthic zone. The benthic is divided into the intertidal or littoral, the sublittoral, bathyal, abyssal, and hadal zones.
- Primary producers, herbivorous consumers, and carnivorous consumers at different levels make up the trophic pyramid. Microorganisms play an important level at each trophic level as decomposers and parasites. Productivity of an ecosystem is considered by taking into account all trophic levels.
- Oceans contribute about 50% of global primary production. Net Primary Production (NPP) by primary producers in the oceans roughly amounts to 53 trillion metric tons of carbon per. Part of the NPP is grazed by herbivores and enters the trophic chain, while the rest undergoes death and enters the microbial decomposition pathway. Recalcitrant organic matter is stored in marine sediments.
- Coastal oceans are biologically much more productive than the open oceans. The estuarine environment, coastal pelagic, mangroves, salt marshes, seagrasses, coral reefs, and algal beds are some of the important habitats of the coastal environment.
- Single-celled phytoplankton are the major primary producers in coastal as well as pelagic ecosystems. Vascular plants and macroalgae are also important primary producers in coastal ecosystems.
- Most of ocean's primary production comes from the vast oceanic pelagic waters.
- Most of the benthic animals as well as microorganisms depend upon the supply of organic material falling from the water column.

(continued)

- The "biological pump" supplies food to animals in the deep-sea benthic in the form of dead organic particles from the euphotic, pelagic waters.
- Deep-sea sediments and hydrothermal vents harbor unique ecosystems, the latter being supported by chemosynthetic bacteria.

## 2.4 Fungi in Marine Ecosystem Processes

***Fungi are associated with organisms of all trophic levels of the ecosystem in the form of (1) Symbiosis and (2) Saprotrophy*** (Cooke and Rayner 1984; Dix and Webster 1995; Kendrick 2000).

### *2.4.1 Symbiosis*

Fungi may live as ectobionts on the surfaces of other organisms, or as endobionts, living within them. This association, **symbiosis, may be in the form of a commensal, a mutualist, or a parasite.**

- Commensals are casual residents within or upon other organisms. They neither cause any harm to the host nor do they obtain any particular benefits from them.
- Mutualists are those that confer benefits to their hosts and also derive advantages from them.
- Parasites obtain benefits from their hosts but are not useful to them. Parasites that cause harm to the health of their hosts are pathogens. Microbial parasites and pathogens are associated with organisms of all trophic levels. They play an important role in ecosystem functioning, similar to animal predators (Gachon et al. 2010).
  - Parasites prevent domination by any single species and maintain a balance in diversity and ecosystem functioning. Thus, they may confer resilience to the ecosystem against perturbation.
  - Pathogenic fungi may devastate the population of their hosts under exceptional circumstances and disturb ecosystem functioning or cause economic losses to human society.
  - Biotrophic fungi are obligate parasites. Biotrophs cannot live in the absence of their living host. Therefore, they rarely cause devastations of their host populations by becoming pathogenic. Facultative parasites are those that may live as commensals.
  - Pathogenic fungi may cause an accelerated leaching out of Dissolved Organic Matter (DOM) from the host, which is then available to other microorganisms for their growth. Parasites can cause the death of their hosts. These are then

available to saprotrophs even earlier than the natural death of the hosts. Thus, parasites may be responsible for an accelerated recycling of organic matter.

Fungi are extremely common in marine organisms as parasites and pathogens. However, their dynamics in regulating natural populations is not sufficiently understood.

- **Endophytes are fungi that live as endobiontic symbionts within an algal or plant tissue without causing any external symptoms.** Endophytic fungi may be commensalistic, mutualistic, or weakly parasitic.

## *2.4.2 Saprotrophy*

Saprotrophic microorganisms utilize dead organic matter for food. All living organisms eventually die and undergo microbial decomposition. **Saprotrophic fungi and bacteria play a key role in the trophic web by organic matter degradation and nutrient recycling. Dead particulate organic matter (POM) with its associated saprobic microorganisms is termed detritus.** Dissolved Organic Matter (DOM) generated from biotic and abiotic activities are also substrates for saprotrophic microorganisms. Since carbon is the major element in POM and DOM, these are often discussed in terms of Particulate Organic Carbon (POC) and Dissolved Organic Carbon (DOC). Saprobic fungi play three major roles.

**Saprotrophic fungi have a greater advantage compared to bacteria in decomposition of particulate organic matter large enough to support endobiontic growth.** Bacteria grow mostly on the surfaces of organic particles. Bacterial degradative activity is mostly confined to the surfaces of plant tissues. Fungi, on the other hand, are capable of penetrating solid substrates. This may be accomplished by hyphae, rhizoids, or ectoplasmic nets (Sect. 1.1). Hence, their degradative ability also prevails in the interior of particles. Many mycetaen fungi are excellent degraders of lignocellulose and therefore are commonly associated with decomposition of vascular plants. They can also efficiently withstand periodic wetting and drying. On the other hand, bacteria are more efficient in utilizing small particles with a large surface area and in using DOM (Torzilli and Andrykovitch 1986; Raghukumar 1990; Newell 1984, 1996a; Newell and Porter 2000).

**Degradation of organic matter by saprotrophic marine microorganisms is characterized by a few key features** (Fig. 2.5).

- **DOM leached out by dead organisms during initial stages is a major nutrient source for saprotrophs.** Microorganisms rapidly take these up through absorption and convert them into their biomass. Conversion of DOM by bacteria and unicellular fungi into their biomass is an extremely important process, since the biomass can be fed upon by phagotrophic protozoans and microzooplankton. The phagotrophs may serve as food for higher level consumers. Thus, the DOM is linked to the food web through a process termed the microbial loop (Raghukumar 2005). In some cases, DOM may first aggregate to

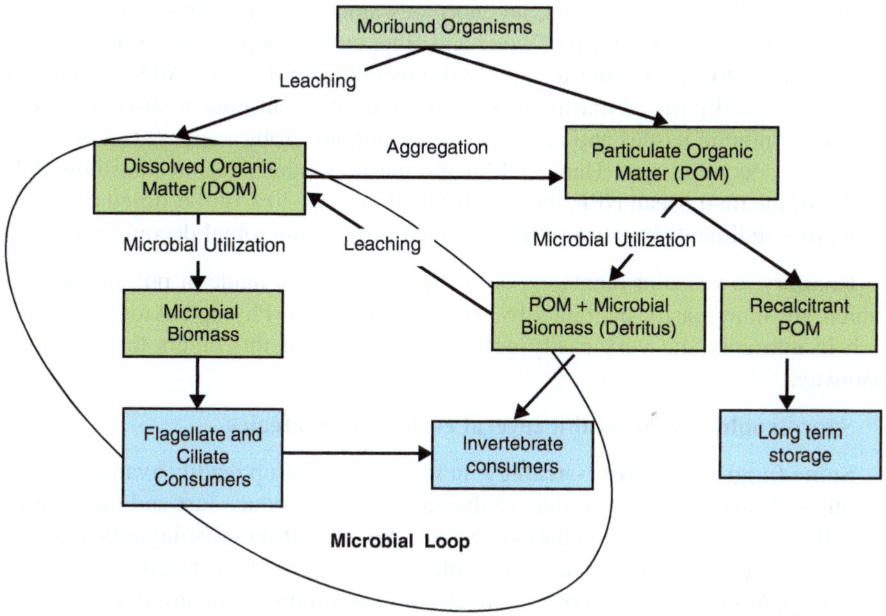


**Fig. 2.5** Key features of organic matter degradation and microbial saprotrophy in marine ecosystems

form flakes. These may subsequently be colonized by microorganisms. These detrital flakes are consumed by larger invertebrates.

- **Saprotrophic bacteria and fungi breakdown complex organic matter of the POM** by use of degradative enzymes and absorb the labile dissolved nutrients thus formed. As a result, organic matter which was originally formed of large and complex molecules is gradually converted into microbial biomass. Growth and enzymatic activity of fungi and bacteria on detritus transforms it biochemically. Detritus is nutritionally enriched because of the microbial activities and their biomass. Detritus forms the food of many detritus-feeding animals or detritivores.
- **Enzymatic degradation of POM by bacteria and fungi results in the release of DOM, as well as inorganic nutrients**. This DOM may again enter the microbial loop.
- **Organic matter that is highly resistant to microbial decay is stored away in marine sediments.** Decay and long-term storage processes are determined to a large degree by the biochemical nature of dead organic matter (Walker and McComb 1988; Harrison 1990; Duarte and Cebrian 1996; Enriquez et al. 1993; Short et al. 1993). For example, structural polysaccharides of vascular plants are composed of lignocellulose, comprising cellulose, hemicellulose, and lignin. Lignocellulose is extremely resistant to decomposition. Hence, decomposition of vascular plant detritus is complex and lasts over a period of several months. In the marine ecosystem, seagrasses, marsh plants, and mangrove detritus are rich

in lignocellulose. In addition, an enormous amount of allochthonous, lignocellulosic material in the form of wood and other terrestrial plant material enters the sea. These too contribute to marine, lignocellulosic detritus. Additionally, the belowground parts of marine macrophytes are nutrient-poor relative to that of photosynthetic tissues and are found in deeper, anaerobic sediments with a low rate of decomposition. Hence, although marine vascular plants contribute only 4% of the total ocean NPP, they contribute to nearly 30% of the stored carbon in marine sediments because of their recalcitrance to microbial decomposition.

Contrary to vascular plants, tissues of macroalgae are made of polysaccharides which are more easily degraded by bacteria and fungi. Phytoplankton detritus is rich in nutrients and very easily decomposed. Most of this enters the consumer pathway.

**Saprotrophic fungi exhibit several ecological strategies.**

- **Some fungi possess an r-strategy**, in which they use up readily available labile nutrients and rapidly colonize a substrate. They reproduce fast and prolifically, following which their population declines. **Several others display a k-strategy**, in which the fungi utilize more complex substrates for their nutrition and grow slowly. Many fungi overcome competition with others by means of antagonism, in which compounds inhibitory to competing organisms are produced. Such strategies define the fungi that colonize various substrates in the sea.
- **Fungal colonization on complex substrates is characterized by a succession of species, the mycosere.** The decomposition of large-sized, macroscopic particulate organic matter takes place over a period of time, depending on its complexity. The decompositional phase of detritus plays an important role in determining the mycosere.
- **Succession of mycetaen fungi in large particulate organic matter is deduced by observing the appearance of sporulating structures, such as ascocarps, basidiocarps, and conidial structures on the surface of the substrate.** A study of these events requires regular observations over a long period of time. Most mycetaen fungi occur in their vegetative, mycelial stage within large particulate matter, such as wood during early stages of colonization and growth. They produce fruiting bodies or sporulating structures on them only during their late stages of colonization and growth. Succession studies based on the appearance of fruiting bodies does not truly represent the process. Culturing method cannot be successfully employed to understand succession because many fungi may not grow in culture. Even if they did, many of them do not sporulate in culture. Therefore, they cannot be identified using microscopic features and can be identified only using molecular methods.

**There are three major phases of decomposition of large particulate detritus in marine environments** (Valiela et al. 1984).

1. **The initial phase of leaching out of labile and soluble compounds:** Coastal plants and algae may be inhabited by endophytic fungi prior to their senescence

and death. Rapid, abiotic leaching of soluble and labile organic matter takes place immediately upon their death. An important consequence of leaching is the removal of inhibitory compounds such as phenolics from decomposing tissues. Some microorganisms colonize detritus during this early decomposition. These too may play a role in breaking down phenolics and enabling further colonization by other microorganisms that would be deterred by high levels of phenolics. Endophytic fungi present earlier in the living tissues often persist for a while after death of the plant or alga. Besides, a few other fungi also colonize the detritus at this stage and initiate decomposition. This leads to further leaching of Dissolved Organic Matter (DOM). Such abiotic and biotically induced leaching causes a rapid loss of weight of detritus. Duration of this phase depends on the host plant and may last a few days to several months. **Fungi with an r-strategy colonize the substrate and rapidly utilize labile, soluble organics during the initial decomposition phase. Diversity during this period is low. A few stress-tolerant species are found during this period.**

2. **The intermediate phase of microbial degradation of refractory components:** A second phase of colonization by fungi takes place after the initial leaching phase and decomposition. Refractory compounds such as lignocelluloses in vascular plants or structural polysaccharides in algae are decomposed by bacteria and fungi. Fungal biomass rapidly builds up during this phase and fungal hyphae pervade the substrate. **Mycosere during this intermediate phase is characterized by colonization from a high diversity of species. These often exhibit a combative strategy and possess antagonistic properties against other fungi, thus preventing them from colonizing.**
3. **The final phase of physical and biological fragmentation of detritus:** The third phase is characterized by declining microbial activities. Biotic activities including microbial degradation, as well as animal activities, combined with abiotic activities such as abrasion result in fragmentation of detritus into fine, particulate detritus. Bacterial biomass is higher on fine particles relative to large POM. Mycosere during this final phase is represented by highly stress-tolerant species with a k-strategy. Diversity during this phase is low. Thus, fungal colonization of substrates takes place by a succession of species with different ecological strategies.

**Degradative activities of bacteria and fungi play a key role in the detrital food web** (Raghukumar 2005).

- **Microbial degradative activities alter the biochemical properties of detritus.** Fungi and bacteria growing in detritus take up inorganic N and P from the environment and incorporate these into microbial protein. This "immobilization" of the inorganic ions alters the chemistry of detritus by enriching it with nitrogen and phosphorus. This results in the lowering of C/N and C/P ratios. Nitrogen-fixing bacteria on detritus may also contribute to increase in nitrogen.

  Easily decomposable organic detritus, such as those from phytoplankton and macroalgae, become highly enriched in total nitrogen owing to buildup of microbial protein.

On the other hand, increase in nitrogen of vascular plant detritus (mangroves, salt marshes and seagrass) takes place through microbial protein, as well as an increase in humic nitrogen (Rice 1982). Microbial degradative activities result in reactive carbohydrates, phenols, small peptides, and amino acids. These react with microbial enzymes and condense to form humifying detritus.

- **Detritus enriched with bacterial and fungal biomass is a major nutrient source to detritivorous animals.** Growth on detritus results in buildup of microbial biomass. The enormous quantity of microbially mediated detritus serves as a major food source to many marine animals, particularly those in coastal environments. These include crabs, gastropods, crustaceans such as prawns, fish, and smaller invertebrates like amphipods and gastropods. Vast amounts of plant and algal detritus from coastal marine vegetations are exported to coastal waters. This process, called "outwelling," is characteristic of coastal wetlands and was first postulated by E. P. Odum and J. M. Teal in the 1960s (Lee 1995). Hence, the importance of detritus as food is not just locally confined to the environment of the coastal vegetation, but extends farther into the coastal waters.
- **Detritus is a nutritious food** for several reasons.
  - Palatability of detritus by detritivores is improved because recalcitrant compounds such as lignocelluloses are degraded or modified.
  - Fungal and bacterial biomass may be substantial in detritus from certain sources, such as salt marsh grass. In such cases, fungi may serve as a large pool of proteins to detritivores.
  - Fungal biomass may provide essential nutrients to detritivores (Mann 1988; Raghukumar 2005). Some of these are amino acids lysine and methionine, various vitamins, polyunsaturated fatty acids, and sterols, an important precursor for the manufacture of cholesterols in marine animals (Phillips 1984). These nutrients are vital for the survival of detritivorous animals, which are unable to synthesize such compounds by themselves. For example, Crustacea require the polyunsaturated fatty acid docosahexaenoic acid for growth, which may be provided by straminipilan fungi (Harrison 1990).
  - Fungi may provide important cues for detritivores to feed on detritus.

**The relationship between a fungus and its host is delicately balanced, based upon the health of the host and environmental conditions. Hence, the gamut of relationships from commensal, mutualist, and saprotroph or parasite is a continuum which can change from one to another.**

Various aspects of marine fungal ecology have been discussed in many reviews (Hyde et al. 1998; Raghukumar 2002, 2008; Hyde 2002; Raghukumar and Damare 2011; Jones and Pang 2012; Nakai and Naganuma 2015).

- Fungi are associated with organisms of all trophic levels in the ecosystem in the form of (1) Symbiosis and (2) Saprotrophy.
- Symbiosis may be in the form of a commensal, a mutualist, or a parasite.

(continued)

- Endophytic fungi live as endobiontic symbionts without causing any external symptoms.
- Saprotrophic fungi and bacteria play a key role in organic matter degradation and nutrient recycling. All organisms in the trophic web undergo microbial decomposition upon death.
- Dead particulate organic matter (POM) with its associated saprobic microorganisms is termed detritus.
- Saprotrophic fungi exhibit r- or k-strategy in colonizing substrates. Growth on detritus results in buildup of fungal biomass.
- Fungal degradative activities alter the biochemical properties of detritus.
- Decomposition of vascular plant detritus in the marine environment consists of (1) the leaching phase, (2) the microbial decomposition phase, and (3) the physical and biological fragmentation phase.
- The enormous quantity of microbially mediated detritus serves as a major food source to many marine animals, particularly those in coastal environments.
- The relationship between a fungus and its host is delicately balanced, based upon the health of the host and environmental conditions. Hence, the gamut of relationships from commensal, mutualist, and saprotroph or parasite is a continnum which can change from one to another.

## 2.5 Diversity of Marine Fungi

**According to popular estimates, about 1,500,000 species of mycetaen fungi inhabit earth.** Studies using sequencing of internal transcribed spacer (ITS) markers from soil DNA samples have even suggested a global total of 3.5–5.1 million species of fungi (O'Brien et al. 2005). Only about 80,000 species of fungi have been described so far.

**The total number of marine fungi has been estimated at about 10,000 species** (Jones et al. 2009, 2015; Jones and Pang 2012).

**There are two major ecological categories of marine fungi.**

- **"Obligate marine fungi"** grow and reproduce exclusively in the marine environment.
- **"Facultative marine fungi"** are terrestrial species, which actively grow and reproduce in marine environments.
- Fungi which are recovered from marine environments through culturing or metagenomic methods, but whose obligate, or facultative marine nature is not certain are called **"marine-derived fungi."**

**Fungi may be detected in marine substrates by two methods** (Fig. 2.6).

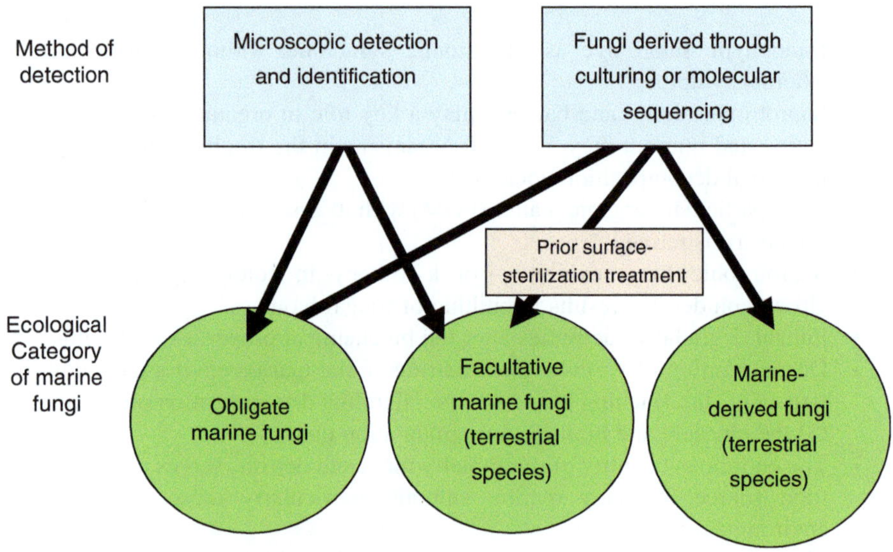


**Fig. 2.6** Methods of detecting different groups of marine fungi

1. **Direct detection of obligate and facultative marine fungi**: Marine substrates are examined microscopically to detect sporulating structures of fungi, followed by identification and isolation in culture. This method unequivocally establishes whether the fungus actively grows in a substrate and whether it is an exclusively marine species or a terrestrial species. However, many fungi may be present within a substrate but may not produce sporulating structures. These may not be detected by this method.
2. **Culturing and metagenomics**: Fungi in marine substrates may be detected by culturing them or by retrieving their molecular sequences. These methods help in discovering the following groups of fungi.

   - **Cultures or sequences may belong to known, or hitherto undescribed obligate marine fungi**.
   - **Facultative marine fungi may be obtained if the substrates are surface sterilized prior to obtaining cultures or sequences.** Terrestrial species of fungi obtained after surface sterilization confirm that they are growing actively in the form of hyphae within the substrate.
   - **"Marine-derived fungi"** are those that are detected by culturing or by detection of their sequences and whose obligate or facultative marine nature is not confirmed.

**Using the above methods, more than 700 species of obligate marine fungi and nearly 550 species of facultative and marine-derived fungi have been described so far** (Jones et al. 2015; Table 2.2). Thus, only a small fraction of nearly 1.5% of fungi from a total of nearly 80,000 species that have been described have been reported till now from marine habitats.

**Table 2.2** Diversity of various groups of marine fungi (Adapted from data by Jones et al. 2015)

| Kingdom | Group | Obligate marine species | Facultative/Marine-derived species |
|---|---|---|---|
| Mycetae | Microsporidia | 6 | – |
| | Cryptomycota | 12 | – |
| | Chytridiomycota | 25 | – |
| | Ascomycota | 531 | 447 |
| | Basidiomycota | 12 | 62 |
| | Asexual fungi of unknown affinities | | 42 |
| Straminipila | Hyphochytriomycetes | 7 | – |
| | Oomycetes | 73 | – |
| | Labyrinthulomycetes | 50 | – |
| Total | | 716 | 551 |

**About 130 species of straminipilan fungi, all of which are obligately marine, are known from the marine environment.**

### *2.5.1 Kingdom Mycetae*

**Members of the Ascomycota are the most frequent among marine mycetaen fungi.** Approximately, 550 obligate marine species and 450 species of facultative or marine-derived ascomycetes have been reported. Nearly 430 of the obligately marine species occur in association with decomposing lignocellulosic materials, such as decaying wood and litter. About a hundred species grow on algae as parasites, saprobes, commensals, and mutualists. Many are associated with animal shells and coral rocks.

**Most of the obligate marine mycetaen fungi belong to Class Sordariomycetes and Class Dothideomycetes of the Ascomycota.**

- Sordariomycetes are characterized by ascocarps which are perithecia or cleistothecia. Asci are single-walled or unitunicate. Asci are arranged basally or peripherally within the ascocarps. Asexual states or anamorphs are produced by many members.
- Dothideomycetes produce perithecia. Asci are double-walled or bitunicate. The inner wall expands and projects out during spore liberation.

**Facultative marine fungi and marine-derived fungi are common in the oceans.** Most of these belong to asexual states of Ascomycota. These occur on lignocellulosic materials as animal and algal parasites and also in the deep-sea sediments. Members of Eurotiomycetes, comprising mostly anamorphic species of the terrestrial genera *Aspergillus* and *Penicillium*, are the most widely represented among these in the sea.

**Basidiomycota are poorly represented among obligate marine fungi.** Thus, only 12 are known to be truly marine. Most of these occur on lignocellulosic materials. Many species of terrestrial Basidiomycota such as mushrooms and bracket fungi have large, macroscopic fruiting bodies or basidiocarps. On the contrary, those of marine species have small basidiocarps, presumably as an adaptation to their aquatic habitat.

Yeasts, which are single-celled, belong to both Ascomycota and Basidiomycota. **More than 200 yeasts have been reported from marine environments** (Fell 2012; Jones et al. 2015). Of these, about 140 are ascomycetes and 70 are basidiomycetes. Most yeasts from marine habitats belong to terrestrial species. Sixteen species of yeasts belonging to Saccharomycetes of the Ascomycota may be truly marine. The ascomycete genus *Metschnikowia* is frequent in oceanic waters and also as parasites of marine animals. Species belonging to *Malassezia* (Exobasidiomycetes, Phylum Ustilaginomycotina of Basidiomycota) are also common in the marine environments, including the deep-sea water column and sediment samples. Marine yeasts may be associated with high nutrient concentrations as occurs in polluted waters, plankton blooms, etc.

**Chytridiomycota are poorly represented in the oceans** and only about 24 species, most of which are probably obligately marine, have been reported so far (Gleason et al. 2012a). However, analyses of environmental samples using molecular techniques have revealed unknown clades of zoosporic true fungi along with other groups of heterotrophic eukaryotic microorganisms in many diverse marine ecosystems.

**Members of Cryptomycota may be important in oceanic systems.** Metagenomic studies have identified several novel groups within Cryptomycota from a variety of marine habitats (Richards et al. 2012; Jebraj et al. 2012).

Six species of Microsporidia parasitic in marine animals and one species of aphelids parasitic in a marine diatom have been described so far (Stentiford et al. 2013; Karpov et al. 2014).

**Obligate marine species are not yet known from many taxonomic groups of fungi belonging to Kingdom Mycetae. Only a few obligately marine mycetaen fungi have been reported from pelagic and benthic habitats of the sea.**

Most of what we know about the diversity of marine mycetaen fungi is based on either direct observation of sporulating structures of fungi on various substrates or culturing of fungi from different habitats. Metagenomic studies based on environmental clone library sequencing has in recent years provided strong evidence that that members of Ascomycota and Basidiomycota are much more diverse and prevalent in oceanic habitats than believed so far.

### 2.5.2 *Kingdom Straminipila*

**About 130 species of marine straminipilan fungi have been reported so far. Oomycetes and Labyrinthulomycetes are important straminipilan fungi in the**

**oceans** (Beakes et al. 2014; Nakai and Naganuma 2015; Laby Base: http://syst.bio.konan-u.ac.jp/labybase/index_en.html).

- Oomycetes are widespread in terrestrial and freshwater environments as saprobes and plant parasites. Nearly 72 species of marine Oomycetes are known, most of which are animal and algal parasites. Only about 7 species of marine Hyphochytriomycetes are known (Marano et al. 2012). Members of the Labyrinthulomycetes are the most widespread and ubiquitous among straminipilan fungi in oceans. A total of 13 genera are known, with approximately 30 described species. Among the three groups of Labyrinthulomycetes, the thraustochytrids appear to be most diverse, comprising 11 genera and 20 species. Aplanochytrids and labyrinthulids are the other two groups of Labyrinthulomycetes. Members of labyrinthulomycetes occur mostly as saprobes on dead organic matter and a few are animal parasites. They are found almost in every marine habitat and may play a major role in marine ecosystem dynamics (Raghukumar 2002; Bongiorni 2012). **Thraustochytrids, belonging to Labyrinthulomycota, are important sources of omega-3 polyunsaturated fatty acids (ω-3 PUFAs), particularly docosahexaenoic acid (DHA), which are required in the diet of many marine animals.**

**Our knowledge of fungal diversity in the marine ecosystem is still very rudimentary.** It is expected that many more marine fungi will be discovered in future.

- According to popular estimates, about 1,500,000 species of mycetaen fungi inhabit earth.
- The total number of marine fungi has been estimated at about 10,000 species.
- There are two major ecological categories of marine fungi, the "Obligate marine fungi" and the "Facultative marine fungi." "Marine-derived fungi" are those whose obligate or facultative nature is not certain.
- Fungi may be detected in marine substrates by direct (microscopic detection) and indirect (culturing and metagenomic) methods.
- Using the above methods, more than 700 species of obligate marine fungi and nearly 550 species of facultative and marine-derived fungi have been described so far.
- About 130 species of stramenipilan fungi, all of which are obligately marine are known from the marine environment.
- Members of the Ascomycota are the most frequent among marine mycetaen fungi. Approximately 550 obligate marine species and 450 species of facultative or marine-derived ascomycetes have been reported. Most of the obligate marine mycetaen fungi belong to Class Sordariomycetes and Class Dothideomycetes of the Ascomycota.

(continued)

Facultative marine fungi and marine-derived fungi are common in the oceans. Basidiomycota are poorly represented among obligate marine fungi. More than 200 yeasts have been reported from marine environments Chytridiomycota are poorly represented in the oceans

- Members of Cryptomycota may be important in oceanic systems.
- Obligate marine species are not yet known from many taxonomic groups of fungi belonging to Kingdom Mycetae. Only a few obligately marine mycetaen fungi have been reported from pelagic and benthic habitats of the sea.
- About 130 species of marine straminipilan fungi have been reported so far. Oomycetes and Labyrinthulomycetes are important straminipilan fungi in the oceans
- Thraustochytrids, belonging to Labyrinthulomycota, are important sources of omega-3 polyunsaturated fatty acids (ω-3 PUFAs), particularly docosahexaenoic acid (DHA), which are required in the diet of many marine animals.
- Our knowledge of fungal diversity in the marine ecosystem is still very rudimentary.

## Future Directions

- Existing culture methods may be inadequate to culture many novel fungi. Novel methods need to be developed to isolate marine fungi from various marine habitats and to understand their role in ecosystem processes.
- Presently used methods of taxonomic identification do not reveal "cryptic species" which may resemble known ones, but which may actually be novel. Further developments in molecular phylogeny and taxonomy may reveal many cryptic species of fungi belong to novel taxa of fungi.
- Many geographical areas have not yet been studied for marine fungi. Future studies in such areas will help provide a more complete picture of the diversity and role of marine fungi.
- Many fungi may have specific substrate and host associations. Studies on benthic fungi and those associated with different plants and animals will help to better elucidate the role of marine fungi.
- Metagenomic studies have revealed the presence of many fungi that have not been cultured and whose role has not been understood. Recent analyses of existing unique fungal SSU rDNA sequence data have suggested that filamentous fungi within Ascomycota and Basidiomycota are more diverse in marine habitats than previously believed. Marine environments may host a significant number of highly novel groups closer to Chytridiomycota. This aspect needs further elucidation and clarification.

# Chapter 3
# History of Marine Mycology

*If I have seen further than others, it is by standing upon the shoulders of giants*

Isaac Newton

Marine mycology has progressed through a series of milestone discoveries by a galaxy of mycologists over the last 160 years.

The first report of a marine fungus was in **1849** by **J.B.H.J Desmaziéres** from France, who reported the discovery of the ascomycete *Phaeosphaeria typharum* from the marshy plant *Typha*. *Phaeosphaeria typharum* is a facultative marine fungus.

Subsequently, in **1856**, **C. Durieu de Maisonneueve and J.F.C. Montagne**, also from France, described the ascomycete *Halotthia posidoniae* in the rhizomes of the seagrass, *Posidonia oceanica*.

The period **1867–1916** witnessed several publications by the **Crouan brothers P.L. Crouan and H.M. Crouan in France, N. Patouillard, M. Reed, and A.D. Cotton. G.K. Sutherland** working from Southampton, southern England, discovered several marine fungi inhabiting cast and living seaweeds. He also described the ascomycete genus *Lulworthia*, which is now known to be ubiquitous in lignocellulosic substrates.

Marine yeasts were reported for the first time in **1894** by **B. Fischer and C. Brebeck**, who described these fungi as a component of North Sea microbial communities. **B.L. Isachenko** in Russia isolated yeasts from seawater from the Barents Sea in **1914**. In **1921–1922, G.A. Nadson** first isolated yeasts from the surface of marine algae and demonstrated that the abundance of yeasts on algae was significantly higher than that in the seawater. The most intensive studies of marine yeasts occurred in the **1940–1970**s, after it became clear that yeasts were an essential component of the seawater microbial community. **D. Ahearn, F.J. Roth, and J.W. Fell** in the USA and **N. van Uden** in Portugal were pioneers in this field. J.W. Fell continues to contribute significantly to marine yeasts.

In **1936**, **Frederick K. Sparrow Jr**., famous for his work on zoosporic fungi, discovered an unusual fungus which had a chytrid-like thallus but produced biflagellate zoospores like in oomycetes. He named this *Thraustochytrium proliferum*.



S. Raghukumar, *Fungi in Coastal and Oceanic Marine Ecosystems*,
DOI 10.1007/978-3-319-54304-8_3

This led to discovery of a number of "thraustochytrids" in later years and laid the foundation for straminipilan fungi belonging to Labyrinthulomycetes (Fig. 3.1).

A significant milestone in the history of marine mycology was the publication by **E.S. Barghoorn and D.H. Linder** in **1944**, who reported a number of indigenous marine fungi that grew on wood immersed in the sea. This generated tremendous excitement among marine mycologists.

**The 1950s and 1960s** witnessed the branching of marine mycology into different disciplines. **I.M. Wilson** in UK **and Samuel P. Meyers** from the University of Miami, USA, continued from where Barghoorn and Linder had left off and made

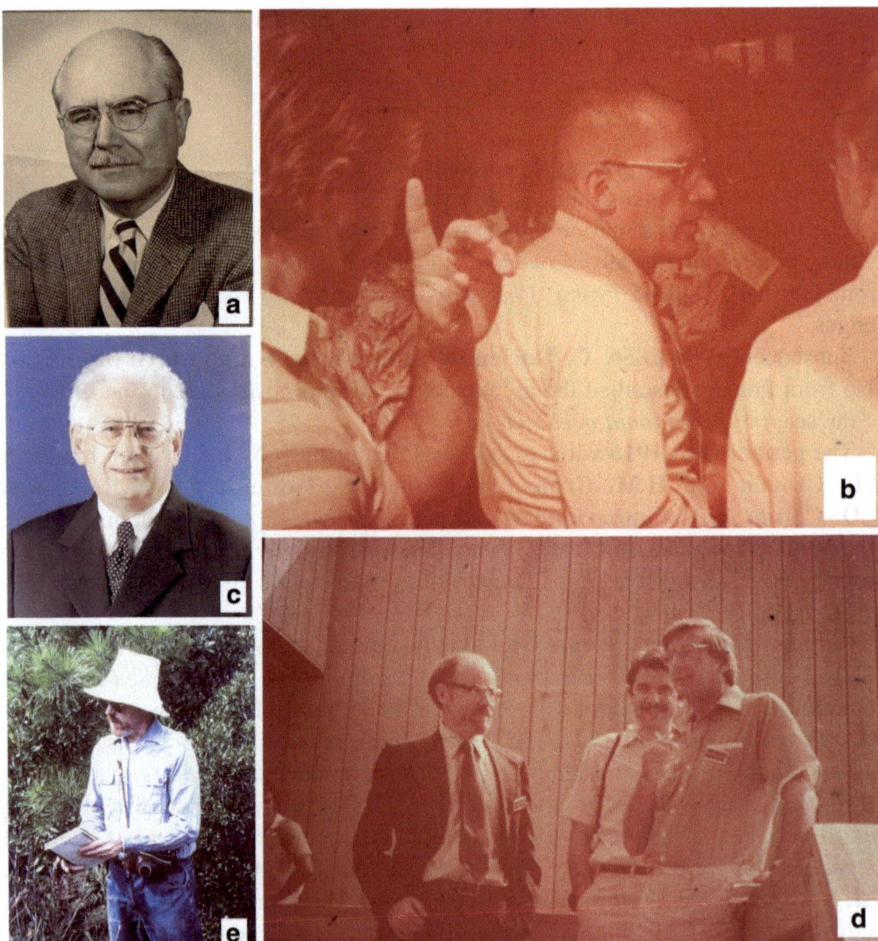


**Fig. 3.1** (**a**) Prof F.K. Sparrow*, Jr. (**b**) Prof. T.W. Johnson* Jr. (**c**) Prof. E.B. Gareth Jones. (**d**) Prof. Jan Kohlmeyer, Dr Steve Newell and Prof. E.B. Gareth Jones at the International Marine Mycology Symposium 1979 held at Tampa, Florida. (**e**) Prof. Jan Kohlmeyer in the field. (*—No more)

some pioneering contributions to the biology of marine lignicolous fungi. S.P. Meyers first brought the attention of marine mycologists to fungi in seagrasses. **T.W. Johnson** (USA), **G. Feldman** and **G. Doguet** (France), and **Willy Höhnk** (Germany) began their contributions to marine lignicolous fungi in the 1950s and 1960s. **A.B. Cribb and J.W. Cribb** (Australia) extended these studies to fungi in mangroves in Australia and described a number of fungi that were specific to mangrove wood, the "manglicolous fungi." Subsequently, **Jan Kohlmeyer** carried out extensive studies on manglicolous fungi from a number of geographical areas. The mantle of I.M. Wilson was passed on to **E.B. Gareth Jones in the 1960s**, who began his work on marine lignicolous fungi from the University of Portsmouth, UK (Fig. 3.1).

The 1950s and 1960s also saw growth in the areas of zoosporic fungi, including chytrids, oomycetes, and thraustochytrids. **Helen S. Vishniac** and subsequently **Solomon Goldstein** from USA carried out pioneering studies on the taxonomy and physiology of thraustochytrids and labyrinthulids in the early **1950s and 1960s**. The first authoritative and comprehensive book on marine fungi was published in the year **1961** by **T.W. Johnson and F.K. Sparrow, Jr.**, titled "Fungi in oceans and estuaries" (Johnson and Sparrow 1961). This book is a classic. The book presented comprehensive information on taxonomy, distribution, physiology, and ecology of marine fungi available at that time. In Japan, **Keisuke Tubaki** was the first to document marine fungi on wood and went on to nurture **Akira Nakagiri**. **R.C. Cavaliere**, a student of T.W. Johnson, contributed to our knowledge of the ascomatal development and structure of marine ascomycetes, with a list of fungi collected in Iceland and on fungi in sea shells.

**Alwin Gaertner** from the Institut für Meereforschung, Bremerhaven, Germany, where W. Höhnk had established research on marine fungi, began his pioneering contributions to the taxonomy, life cycle, and ecology of thraustochytrids in the **1960s** (Fig. 3.2). He brought to light the common occurrence of these fungi in coastal and oceanic waters and sediments and optimized the pine pollen baiting technique.

The growing interest in marine fungi led to the organization of the first Marine Mycology Symposium at Bremerhaven, Germany, by W. Höhnk.

**Jan Kohlmeyer** began his comprehensive studies on marine lignicolous and manglicolous fungi starting from the late **1950s.** His extensive field collections resulted in the discovery of a number of marine lignicolous fungi and led to numerous publications and synoptic plates describing new lignicolous fungi.

A flurry of developments took place in the **1970s**. Detailed studies on marine lignicolous fungi continued from many parts of the world. **C.A. Shearer** from the USA began her contributions on marine lignicolous ascomycetes in the 1970s and continues to do so till date. The second International Marine Mycology Symposium was organized once again in Bremerhaven, Germany, by Alwin Gaertner in 1973. **Frank O. Perkins** from Virginia Institute of Marine Science, USA, and **David Porter** from the University of Georgia began detailed ultrastructural studies on Labyrinthulomycetes (Fig. 3.2). E.B. Gareth Jones, as well as his students and colleagues, namely **Graham Bremer, D.J. Alderman,** and **D.J. Miller,** began to

**Fig. 3.2** (**a**) Dr. Alwin Gaertner*, Dr Annemarie Ulken*, Dr David Porter and Dr Steve Moss* at the International Marine Mycology Symposium held at Tampa, Florida, 1979. (**b**) Dr Alwin Gaertner* in the field. (**c**) Dr David Porter, Dr Chandralata Raghukumar, Dr Karsten Schaumann* and Dr Steve Moss* at the International Marine Mycology Symposium held at Vancouver, Canada, 1994. (**d**) Dr Kevin D. Hyde (**e**) Dr Jack W. Fell. (**f**) Prof. K.R. Sridhar. (**g**) Dr Daiske Honda. (*—No more)

contribute strongly to marine mycology from the laboratory of E.B. Gareth Jones. **Stephen Moss** from there made excellent contributions to the ultrastructure of Labyrinthulomycetes, as well as marine lignicolous fungi (Fig. 3.2). **David Jennings** from England contributed to our understanding of many aspects of fungal physiology.

**The early 1970s** was also the dawn of fungal studies in salt marsh grasses, where they were subsequently shown to play a crucial role in the ecosystem by **R.D. Goos** and **R.V. Gessner** from the University of Rhode Island, USA. This complemented the taxonomic studies of salt marsh grass fungi by Jan Kohlmeyer. **Steve Newell** from the University of Georgia together with **Jack Fell** at the University of Miami, USA, began his studies on marine fungi in the 1970s and subsequently established their important role in decomposition of standing, dead salt marsh grass. This complemented **Jan Kohlmeyer**'s studies on fungi in salt marsh grass. **G.C. Hughes** from Canada contributed to the biogeography of marine, lignicolous fungi. **E.B. Gareth Jones** continued to make important contributions to the taxonomy and physiology of marine lignicolous fungi. Together with **Stephen T. Moss**, he published several papers on electron microscopy of ascospores of these fungi, a work which continued into the 1980s. E.B. Gareth Jones has continued his valuable contribution to the taxonomy of obligate marine fungi to this day and has published several important and comprehensive reviews on the subject.

Many of the important findings regarding marine fungi were compiled in a book in **1976** edited by E.B. Gareth Jones (Jones 1976) titled "Recent advances in aquatic mycology." This was soon followed by the classic book "Marine mycology: The Higher Fungi" by **Jan Kohlmeyer and Erika Kohlmeyer** in **1979** (Kohlmeyer and Kohlmeyer 1979). This book brought together detailed information on the taxonomy, diversity, ecology, and biology of "higher fungi," namely mycetaen fungi in a highly useful manner. Jan Kohlmeyer's subsequent work led to intensive research on salt marsh fungi, fungi in mangroves, coral reefs, and the deep sea. These publications continue to be highly relevant today. Another valuable book that was published in **1986** is "Biology of Marine Fungi" edited by **S.T. Moss.**

The **1980s and 1990s** were periods of consolidation of marine mycology in a number of areas. This was a period when the taxonomy and phylogeny of marine fungi were refined, and an enormous amount of information on the ecology and biotechnology of marine fungi were studied. **Seshagiri Raghukumar** and **Chandralata Raghukumar**, coming from the school of Prof. C.V. Subramanian in India and later groomed by Alwin Gaertner in Germany, began to contribute to marine fungi from the National Institute of Oceanography, India. In addition to taxonomy and ecology of filamentous fungi and thraustochytrids, algal diseases, and biotechnology, they extended the boundaries of marine mycology to the oceans. The Hong Kong University became a hub of studies on marine fungi. **Kevin Hyde**, from the school of **E.B. Gareth Jones,** who started with marine fungi from Seychelles began contributing to mangrove fungi in the 1980s (Fig. 3.2). He established the Hong Kong University as a center of marine mycology studies, which he continues to pursue now in Thailand. **Lillian Vrijmoed** and **S.B. Pointing** from Hong Kong contributed to ecology, diversity, and physiology of marine fungi.

**Kishio Hatai** from Japan began working extensively on oomycetan parasites of marine animals during this period.

The **1990s** saw the commercialization of the omega-3 polyunsaturated fatty acid production from thraustochytrids by **William Barclay** of OmegaTech, USA. A number of researchers, particularly from Martek Biosciences in the USA (now part of DSM) and Japan, have been contributing to this field. **Daiske Honda**, Japan, began detailed studies on the taxonomy and phylogeny of thraustochytrids and redefined the species and generic circumscriptions of the group. Honda has established an excellent center dealing with these fungi at Konan University, Japan (Fig. 3.2).

**The twenty first century** has seen the blossoming of interest in marine fungi in extreme environments and as a source of novel drugs, a trend that started in the 1990s. Increasing interest in the diversity of marine fungi led to the search of a variety of several marine habitats, particularly the deep sea and hydrothermal vents, resulting in the discovery of a number of novel fungal phylotypes in the sea. **Chandralata Raghukumar** and her colleagues in India began their studies on deep-sea fungi. **Takahiko Nagahama** and others from Japan have been discovering several novel and interesting yeasts from deep-sea environments. **Nina Gunde-Cimerman** from University of Ljubljana, Slovenia, has been carrying out research on fungi in hypersaline environments. **Frithjof C. Kuepper** from the University of Aberdeen has been studying fungal parasites of algae. Natural product chemists, such as **Rainer Ebel** from the University of Aberdeen, Scotland, are contributing enormously to the discovery of novel secondary metabolites from marine fungi. Progress in methods in metagenomics has encouraged a number of marine biologists, microbiologists, and mycologists to study fungal diversity. This is leading to a greater realization of the presence and role of fungi in a number of exotic marine habitats such as the deep sea. Many of the recent developments have been compiled in "Marine Mycology—A Practical Approach" edited by Kevin D. Hyde and Stephen B. Pointing (2000), "Biology of Marine Fungi" edited by Chandralata Raghukumar (2012), and "Marine Fungi and Fungal-like organisms" edited by E.B. Gareth Jones and Ka-Lai Pang (2012).

Several strong centers of marine mycology have now emerged. Some of these are Hawaii, USA (Guang-Yi Wang), Taiwan (Ka-Lai Pang), Malaysia (S.A. Alias), Thailand (Kevin Hyde, J. Sakayaroj, S. Suetrong), India (K.R. Sridhar, S. Raghukumar, C. Raghukumar, Samir Damare, Cathrine Manohar, B.D. Borse, V.V. Sarma), Mexico (M. González), Saudi Arabia (M. A. Abdel-Wahab ), Portugal (M. Barata), Philippines (E. Leaño, T. E. dela Cruz ), China (Song, J. Jin), Japan (Daiske Honda), and Thailand (Kevin Hyde).

The future of marine mycology appears exciting.

# Chapter 4
# Allochthonous Wood in Coastal Waters

*The world depends on fungi, because they are major players in the cycling of materials and energy around the world.*
E. O. Wilson

A major milestone in marine mycology was the publication by E.S. Barghoorn and D.H. Linder in 1944, describing an exciting, special group of fungi that grow on wood submerged in the sea (Barghoorn and Linder 1944). Subsequently, a number of mycologists contributed enormously to our understanding of the taxonomy, distribution, and physiology of these "marine lignicolous fungi." These include the eminent marine mycologists A.B. Cribb and J.W. Cribb, Samuel P. Meyers, Jan Kohlmeyer, E.B. Gareth Jones, and Kevin Hyde, as well as a number of others. Important literature on these fungi can be found in Kohlmeyer and Kohlmeyer (1979), Kohlmeyer and Volkmann-Kohlmeyer (1991), Kohlmeyer (1984), Hyde and Sarma (2000), and Jones et al. (2009).

Wood is an important substrate for marine fungi in the sea. **Enormous amounts of allochthonous wood from land plants constantly enter the marine environment in various ways** (Fig. 4.1).

- Parts of dead trees may drift off into estuaries and coastal seas as **driftwood** (Fig. 4.1a, b). These may be deposited in intertidal beaches, float in the water, or eventually become submerged in the sea. Wood under late decompositional stages can be recognized by the presence of tunnels made by wood borers (Fig. 4.1b, c).
- **Man-made constructions** introduce wood into the marine environment in the form of jetties, pilings, and wooden boats (Fig. 4.4d, e).
- Lignocellulosic **cordage** such as the manila cordage made of hemp or those made of coconut coirs are often used to tether boats and other objects.

Occasionally, catastrophic events may introduce large amounts of wood into the marine environment. A typhoon nicknamed "Morakot" disastrously affected



S. Raghukumar, *Fungi in Coastal and Oceanic Marine Ecosystems*,
DOI 10.1007/978-3-319-54304-8_4

**Fig. 4.1** Woody lignocellulosic material in the sea are substrates for marine lignicolous fungi. (**a**) An uprooted tree deposited in the sea. (S. Raghukumar). (**b**) A piece of driftwood (S. Raghukumar). (**c**) A piece of allochthonous wood in the sea riddled with borer tunnels (Courtesy: Scott Frazier). (**d**) Wooden stakes in the intertidal for tying nets (S. Raghukumar). (**e**) Wooden pilings attacked by wood borer and microbial attack (Courtesy: Tom Reis, Substructure, Inc.)

Taiwan during 2009, as a result of which nearly $8.4 \times 10^9$ kg of woody material was carried into the adjacent oceans off Asia (West et al. 2011).

In addition to allochthonous wood, coastal mangrove trees contribute autochthonous woody material to the marine environment. These are discussed later (see Chap. 5). The present chapter is confined to allochthonous woody material.

**More than 80% of wood is comprised of lignocellulose, which is a complex of cellulose, hemicellulose, and lignin** (Fig. 4.2). **Lignocellullose constitutes the most important structural matter of vascular plants** (Motta et al. 2011; Sajith et al. 2016)**.**

- Cellulose is a high molecular weight, homopolymer polysaccharide consisting of a linear chain of hundreds to thousands of (1,4)-D-glucopyranose units joined by β-1,4 linkages.

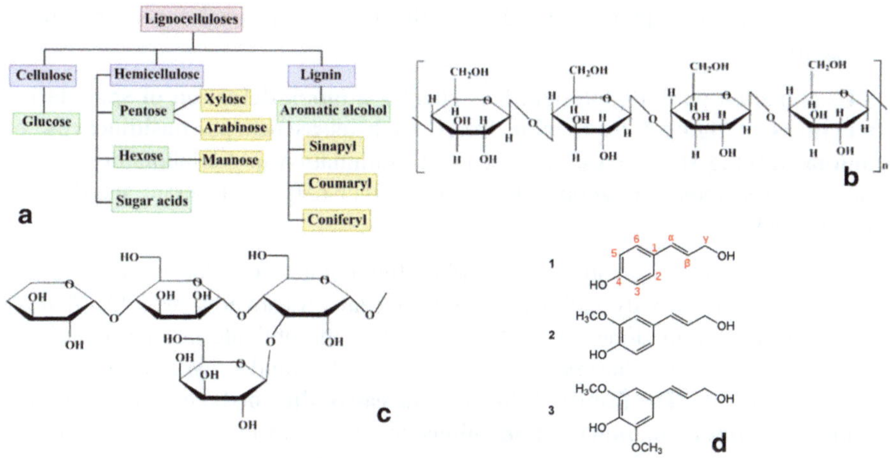


**Fig. 4.2** The lignocelluloses complex. (**a**) General composition of lignocelluloses. (**b**) Structure of cellulose. (**c**) Structure of hemicelluloses. (**d**) Monolignols, the common components of lignin, paracoumaryl alcohol (1), coniferyl alcohol (2) and sinapyl alcohol (3) (Source: **a–c** Sajith S, et al. (2016) J Nutr Food Sci 6:461. **d** Wikipedia Commons)

- Hemicelluloses are highly branched heterogeneous polymers composed of sugar acids; the pentoses arabinose and xylose; and the hexoses glucose, galactose, and mannose. Of these, xylans are the most abundant and are categorized as linear homoxylan, arabinoxylan, glucuronoxylan, or glucuronoarabinoxylan based on the common substituents found on the backbone.
- Lignin is an amorphous, complex molecule made up of phenylpropane units such as coumaryl alcohol, coniferyl alcohol, and sinapyl alcohol. **Lignocellulose, particularly lignin, is highly resistant to microbial decomposition and has little nutritive value for animals.** Many terrestrial, mycetaen fungi, particularly the white-rot fungi belonging to Basidiomycota, are extremely efficient lignocclluloses degraders.

**All lignicolous fungi described so far are mycetaen fungi,** which are the most important colonizers and degraders of wood submerged in the sea. Therefore, the term "lignicolous fungi" in this book refers exclusively to such fungi.

## 4.1 Diversity of Fungi on Allochthonous Wood in the Sea

One may walk along a beach looking out for driftwood and collect these to study lignicolous marine fungi. An easy and reliable way to study them is to submerge test panels of wood in the sea, recover them periodically and examine them. A number of mycologists used the latter technique extensively in the immediate "post-Barghoorn and Linder era." Most of our information on diversity and colonization of allochthonous wood by marine lignicolous fungi comes from these two sources.

Examination with a good field lens enables **a quick detection of sporulating structures of lignicolous marine fungi, such as ascocarps, basidiocarps, or conidia-bearing structures.** A subsequent examination in the laboratory using a good set of stereoscopic zoom microscope further helps in accurate and detailed observations.

- About 190 species of obligately marine, lignicolous fungi have been found in allochthonous woody substrates (http://ocean.otr.usm.edu/~w529014/index_files/Page319.htm; Jones et al. 2009). Most of the obligate marine lignicolous fungi belong to the Classes Sordariomycetes and Dothideomycetes (Sect. 2.4.1) of Ascomycota. **The family Halosphaeriaceae of the Sordariomycetes has the largest number of marine lignicolous species**, with more than 50 genera in 130 species. Only 12 members of marine Basidiomycota are known from wood. Many species of ascomycetes and basidiomycetes exist only in their asexual or anamorphic states.
- Culture-dependent and culture-independent metagenomic studies have shown that the diversity of fungi in allochthonous wood in the sea is much higher than that what is now known. For example, 577 cultures were obtained from North Atlantic driftwood samples. Out of 125 OTUs (Opertional Taxonomic Unit) that were detected, 67 OTUs corresponded to marine species and 58 corresponded to terrestrial species. Nine could not be assigned, being probably novel (Rämä et al. 2014). Metagenomic studies on these drifrwood samples yielded 807 fungal OTUs, out of which 74% were ascomycetes and 20% were basidiomycetes. Most were marine-derived fungi, belonging to terrestrial species. At the same time, one-fourth of these appeared to be novel, undescribed species (Rämä et al. 2016).

**Spores of marine lignicolous fungi possess several structural adaptations that are useful for an aquatic life.**

- Among terrestrial species of ascomycetes, the asci mostly eject their ascospores forcibly into air. On the contrary, **asci in the family Halosphaeriaceae disintegrate at maturity, releasing the ascospores into the cavity of the perithecium, the ascocarp**. The spores then pass out through the opening of the perithecium, the ostiole into the aqueous environment. Lignicolous marine fungi belonging to the Dothideomycetes mostly inhabit wood that is present in the intertidal region and wetted only periodically. Asci of Dothideomycetes possess double-walled "bitunicate" asci with a special mode of forcibly ejecting ascospores out.
- **Spores of obligate marine lignicolous fungi show special adaptations that enable them to float in water and attach to substrates. Ascospores of most marine ascomycetes possess gelatinous, hair-like, or cap-like sticky appendages** (Hyde and Jones 1989). Some of the different forms of appendage forms are

as follows. (1) Release of a drop of mucilage from polar end chambers of the spores; (2) hamate or cap-like appendages which uncoil to form viscous threads; (3) spores with a sticky mucilaginous sheath; (4) disc or pad-like attachment; (5) sticky vermiculate appendages surrounding the spore; (6) ribbon-like appendages; (7) tufts of fibrillar appendages; (8) irregular amorphous appendages with or without a fibrillar component, and (9) adhesive spore wall. Basidiomycetes or anamorphic fungi often possess branched or tetraradiate spores that increase their surface areas and enable them to float. Examples of some of these are found in Fig. 4.3.

The morphology of spores and appendages are key features for taxonomic identification of obligately marine, lignicolous fungi (Hyde and Sarma 2000; Jones et al. 2009).

**A number of factors determine the diversity of lignicolous marine fungi** (Jones 2000).

- **The type of wood strongly influences the diversity of obligate marine fungi.**
  - For example, *Halosphaeria appendiculata* Linder, *Nautosphaeria cristaminuta* E.B.G. Jones, *Halosarpheia hamata, Marinospora calyptrata* (Kohlm.) A.R. Caval., and *M. longissima* (Kohlm.) A.R. Caval. have been shown to prefer *Fagus sylvatica* (beechwood), while *Lautisporopsis circumvestita* (Kohlm.) E.B.G. Jones, Yusoff & S.T. Moss and *Cirrenalia macrocephala* (Kohlm.) Meyers & R.T. preferred *Pinus sylvestris* (pinewood) (Jones 2000).
  - In Hong Kong waters, the fungi *Ceriosporopsis halima* Linder and *Cirrenalia macrocephala* did not colonize teak, but grew on pine. *Cytospora* sp. did not grow on pine, but did so on teak (Vrijmoed et al. 1986).
  - A metagenomic study of fungi in driftwood of the Northern Atlantic showed that fungal communities in conifer logs were different, compared to the deciduous ones (Rämä et al. 2016).
- **"Arenicolous fungi" live among or on sand particles in intertidal beaches, while deriving their nutrition from particulate organic matter, such as decomposing wood** (Kohlmeyer and Kohlmeyer 1979). They produce thick-walled, hard, and carbonaceous ascocarps firmly attached to the sand grains. Examples are species of *Corollospora* and *Carbosphaerella*. Spores of such sand-inhabiting, arenicolous fungi often are trapped in sea foam. Microscopic examination of sea foam is a convenient and exciting way of discovering spores of marine lignicolous fungi. Such a method is also adapted by mycologists who study fresh water aquatic or Ingoldian fungi.
- **Lignicolous marine fungi display a distinct geographical distribution in tropics, temperate and polar.** Thus, temperature plays a major role in determining their diversity. Several species are restricted to particular temperature

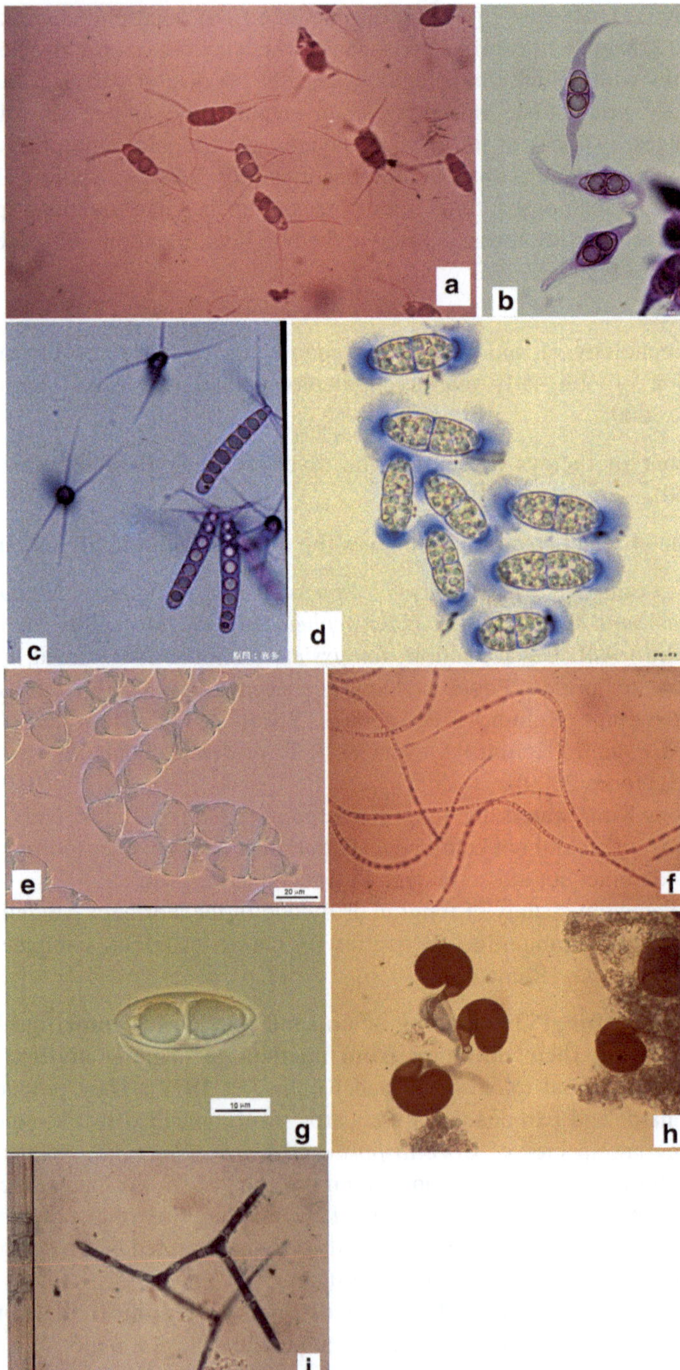


**Fig. 4.3** (**a–g**) Ascospores of marine lignicolous fungi with various types of sticky appendages. (**a**) *Antennospora quadricornuta;* (**b**) *Ceriosporopsis halima;* (**c**) *Torpedospora radiata;* (**d**) *Remispora gallerita.* (**e**) *Saagaromyces abonnis.* (**f**) *Lulworthia sp.;* (**g**) *Aniptodera*

regimes, while some are cosmopolitan (Table 4.1). Culturable fungi in driftwood showed that the western and eastern part of the Norwegian Barents Sea coast hosted different communities (Rämä et al. 2014).

## 4.2 Growth and Degradative Activities of Marine Lignicolous Fungi

There are three important aspects of growth of fungi in wood submerged in the sea

- Colonization and growth
- Enzymatic degradation of wood
- Interaction with wood borers

### 4.2.1 Colonization and Growth

Spores of marine lignicolous fungi have a large surface-to-volume ratio and easily float passively in water. **Sticky appendages of floating ascospores help them to attach to a woody substrate immediately upon coming into contact with them, while branched spores become entrapped to rough surfaces of wood.** The spores rapidly germinate to form a germ tube and produce an extracellular mucus which helps to anchor them even more firmly to wood. The germ tube then produces an appressorium, a swollen tip of the hypha, from where the fungus penetrates the woody substrate through a penetration hypha. This process of attachment, germination, and growth within the wood results in colonization of the substrate.

**Fungi colonize wood submerged in the sea in different phases, in the form of a succession of species** (Jones and Hyde 2002). We can experimentally deduce the successional pattern of fungi in wood submerged in the sea by suspending wood panels in the sea, retrieving samples periodically, and examining them for presence of fungal fruiting bodies. Such experiments have provided us an insight into fungal colonization of wood in the sea.

- **Certain fungi may require conditioning of the wood in the sea before they can pervade and successfully colonize them** (Jones 2000). Wood immersed in the sea undergoes biotic and abiotic changes that alter its chemistry. Fungal

**Fig. 4.3** (continued) chesapeakensis; (**h**) Conidia of *Hydea pygmaea;* (**i**) Tetraradiate conidia of *Varicosporina ramulosa.* (**a**, **f**, **h** and **i** S.Raghukumar; **b–d** Courtesy "'Marine fungi' by HARC/NARO available at http://togodb.biosciencedbc.jp/togodb/view/marine_fungi under https://creativecommons.org/licenses/bysa/2.1/jp/deed.en". **e** and **g** (Photographs: V.V. Sarma and S. Raghukumar. Marine Lignicolous fungi. A Database on taxonomy. With kind permission of CSIR-National Institute of Oceanography, Goa, India. Under Creative Commons 4.0. http://www.niobioinformatics.in/fungi/index.htm)

**Table 4.1** Geographical distribution of some lignicolous marine fungi

| Tropical | Temperate/Arctic | Cosmopolitan |
|---|---|---|
| *Antennospora quadricornuta* (Cribb & J.W. Cribb) T.W. Johnson | *Asteromyces cruciatus* Moreau & M. Moreau ex. Hennebert | *Arenariomyces trifurcatus* Höhnk |
| *Arenariomyces triseplatus* Kohlm. | *Lautisporopsis circumvestita* (Kohlm.) E.B.G. Jones, Yusoff & S.T. Moss | *Ceriosporopsis halima* Linder |
| *Dactylospora haliotrepha* (Kohlm. & E. Kohlm.) Hafellner | *Ceriosporopsis tubulifera* (Kohlm.) P.W. Kirk ex Kohlm. | *Corollospora maritima* Werderm. |
| *Verruculina enalia* (Kohlm.) Kohlm. & Volkm. Kohlm. | *Cirrenalia macrocephala* (Kohlm.) Meyers & R.T. Moore, | *Corollospora pulchella* Kohlm., I. Schmidt & N.B. Nair |
| *Saagaromyces abonnis* (Kohlm.) K.L. Pang & E.B.G. Jones | *Dendryphiella salina* (G.K. Sutherl.) G.J.F. Pugh & Nicot | *Halosphaeriopsis mediosetigera* (Cribb & J.W. Cribb) T.W. Johnson |
| *Bathyascus tropicalis* Kohlm. | *Digitatispora marina* Doguet | *Lignincola laevis* Höhnk |
| *Cirrenalia tropicalis* Kohlm. | *Halosarpheia trullifera* (Kohlm.) E.B.G. Jones | *Monodictys pelagica* (T.W. Johnson) E.B.G. Jones |
| *Halosarpheia fibrosa* Kohlm. & E. Kohlm. | *Havispora longyearbyenensis* K.L. Pang & Vrijmoed | *Torpedospora radiata* Meyers |
| *H. marina* (Cribb & J.W. Cribb) Kohlm. | *Trichocladium alopallonellum* (Meyers & R.T. Moore) Kohlm. & Volkm.-Kohlm. | *Leptosphaeria australiensis* (Cribb & J.W. Cribb) G.C. Hughes |
| *Lulworthia grandispora* Meyers | *Leptosphaeria pelagica* E.B.G. Jones | |
| *Natantispora retorquens* (Shearer & J.L. Crane) J. Campb. | *Monodictys pelagica* (T.W. Johnson) E.B.G. Jones | |
| *Halocyphina villosa* Kohlm. & E. Kohlm. | *Nais inornata* Kohlm. | |
| *Periconia prolifica* Anastasiou | *Ocostaspora apilongissima* E.B.G. Jones, R.G. Johnson & S.T. Moss | |
| *Saagaromyces ratnagiriensis* (S.D. Patil & Borse) K.L. Pang & E.B.G. Jones | *Ondiniella torquata* (Kohlm.) E.B.G. Jones, R.G. Johnson & S.T. Moss | |
| | *Zalerion maritimum* (Linder) Anastasiou | |

colonization also depends on the type of wood and environmental conditions. "Early colonizers" are able to pervade wood during the initial period. Wood submerged in temperate waters is slow to be colonized by fungi. E.B. Gareth Jones noticed that fungal colonization of wood in British waters started by about 12 weeks. On the contrary, wood submerged in subtropical Hong Kong waters and the tropical waters of Goa, India, were colonized even by 4 weeks (Vrijmoed et al. 1986; Viswakiran et al. 2001). In Hong Kong waters, *Periconia prolifica* colonized both pine and teak wood by 4–5 weeks (Vrijmoed et al. 1986). Other early colonizers in these waters were *Trichocladium achrasporum* and *Lulworthia* sp. *Antennospora quadricornuta* was an early colonizer on pine wood, but a late colonizer on teak. *Monodictys pelagica* showed the opposite trend. With regard to environmental conditions, the frequency of wood colonization by *Antennospora quadricornuta* increased from a low salinity site of 12.5 ppt to that of a higher salinity of 27.5 ppt. Fungal succession depends upon the kind of wood, the time of submergence, environmental conditions, and the geographical location.

- **Succession may often result from a particular species competing and preventing another to colonize** (Jones 2000). For example, it has been observed that *Lulworthia* sp. produced more number of ascocarps on wood when no other fungus was present in it, as judged by incubation. However, fewer ascocarps of *Lulworthia* were found in wood where the fungi *Ceriosporopsis halima* or *Amylocarpus encephaloides* were also present. When test blocks that were colonized by *C. halima*, *Corollospora maritima*, and *Halosphaeriopsis mediosetigera* in the laboratory were immersed in the field, no other fungus was able to colonize them, suggesting the presence of antagonistic metabolites. Study of such antagonistic behavior provides important clues to discover antifungal metabolites. Thus, sporulation of the fungus *Lignicola laevis* on wood is affected by the presence of *Aigialus parvus* and *Verruculina enalia*. Subsequently, *Aigialus parvus* was found to produce a number of bioactive compounds. Another lignicolous marine fungus, *Leptosphaeria oraemaris*, produces an antifungal sesquiterpene called culmorin in the laboratory.
- **Within a single phase of succession, several species may simultaneously colonize wood, but form their reproductive, fruiting bodies at different time intervals.** When a sample of decaying wood from the sea is brought to the laboratory and examined, the presence of some fungi may be recognized immediately upon recovery through the presence of their fruiting bodies. However, upon further incubation, ascocarps, basidiocarps, or conidial bearing structures of other species may appear at different periods. This demonstrates that even when fungi colonize wood at the same time, the time to produce their reproductive structures varies. For example, in one set of observations on driftwood samples from the west coast of India, Prasannarai and Sridhar (2001) observed that *Hypoxylon oceanicum* and *Lophiostoma mangrovei* were present initially, but disappeared upon later incubation. Fruiting bodies of many fungi appeared only after incubation for 6 months. *Corollospora* sp. 1 and *Dactylospora heliotrepha* were found after 12 months, whereas *Corollospora colossa*, *Corollospora* sp. 2, and *Lulworthia* sp. were encountered after 18 months incubation. Besides, frequency of many species, such as *Aniptodera chesapeakensis*, *Crinigera maritima*, *Torpedospora radiata*, and *Calathella mangrovei* increased with extended incubation periods. Up to 21 different fungi colonized mango wood along tropical coastal waters of Goa, India, within 1 month. As the wood aged to 4 months, only a maximum of 7 persisted. The most persistent were the ubiquitous *Antennospora quadricornuta*, *Lulworthia* sp., *Periconia prolifica*, and *Zalerion varium*. It is, therefore, important to incubate wood samples for a long period of even up to a year and examine them periodically to detect fungi.

### 4.2.2 Enzymatic Degradation of Wood

**Wood submerged in the sea is degraded by the combined action of wood borers, fungi, and bacteria, resulting in loss of weight.** It is difficult to quantify the individual roles of each of these in weight loss of wood. However, many controlled studies using wood blocks inoculated with single species of fungi have shown that fungi cause loss of weight in wood owing to their degradative activity. The role of fungi in degradation of wood in the sea has been ascertained by (1) their capability to produce various lignocellulose degrading enzymes (Hyde et al. 1998) and (2) their ability in the laboratory to induce weight loss of wood on which they grow.

**Cellulose, xylan, and lignin components are degraded by the lignocellulase enzyme complex that includes cellulases, xylanases, and lignin-degrading enzymes** (see Chapter 13).

- The cellulase system consists of endocellulase, cellobiohydrolase, and beta-glucosidase. Endocellulase randomly cleaves internal bonds at amorphous sites and creates new chain ends. Cellobiohydrolase cleaves two to four units from the ends of the exposed chains produced by endocellulase and generates tetrasaccharides or disaccharides. Beta-glucosidase hydrolyses the tetra- and disaccharides into individual glucose units. Endocellulase is extremely common among marine lignicolous fungi. Cellobiohydrolase is also present in several lignicolous fungi. Beta-glucosidase has been demonstrated in a few. Xylanases include a large repertoire of enzymes. Lignin-degrading enzymes (LDEs) comprise Lignin peroxidase (LiP), Manganese peroxidase (MNP), versatile peroxidase (VP), and laccase.

**Fungal hyphae penetrate wood through their apical growth. Production of lignocellulases subsequently allows the lignicolous fungi to digest cell walls of wood fibers, penetrate them, and grow into the lumen of the wood cells.** A successional series of this process helps them to permeate the interior of the wood. The bore hole that passes through the cell wall is typically narrow, and the hypha is extremely constricted as it penetrates through the cell wall (Fig. 4.4a).

**Production of cellulases and xylanases is widespread among lignicolous marine fungi** (Table 4.2; Gessner 1980; Leightley 1980; Rohrmann and Molitoris 1992; Raghukumar et al 1994a; Pointing and Hyde 2000; Bucher et al. 2004). **Several lignicolous marine fungi, such as *Corollospora maritima*, *Monodictys pelagica*, *Lignicola laevis*, *Nia vibrissae*, and *Stagonospora* sp., have been shown in the laboratory to utilize and grow on cellulose.** Many fungi also produce oxidative enzymes which help in cellulose utilization.

**Hemicellulolytic enzymes appear to be less frequent among marine lignicolous fungi, compared to cellulolytic and lignin-degrading enzymes.** Many marine fungi can decolorize lignin model compounds, such as the aromatic polymeric dyes PolyR-478 and Azure B (Raghukumar et al 1994a; Pointing et al. 1998). This demonstrates their capability to degrade lignin. The capability to mineralize lignin to $CO_2$ has been demonstrated in several marine fungi (Sutherland et al. 1982).

**Growth of marine fungi in wood results in wood rot.** There are three types of wood rot, namely, "soft rot," "white rot," and "brown rot." Marine fungi cause the "soft rot" and "white rot."

1. **Soft rot**: In this mode of wood rot, caused by most marine lignicolous fungi, the fungi grow within the secondary walls of the cell, forming fine T-shaped branches. Cellulolytic activity results in rhomboid cavities within the secondary wall (Fig. 4.4b). The hypha within these cavities are broader than the penetration hyphae. Alternate stop-and-grow pattern results in a series of rhomboid cavities. Surface softening of the wood is characteristic of soft rot. Ligninolytic activities of soft-rot fungi are generally low, while cellulolytic and hemicellulolytic activities are high.
2. **White rot**: This is found in fungi belonging to basidiomycetes such as *Digitatispora marina*, *Halocyphina villosa*, and *Nia vibrissae*. The fungi produce cellulolytic, hemicellulolytic, and ligninolytic enzymes. The hyphae grow along the walls within the lumen of the vessels and erode the cell walls by enzymatic means. They do not penetrate into the secondary wall of the vessels as with "soft rot." Typically, white rot fungi utilize both lignin and cellulose. White color of the decomposed wood owing to loss of lignin is characteristic of white rot.
3. **Brown rot:** Fungi causing brown rot rapidly utilize cellulose and hemicelluloses, with very little lignin degradation, resulting in dark brown, dry, and cracked wood. This type is rare or absent in the marine environment.

The fact that most marine fungi cause soft rot that requires high cellulase and hemicellulase activity but little or no lignin-degrading activity appears to be in congruence with experimental results on enzyme production (Table 4.2). Experimental work has shown that most lignicolous marine fungi produce lignin-degrading enzymes, but scant hemicellulases. Clearly, we need to understand more regarding the actual patterns of enzyme production and wood decay in the sea.

**Growth of marine lignicolous fungi causes weight loss of the wood over a period of time. Weight loss depends on the kind of wood and temperature. The process is greater when the wood is incubated under moist conditions compared to submerged wood.** Hard wood *Fagus sylvatica* is degraded much faster than *Pinus sylvestris*, which is a softwood. Among various species tested, *Humicola alopallonella*, *Corollospora maritima*, *Savoryella lignicola*, *Lulworthia* sp., and *Nais inornata* have been shown to cause substantial weight loss of nearly 10% or more by 18 weeks. Some ascomycetes caused very high mass loss of up to 20.1% (Bucher et al. 2004). Warmer temperatures cause a greater loss of wood. Warm water and cold water species may prefer different temperature regimes for growth and wood degradation. Two warm-water species, *Nia vibrissa* and *Halocyphina villosa*, caused greater weight loss of *Ochroma lagopus* (balsa) at 22 °C than at 10 °C (Mouzouras 1986). On the contrary, the cold-water marine fungus,

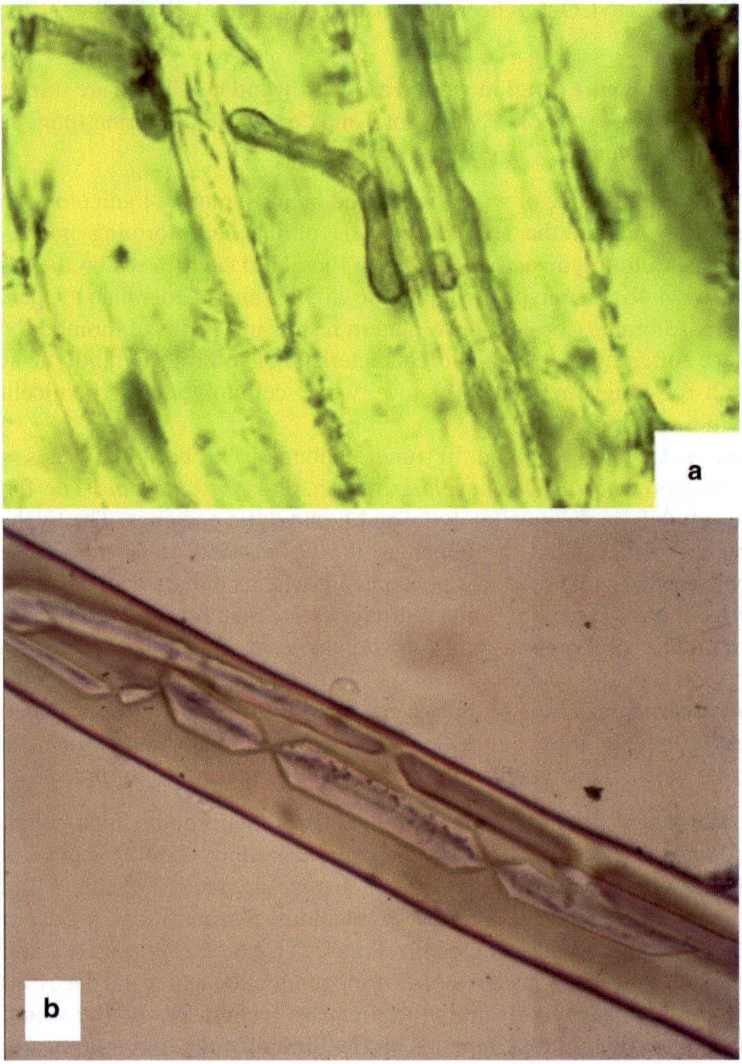

**Fig. 4.4** (**a**) Hyphae of a marine lignicolous fungus grows through the cell wall of wood fibres in the form of narrow penetration hyphae and enters the lumen of the wood cell where it assumes its normal width (Source: Raghukumar (2008). Marine fungal biotechnology: an ecological perspective. Fungal Diversity 31: 19–35. Kind permission of Dr Kevin Hyde, Fungal Diversity). (**b**) Hyphae of a marine fungus within the secondary fibre of wood. Note the rhomboidal cavities with hyphae stained blue

*Digitatispora marina,* caused weight losses of 8.06 and 14.33% after 8 and 24 weeks incubation at 10 °C. When incubated at 22 °C, the same fungus caused only 1.06 and 4.94% weight loss, respectively.

**Table 4.2** Examples of lignocellulolytic enzymes produced by some marine fungi (adapted from Pointing and Hyde 2000)

| Fungus | Cellulase | Hemicellulase | Lignin-degrading enzymes |
|---|---|---|---|
| *Nia vibrissa* Moore *et* Meyers | + | + | + |
| *Amylocarpus encephaloides* Curre | + | – | + |
| *Arenariomyces trifurcata* Höhnk | – | – | + |
| *Ceriosporopsis halima* Linder | – | – | + |
| *Corollospora intermedia* I. Schmid | + | – | + |
| *Corollospora lacera* (Linder in Barghoom *et* Linder) Kohlm | + | – | + |
| *Corollospora maritima* Werderman | + | – | + |
| *Lignincola laevis* Höhnk | + | – | + |
| *Lulworthia lignoarenaria* J. Koch *et* E.B.G. Jones | + | – | + |
| *Lulworthia sp* | + | – | + |
| *Marinospora longissima* (Kohlm.) Cavaliere | – | – | + |
| *Nautosphaeria cristaminuta* Jones | – | – | + |
| *Pleospora pelagica* T.W. Johnson | + | – | + |
| *Remispora stellata* Kohlm. | – | – | + |
| *Savoryella lignicola* Jones *et* Eaton | + | + | + |
| *Cirrenalia macrocephala* (Kohlm.) | + | + | + |
| *Hydea pygmea* (Kohlm.) K.L. Pang & E.B.G. Jones | – | – | + |
| *Cirrenalia tropicalis* Kohlm. | + | – | + |
| *Culcitalna achraspora* Meyers *et* Moore | + | + | – |
| *Dendryphiella salina* (G.K. Sutherl.) | + | – | + |
| *Humicola alopallonella* Meyers *et* Moore | + | – | + |
| *Mondictys pelagica* (Johnson) Jones | + | – | + |
| *Varicosporina ramulosa* Meyers *et* Kohlm. | + | – | + |
| *Zalerion maritimum* (Linder) Anastasiou | + | – | + |
| *Zalerion varium* Anastasiou | – | + | + |

### 4.2.3 Relationship of Lignicolous Fungi with Wood Borers and Bacteria

**Wood borers are terrible pests and have been responsible for incalculable economic losses in earlier times.** When ships were made of wood, many a vessel perished in the sea by wood borers, the enemy invisible to the eye. Lignicolous fungi and bacteria are always found in wood along with marine wood borers.

**There are two major groups of marine wood borers. One of these belongs to bivalve molluscs and the other to crustaceans.** Wood-boring molluscs fall under

Teredinidae (shipworms) and Pholadidae (pholads or piddocks). Wood-boring crustaceans belong to the isopod genus *Limnoria* (gribbles) and the amphipod genus *Sphaeroma* (pill-bug).

Wood borers cannot withstand exposure to air and drying. **Therefore, while wood borers are the major degraders of submerged wood, fungi are more important on wood present in the intertidal region.**

Marine lignicolous fungi interact closely with marine wood borers during degradation (Kohlmeyer and Kohlmeyer 1979).

**There is strong evidence that fungi may play an important role in promoting the settlement of larvae belonging to shipworms and pholads. Fungi that colonize wood soften the wood surface by their degradative activity. This "conditioned" wood might be preferred by larvae for settling, rather than on wood which has not been thus modified** (Kohlmeyer and Kohlmeyer 1979).

Boring crustaceans are also aided by lignicolous fungi. Experiments have shown that the **lifespan of crustaceans is longer when they settle on fungus-infested wood**. A diet of wood containing lignicolous fungi seems to be essential to the limnorids for their reproduction.

Diverse lignocellulolytic bacteria inhabit wood in the sea, along with fungi. It is believed that bacterial wood decay in marine environments is relatively slow and superficial (Singh and Butcher 1991). However, the biomass of fungi and bacteria in decaying wood has not been quantitatively evaluated (Pointing and Hyde 2000).

**Marine wooden structures are coated with various preservatives against wood borers.** The major one is CCA (copper-chrome-arsenate). Although the preservative may contain wood borers and marine fungi for a while, both these groups of organism are eventually capable of overcoming the preservative qualities of CCA and colonizing wood treated with it (Eaton 1985).

- Enormous amounts of allochthonous wood from land plants enter the marine environment in the form of driftwood, man-made constructions, and cordage.
- Wood from vascular plants is made up of lignocellulose, which is a complex of cellulose, hemicelluloses, and lignin. It is highly resistant to microbial decomposition and has little nutritive value for animals.
- "Lignicolous fungi" grow by utilizing lignocellulosic components of wood. Among these, ascomycetes and basidiomycetes are the best studied and the most important colonizers and degraders of wood submerged in the sea.
- Obligate marine lignicolous fungi are recognized by their sporulating structures such as ascocarps, basidiocarps, or conidia-bearing structures on the wood surface. The family Halosphaeriaceae of the Sordariomycetes has the largest number of marine lignicolous species. Asci in the family Halosphaeriaceae disintegrate at maturity, releasing the ascospores into the cavity of the perithecium, the ascocarp.

(continued)

- Spores of obligate marine lignicolous fungi show special structural adaptations to enable their flotation in water and attachment to substrates. Ascospores of most marine ascomycetes possess gelatinous, hair-like, or cap-like sticky appendages. Conidia of anamorphic lignicolous fungi are often tetraradiate. Spore morphology is an important characteristic for identification of marine lignicolous fungi.
- Diversity of obligate marine lignicolous fungi is determined by the water temperature and the type of wood.
- "Arenicolous fungi" live among or on sand particles in intertidal beaches, while deriving their nutrition from particulate organic matter, such as decomposing wood.
- The sticky appendages of many ascospores help them to attach to wood, followed by germination and growth. Branched spores become entrapped to rough surfaces of wood. The spores germinate immediately upon attachment.
- Fungi colonize wood submerged in the sea in a succession of species.
- Several fungi may colonize wood simultaneously, but each becomes visible on the wood surface through their reproductive, fruiting bodies at different time intervals.
- Marine lignicolous fungi might compete with each other while colonizing wood.
- Lignicolous fungi produce lignocellulose degradative enzymes cellulases, hemicellulases, and lignin-degrading enzymes that help them to penetrate and grow into the lumen of the wood cells.
- Most marine fungi cause the "soft rot," while a few are responsible for "white rot."
- Allochthonous wood in the sea is degraded by the combined action of wood borers, fungi, and bacteria, resulting in loss of weight.
- The shipworms and pholads, which are bivalve molluscs, and the gribbles and the pill bug, which are crustaceans, are the two major groups of wood borers that inhabit wood in seawater.
- Fungi may play an important role in promoting the settlement of larvae belonging to shipworms and pholads. Wood "conditioned" by fungal growth might be preferred by larvae for settling.
- Lifespan of crustaceans is longer in fungus-infested wood.
- Lignicolous fungi are capable of colonizing and growing on wood treated with wood preservatives.

## Future Directions

1. Most studies on marine lignicolous fungi have been confined to obligate marine species. The role of facultative marine fungi has not been studied in detail.

2. Colonization and successional patterns have been deduced by direct observations to detect sporulating structures. Many fungi may colonize, but not sporulate successfully. Accurate information on colonization by different fungi can be obtained from metagenomic methods.
3. Mycetaen fungi are easy to detect on wood because of the sporulating structures, and hence these have been studied in detail. We have little information on straminipilan fungi, particularly those belonging to thraustochytrids, which are common in the sea and may play a role in lignocellulose degradation.
4. It can be presumed that both DOM and fine POM containing fungi and bacteria (detritus) are released from decomposing wood. The role of woody detritus in marine food web is not known. How much DOM is generated? Does decaying wood support detritivores other than wood borers?
5. Although earlier studies provided several clues on the relationships between marine wood borers and lignicolous fungi, details of this relationship have not been elucidated.

# Chapter 5
# The Mangrove Ecosystem

*If there are no mangroves, then the sea will have no meaning.*
*It is like having a tree without roots, for the mangroves are the roots of the sea...*
*Words of a Thai fisher from the Andaman Coast*

The hundreds of islands with mangrove vegetation that constitute Sunderbans in the Gangetic delta region of India, the home of the Bengal tiger, traversed by innumerable creeks and which sustain a million lives is a natural wonder. Equally important and wonderful are the numerous coastal mangrove vegetations and forests in other tropical and subtropical coasts.

Mangroves are autochthonous marine vegetations. They are found along marshy tidal areas of estuaries and coastal shorelines of the tropics and subtropics, between latitudes of 25°N and 25°S. Mangrove vegetations are also called mangal. Mangrove trees are adapted to extreme, fluctuating conditions of salinity and temperature and to the swampy sediments with low oxygen conditions. The total global coverage of mangroves is approximately 24 million hectares of area.

There are around 30 "typical mangrove plants" and more or less equal number of "mangrove associated plants" that occur in mangroves forests around the world (Hutchinson et al. 2014). Mangrove trees have several interesting ecological adaptations. Species of *Rhizophora* support themselves in the marshy soil by producing prop roots. This genus also produces viviparous seedlings in which germination takes place while the seeds are still attached to the trees. Species of *Avicennia* have characteristic "breathing roots" called pneumatophores.

Mangroves are a treasure house for marine mycologists. The enormous amount of decomposing litter generated by the dense vegetation of a variety of mangrove trees, the marshy sediments, and the winding creeks of a mangrove forest are all excellent habitats for fungi (Fig. 5.1)

**Net primary production (NPP) of mangrove trees results in the woody trunk, the dense foliage of leaves, underground roots, prop roots of many species that help in supporting the tree in marshy conditions, the aerial roots or pneumatophores of several species, and the viviparous seedlings of species such as *Rhizophora*.** The global net primary production (NPP) of mangrove trees



S. Raghukumar, *Fungi in Coastal and Oceanic Marine Ecosystems*,
DOI 10.1007/978-3-319-54304-8_5

**Fig. 5.1** A mangrove forest with *Rhizophora* trees supported by prop roots, the creek, and marshy sediments (Courtesy: Pieter van Eijk, Wetlands International)

through wood, leaves, and roots is 218 × $10^6$ metric tons of C (Fig. 5.2; Bouillon et al. 2008).

**The enormous quantity of detritus generated by mangroves serves as a major food source to many estuarine and coastal animals, such as prawns and fish.** One of the important aspects of mangroves, therefore, lies in their export of organic matter in the form of detritus to adjacent coastal waters.

- **Fungi play an important role in decomposition of dead lignocellulosic material.** A number of mycetaen and straminipilan fungi carry out this task in the mangrove ecosystem. Most of our knowledge on mangrove fungi comes from those that inhabit dead mangrove wood, prop roots, leaves, and viviparous seedlings. Fungal dynamics of these depends on the physical and chemical nature of the substrates. **Mangrove wood is more recalcitrant to degradation compared to leaves and viviparous seedlings because of several significant differences in their structural and chemical composition** (Marchand et al. 2005). Fungi which grow on wood have a "k" strategy for growth and survival in that they grow slowly, face little competition from other microbes, and break down recalcitrant compounds.

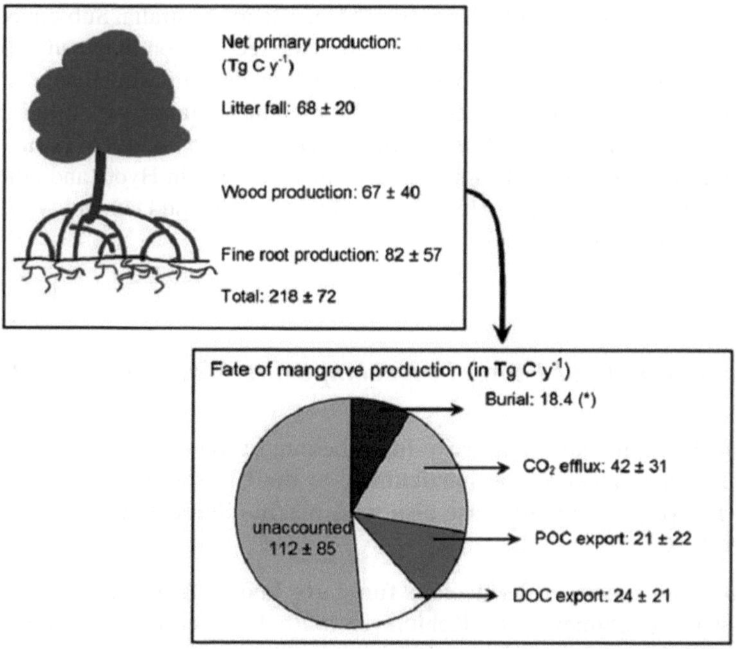


**Fig. 5.2** The fate of mangrove primary production (Copyright (2008) Wiley. Used with permission from Bouillon, S., et al. (2008) Mangrove production and carbon sinks: A revision of global budget estimates, Global Biogeochem. Cycles, 22, GB2013. doi: 10.1029/2007GB003052. John Wiley and Sons.)

These features lead to differences in fungal dynamics in leaves and wood.

1. The three distinct phases of decomposition (Chap. 2), namely, leaching, decomposition, and fragmentation phases, are more clearly seen in the case of mangrove leaves than in case of wood.
2. The zoosporic, straminipilan fungi follow the substrate capture or r-strategy in the case of decaying mangrove leaves (Newell 1996a). On the other hand, obligate, marine, lignicolous mycetaen fungi adopt a k-strategy to colonize and exploit the highly recalcitrant nature of decomposing woody and lignocellulosic materials.
3. Leaves are converted to fine POM in a matter of a few weeks, while wood takes several months to totally degrade.

## 5.1 Fungi in Woody Detritus of Mangroves

The estimated global production of mangrove wood is $67 \times 10^6$ metric tons per year (Bouillon et al. 2008). Lignocellulose-degrading, lignicolous fungi are extremely common on decomposing mangrove wood. A.B. Cribb and J.W. Cribb described the first lignicolous mangrove fungi from Queensland, Australia. Subsequently, the research of Jan Kohlmeyer in the 1960s led to the recognition that many obligately marine lignicolous fungi were found exclusively on decomposing mangrove wood, prop roots, or viviparous seedlings. **Lignicolous mangrove fungi include "manglicolous fungi" which are found exclusively in mangroves, as well as other marine, lignicolous fungi.** E.B. Gareth Jones, Kevin Hyde, and others have contributed enormously to our knowledge of the diversity and taxonomy of fungi in woody mangrove substrates.

### *5.1.1 Diversity of Fungi in Decomposing Mangrove Wood*

**Most of the lignicolous mangrove fungi belong to the division Pezizomycotina of the Phylum Ascomycota, particularly to the Classes Dothideomycetes and Sordariomycetes, as is the case also with marine lignicolous fungi in general** (Sect. 1.2.1).

**About 300 species of lignicolous fungi are known from decomposing mangrove wood** (Kohlmeyer and Kohlmeyer 1979; Jones et al. 2009; Sridhar et al. 2012a). While members of Halosphaeriaceae with unitunicate asci are dominant in allochthonous wood, members of both Halosphaeriaceae and also those of the Class Dothideomycetes with bitunicate asci are common on mangrove wood, prop roots, and seedlings (Sarma 2012).

- **Nearly 80 species form the core group of lignicolous mangrove fungi that are frequent on mangrove wood.** Some of these are shown in Fig. 5.3.
- **Fungi belonging to the Dothideomycetes, which are characterized by bitunicate asci, are important inhabitants of mangrove wood in the intertidal region.** Nearly 110 species of marine Dothideomycetes, belonging to 64 genera, are known. Nearly all of these are intertidal species (Suetrong et al. 2009). Bitunicate asci of Dothideomycetes have a "jack-in-the-box" mechanism, wherein the inner wall of the two-walled asci is capable of expanding under moist conditions when mature. It expands beyond the outer wall and discharges the ascospores forcibly. This mechanism is well adapted for intertidal conditions. These fungi do not have the elaborate appendages found in marine lignicolous fungi belonging to the Halosphaeriaceae (Chapt. 4). However, ascospores of many of their species have a mucilaginous sheath, often in the form of polar appendages (Jones 2006).
- Some of the common manglicolous ascomycete fungi are:

  - *Aigialus mangrovei* Borse.
  - *A. grandis* Kohlm. & S. Schatz.
  - *Dactylospora haliotrepha* Kohlm. & E. Kohlm.) Hafellner.
  - *Halorosellinia oceanica* (S. Schatz) Whalley, E.B.G. Jones, K.D. Hyde & Lassøe)
  - *Halosarpheia marina* (Cribb & J.W. Cribb) Kohlm.

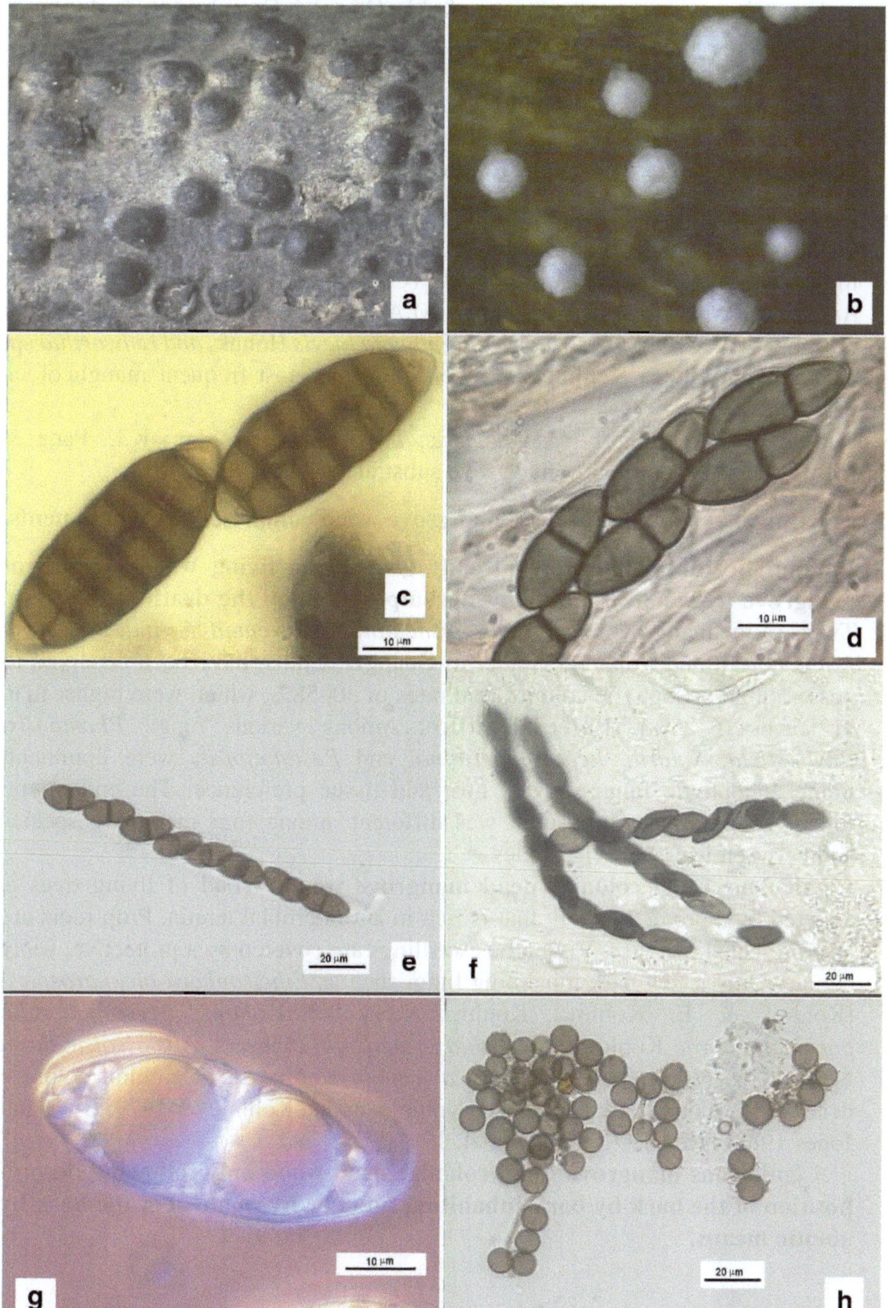


**Fig. 5.3** Some common lignicolous marine fungi of decomposing mangrove wood. (**a**) Ascocarps of *Halorosellinia oceanica*. (**b**) Basidiocarps of *Halocyphina villosa*. (**c**) Ascospores of *Aigialus mangrovei*. (**d**) Ascosores of *Dactylospora haliotrepha*. (**e**) Ascospores of *Verruculina enalia*. (**f**) Ascospores of *Halorosellinia oceanica*. (**g**) Ascospores of *Saagaromyces ratnagiriensis*.

- *Natantispora retorquens* (Shearer & J.L. Crane) J. Campb., J.L. Anderson & Shearer.
- *Quintaria lignatilis* (Kohlm.) Kohlm. & Volkm.- Kohlm.
- *Saagaromyces abonnis* (Kohlm.) K.L. Pang & E.B.G. Jones.
- *S. ratnagiriensis (*S.D. Patil & Borse) K.L. Pang & E.B.G. Jones.
- *Verruculina enalia* (Kohlm.) Kohlm. & Volkm.- Kohlm.

- Several marine lignicolous fungi which are not confined to mangroves and which are also found on allochthonous wood grow on dead mangrove wood. Some of these are *Antennospora quadricornuta* (Cribb & J.W. Cribb) T.W. Johnson, *Aniptodera chesapeakensis* Shearer & M.A. Mill., *Leptosphaeria australiensis* (Cribb & J.W. Cribb) G.C. Hughes, *Lignincola laevis* Höhnk, *and Lulworthia* sp.
- Halocyphina villosa Kohlm. & E. Kohlm. is the most frequent manglicolous basidiomycete.
- The anamorphic fungus *Hydea (Cirrenalia) pygmea* (Kohlm.) K.L. Pang & E.B.G. Jones is common in mangrove substrates.

**The fungal assemblage of dead mangrove wood comprises two components.**

1. **Endophytic fungi that were already growing in living woody trunks of mangrove trees may continue as decomposers upon the death of the trees.** Twigs of the mangroves *Aegiceras corniculatum, Avicennia marina, Bruguiera gymnorrhiza*, and *Kandelia candel* in Guangxi, China, have been observed to possess high endophytic colonization rates of 30–58%, which were higher than in leaves (6–25%) (Li et al. 2016). Among various fungi, *Phomopsis, Phyllosticta, Xylaria, Leptosphaerulina,* and *Pestalotiopsis* were dominant. Some endophytic fungi showed host and tissue preference. The endophytic fungal community composition was different among four mangrove species and between leaf and twig tissues.
2. **Lignicolous fungi colonize dead mangrove wood.** Wood of living trees is covered by a protective bark that is rich in antimicrobial tannin. Prop roots are also protected likewise. Viviparous seedlings are covered by a protective, waxy cuticle. Only a few bark-inhabiting fungi such as *Etheirophora blepharospora* (Kohlm. & E. Kohlm.) Kohlm. & Volkm.-Kohlm., *Mycosphaerella pneumatophorae* Kohlm., *Lulworthia grandispora* Meyers, *Alisia salsuginosa* Nakagiri & Tad, and *Rhabdospora avicennia* Kohlm. & E. Kohlm. colonize dead wood where a bark is retained (Kohlmeyer and Kohlmeyer 1979; Hyde and Jones 1988; Hyde 1991; Leong et al. 1991).

   **Lignicolous mangrove fungi colonize dead wood only following decomposition of the bark by bark-inhabiting fungi or by removal of the bark by abiotic means.**

---

**Fig. 5.3** (continued) (**h**) Conidia of *Periconia prolifica* (Photographs: V.V. Sarma and S. Raghukumar. Marine Lignicolous fungi. A Database on taxonomy. With kind permission of CSIR-National Institute of Oceanography, Goa, India. Under Creative Commons 4.0. http://www.niobioinformatics.in/fungi/index.htm)

Ascospores, basidiospores, and conidia become attached to mangrove wood and other lignocellulosic material in a passive manner, through impaction, entrapment, and adhesion owing to their sticky ascospores or appendages. Attachment of ascospores of many mangrove dothideomycete fungi is aided by the mucilaginous sheath present on them. Many fungi with unitunicate asci possess ascospores with various forms of sticky appendages which too are of importance in their adhesion to wood. Further germination and colonization takes place as with other marine lignicolous fungi (Sect. 4.2.1).

**Species assemblage and diversity of lignicolous fungi in mangrove wood is determined by various factors.**

1. **Many mangrove fungi display specific host preferences or even specificity towards wood from particular mangrove tree species** (Table 5.1). Mangrove trees differ in terms of their wood anatomy. They may also differ with reference to their chemistry such as the presence of tannins and other inhibitory compounds and the composition of lignocelluloses. These factors are probably important determinants of the preference of individual species to wood of different mangrove trees.

   Wood of mangrove species belonging to *Rhizophora* is a highly favorable substrate for mangrove fungi. Along the mangroves of the west coast of India, *Rhizophora apiculata* Blume has the highest diversity, followed by *Avicennia officinalis* L. and *A. marina* (Forssk.) Vierh. Decomposing prop roots of *R. apiculata*, in particular, are excellent substrates for growth of mangrove fungi (Sarma and Vittal 2000). More than 201 lignicolous fungi belonging to 139 genera are known from *Rhizophora* spp. Thus, approximately two-thirds of total number of lignicolous marine fungi are known to occur on this host. Some species prefer substrates belonging to *Rhizophora* sp., while others prefer those of *Avicennia*. Many others are known to prefer the fronds of the mangrove palm *Nypa fruticans*. Several fungi show no host preference and colonize lignocellulosic material from a wide variety of mangrove species.
2. **Several species of mangrove fungi display distinct geographical distributions** (Table 5.2). Lignicolous fungi have been studied from a wide range of tropical and subtropical mangrove ecosystems of the Indian (India, Malaysia, Thailand, Philippines, Hong Kong, Seychelles, New Zealand, and Australia), Atlantic (Sierra Leone in West Africa, Florida in USA, Baja California in Mexico, Bermuda, Brazil, South Africa, and Egypt), and the Pacific Oceans (Peru, Hawaii, Fiji). Diversity of fungi is greater in tropical mangroves, particularly in the Indian Ocean, than in subtropical ones. This might correspond with the greater mangrove tree diversity of tropics. Although similar fungi are found both in tropical and subtropical mangroves, frequency of certain species differs in the two (Abdel-Waheb and El-Sharouny 2002).
3. **Location of the wood in the intertidal zone is an important determinant of the diversity present in it** (Table 5.3). Mangroves are subject to strong tidal variations. Hence, different parts of mangroves occurring at different zones from the highest high tide to the lowest low tide are inundated by seawater or exposed

**Table 5.1** Host preference of lignicolous mangrove fungi (Gratefully acknowledge: Md. Farid Hossain, Md. Anwarul Islam. Utilization of Mangrove Forest Plant: Nipa Palm (*Nypa fruticans* Wurmb.) American Journal of Agriculture and Forestry 2015; 3(4): 156–160 for the photograph of *Nypa fruticans*)

| | | |
|---|---|---|
|  *Rhizophora apiculata* | *Caryosporella rhizophorae*, *Cryptosporella mangrovei*, *Etheirophora blepharospora*, *Hypophloeda rhizospora*, *Lophiostoma rhizophorae*, *Pedumispora rhizophorae*, *Rhizophila marina*, *Sacardoella rhizophorae*, *Trematosphaeria mangrovei*, *Phomopsis mangrovei*, *Robillarda rhizophorae*, *Xylomyces rhizophorae* | *Leptosphaeria australiensis*, *Savoryella lignicola*, *Verruculina enalia*, *Halorosellinia oceanica*, *Hypoxylon* sp., *Lophiostoma mangrovei*, *Lulworthia* sp., *Lulworthia grandispora*, *Trichocladium achrasporum* |
|  *Avicennia marina* | *Adomia avicenniae*, *Camarosporium roumeguerii*, *Bathyascus avicenniae*, *Cryptovalsa* sp, *Didymella avicenniae*, *Eutypa bathurstensis*, *Eutypella naqsii*, *Julella avicenniae*, *Leptosphaeria avicenniae*, *Mycosphaerella pneumatophorae*, *Rhabdospora avicenniae*, *Zopfiella latipes*, *Z. marina* | |
|  *Nypa fruticans* | *Aniptodera nypae*, *A. intermedia*, *Anthostomella nypae*, *Astrosphaeriella striatispora*, *Fasciatispora nypae*, *Helicascus nypae*, *Helicorhoidion nypicola*,, *Lignincola nypae*, *Linocarpon appendiculatum*, *L. bipolaris*, *L. angustatum*, *Oxydothis nypae*, *Tirisporella beccariana*, *Trichocladium nypae* | |

**Table 5.2** Preference of lignicolous mangrove fungi to tropical and subtropical regions

| Tropical mangroves | Subtropical mangroves |
|---|---|
| *Antennospora quadricornuta* | *Cirrenalia macrocephala* |
| *Aigialus grandis* | *Dactylospora haliotrepha* |
| *Dactylospora haliotrepha* | *Halosarpheia fibrosa* |
| *Caryosphaerella rhizophorae* | *Leptosphaeria australiensis* |
| *Aniptodera mangrovei* | *Lignincola laevis* |
| *Halocyphina villosa* | *Periconia prolifica* |
| *Kallichroma tethys* | *Phomopsis* sp. |
| *Lulworthia grandispora* | *Zalerion varium* |
| *Massarina velatospora* | |
| *Halorosellinia oceanica* | |
| *Halosarpheia marina* | |
| *Antennospora quadricornuta* | |

**Table 5.3** Vertical distribution of lignicolous mangrove fungi in the littoral zone

| Lower littoral or tidal | Mid-littoral or intertidal zone | Upper littoral zone |
|---|---|---|
| Ascomycetes: *Antennospora quadricornuta, Thalassogena sphaerica, Acrocordiopsis patilii* | Dothideomycetes (bitunicate ascomycetes): *Aigialus grandis, Caryosporella rhizophorae, Pyrenographa xylographoides, Julella avicenniae, Massarina velatospora, M. ramunculicola* | *Cytospora rhizophorae, Halorosellinia oceanica, Savoryella lignicola* |
| Anamorphic fungi: *Trichocladium achrasporum, Trichocladium alopallonellum* | Unitunicate ascomycetes: *Cucullosporella mangrovei, Marinosphaera mangrovei, Lulworthia* sp., | |
| Basidiomycete: *Nia vibrissa* | | |
| *Cryptovalsa* sp., *Lulworthia* sp., *Leptosphaeria australiensis* and the anamorphic fungus *Hydea pygmea* | | |

to air for different periods. Objects in upper littoral or intertidal region are wetted only briefly and remain exposed to air for a large part of the tidal cycle. Those at the lower littoral region are mostly submerged in water and are exposed to air only for a short period. These variations strongly influence the distribution of species in mangrove wood, prop roots, and seedlings. Distinct species assemblages of fungi occur in wood of *Rhizophora apiculata, R. mucronata, Sonneratia griffithii, and Xylocarpus granatum* at different zones of the tidal region (Hyde 1988; Hyde et al. 1993; Hyde and Lee 1995; Sridhar et al. 2012a). On the other hand, no evidence of vertical distribution of fungi has been found on the palm *Nypa fruticans* (Besitulo et al. 2010).

a. **Wood in the lower littoral close to subtidal area is frequently colonized by fungi with unitunicate asci of the family Halosphaeriaceae** and other ascomycete families. Such fungi are also found on allochthonous wood (Chap. 4). Fungi with ascomata having membranous walls and immersed ascomata appear to be common on wood below mean tide level. Ascospores of these fungi have typical appendages for floatation and attachment.
b. **Wood in the mid-littoral or intertidal zone has the greatest diversity of fungi** (Hyde 1990; Alias and Jones 2000). Wood at this zone is inundated or exposed to air at regular and almost equal intervals within each day. Fungi belonging to Dothideomycetes are common in this zone. Adaptations to this intermittently wet and dry zone appear to be superficial, carbonaceous ascomata, bitunicate asci with an active spore dispersal mechanism and colored or ornamented ascospores, often with a mucilaginous wall. Other ascomycetes characterized by unitunicate asci and passive dispersal of ascospores are also found. *Halocyphina villosa* J. & E. Kohlm. is the most frequent basidiomycete on wood in the intertidal. Several anamorphic fungi are common inhabitants. Most of these fungi are not found on allochthonous driftwood samples. They are adapted to withstand exposure to sunlight, desiccation, and dryness.
c. **Wood in the upper littoral zone** also has its characteristic fungal diversity.
d. **Some fungi are versatile and are found on wood throughout the littoral zone.** In general, these versatile fungi possess passive spore dispersal and have immersed ascocarps with membranous walls.
e. **Several facultative marine fungi**, such as the ascomycetes *Anthostomella* sp., *Hysterium* sp., and *Lecanidion atratum,* as well as anamorphic fungi, such as *Epicoccum purpurascens, Trimmatostroma* sp., *Helicoma hyalonema, Helicosporium pannosum, H. hongkongense*, and *Thozetella nivea,* are also common in intertidal mangrove wood and fronds of the mangrove palm *Nypa* (Sarma and Vittal 2001; Loilong et al. 2012).

4. **Salinity is an important determinant of species diversity in mangrove wood** (Sridhar et al. 2012a). Salinity gradients may exist in mangrove ecosystem depending on proximity to the sea or salinity changes may take place depending on the season. Salinity closer to freshwater is found in the upper reaches of an estuary. Strong monsoonal rains lead to rain and freshwater inflow from land and lower salinities in mangroves along the west and east coast of India. Compared to allochthonous wood immersed in the sea, mangroves are exposed to a wide range of salinities during various seasons.

   a. **Low salinities promote colonization of mangrove wood by facultative marine fungi** belonging to terrestrial as well as freshwater species. Thus, species of *Alternaria, Arthrobotrys, Aspergillus, Penicillium, Phoma*, and

*Tetracrium*, as well as several freshwater and aero-aquatic fungi, colonize mangrove wood during monsoon periods along the Indian coasts. The mangrove palm *Nypa fruticans* is found along a large salinity gradient ranging from freshwater to seawater conditions. Fronds of this palm along low salinity reaches are colonized by facultative marine fungi such as freshwater species belonging to *Acrogenospora, Helicoma*, and *Helicosporium* spp. and the basidiomycete *Grammothele fuligo.*

b. **Salinities approaching that of seawater promote colonization of mangrove wood by typical manglicolous fungi**. Thus, summer months along Indian coasts accompanied by higher salinities promoted colonization of typical mangrove fungi (Sridhar and Maria 2006). Diversity of fungi as judged by sporulating structures was also higher during summer (Ananda and Sridhar 2004). Dead fronds of *Nypa fruticans* growing at higher salinity reaches of 17–26 ppt are colonized by species belonging to *Halosarpheia, Helicascus, Lulworthia, Saagaromyces*, and *Savoryella.* (Sridhar et al. 2012a).
c. **Many manglicolous fungi show a broad salinity tolerance.** Thus, *Aigialus mangrovei, Halosarpheia cincinnatula, Hydea pygmea, Lignincola laevis, Matsusporium tropicale, Oceanitis cincinnatula, Lulworthia grandispora, Passeriniella mangrovei, Savoryella lignicola, Verruculina enalia, Bactrodesmium linderi*, and *Zalerion maritimum* are found both during monsoon and summer periods in mangroves of Indian coasts.

**Fungi colonize decomposing wood of any given mangrove tree in a succession of species** (Table 5.4).

- Fungal succession is determined by two factors.
  - Fungi that colonize wood may alter the chemistry in such a way that after a while it becomes unsuitable for them. Other fungi which can optimally utilize the wood with altered chemistry then colonize and grow.
  - Fungi may exhibit antagonism against each other. A fungus may not colonize till the presence of an antagonistic fungus declines. The slow growing lignin-degrading fungi colonizing woody substrates in the marine environment are probably k-strategists and may be a good source of antibiotics (Strongman et al. 1987).

- **Fungal succession in wood is deduced by observing the appearance of fruiting bodies of the fungi on the surface of the substrate** (Sect. 2.4.2). We have a fair idea of these processes in mangrove wood through long-term moist chamber incubations and periodic observations (Table 5.4; Alias and Jones 2000; Maria and Sridhar 2003).

  (i) **Wood is initially colonized by a few fungi.**
  (ii) **Diversity of fungi increases with progressive decomposition of wood.**

**Table 5.4** Some examples of successional patterns of lignicolous fungi at different stages of decomposition in mangrove wood

| Host | Early colonizer | Intermediate colonizer | Late colonizer | References |
|---|---|---|---|---|
| *Avicennia marina* and *Bruguiera parviflora* at Kuala Selangor mangrove stand in Malaysia | *6–18 weeks Halosarpheia marina, Halosarpheia retorquens, Lignincola laevis, Lignincola longirostris*, and *Lulworthia grandispora* | 26–54 weeks: *Dictyosporium pelagicum, Halocyphina villosa, Halosarpheia ratnagiriensis, Periconia prolifica, Savoryella lignicola, Trichocladium achrasporum, Trichocladium alopallonellum*, and *Verruculina enalia* | *60–96 weeks: Aigialus parvus, Leptosphaeria australiensis, Nais glitra, Quintaria lignatilis, Saccardoella marinospora*, and *Tirispora unicaudata* | Alias and Jones (2000) |
| Split wood blocks of *Bruguiera gymnorrhiza* and *Rhizophora muronata* at the intertidal zone at five mangrove sites in Mauritius | *Cumulospora marina, Hydea pygmea*, and *Lulworthia* sp. *Periconia prolifica* | *Lignincola laevis Periconia prolifica* as an early to intermediate colonizer | *Periconia prolifica, Dactylospora haliotrepha*, and a pycnidial fungus as late colonizers | Poonyth et al. (2001) |
| *Avicennia officinalis* | *4–12 weeks: Hydea pygmea, Lignincola laevis, Lulworthia sp., Trichocladium achrasporum* | *16–32 weeks: Hydea pygmea, Lignincola laevis, Lulworthia* sp., *Tirispora* sp. | *36–72 weeks: Lulworthia* sp., *Tirispora* sp., *Verruculina enalia, Savoryella lignicola* | Maria and Sridhar (2003) |
| *Avicennia alba* | *6–18 weeks: Halosarpheia marina* | 22–32 weeks: *Lignincola laevis, Lulworthia* sp., *Verruculina enalia, Payosphaeria minuta* | 37–60 weeks: *Halocyphina villosa, Lulworthia* sp., *Verruculina enalia, Payosphaeria minuta* | Leong et al. (1991) |

(iii) **Diversity of fungi declines with late stages of decomposition.**
(iv) **Some species colonize wood at late stages of decomposition.**
(v) **A few fungi occur throughout the decomposition, while others are restricted to different stages of decomposition.**

### 5.1.2 *Enzymatic Degradation of Lignocellulosic Material by Mangrove Fungi*

**Mangrove fungi produce cellulases, hemicellulases, and lignin-degrading enzymes (LDEs) upon colonization of the substrate** (Sect. 4.3.2). **These enzymes break down the substrate and enable them to grow.**

**A number of obligate as well as facultative mangrove fungi have been shown to be capable of degrading cellulose and hemicellulose** (Raghukumar et al. 1994a; Rohrmann and Molitoris 1992; Pointing et al. 1998; Bucher et al. 2004). Out of 45 marine fungi, 89% showed cellulolytic and 84% showed xylanolytic activities under *in vitro* conditions (Bucher et al. 2004). The lignicolous mangrove fungi *Corollospora maritima, Julella avicenniae, Lignincola laevis, Monodictys pelagica, Nia vibrissa,* and *Stagonospora* sp. can utilize cellulose at salinities of 0–34 ppt. Others such as *Aigialus mangrovei, Gongronella* sp., *Lophiostoma mangrovei*, and *Halorosellinia oceanica* produce high levels of xylanase.

**Many manglicolous fungi produce various LDEs** such as lignin peroxidise, manganese peroxidase, and laccase. *Julella avicenniae, Lignincola laevis, Nia vibrissa,* and *Stagonospora* sp. utilize glucose or cellulose for growth, while simultaneously decolourizing polymeric dyes Poly R-478 and Azure B and oxidation of syringaldazine. This indicates their lignin-degrading capability. *Nia vibrissa* is an excellent lignin degrader. This fungus as well as the anamorphic fungus *Monodictys pelagica* can mineralize lignin to $CO_2$. *Ascocratera manglicola, Astrosphaeriella striatispora, Cryptovalsa halosarceicola, Linocarpon bipolaris,* and *Rhizophila marina* possess excellent lignin solubilizing abilities and may be comparable to terrestrial white-rot basidiomycetes. An isolate of the facultative marine fungus, *Cerrena unicolor* has been shown to decolorize a number of lignin model compounds and produce extremely high levels of laccase (Raghukumar 2008).

**Enzymatic degradation of wood by fungi results in weight loss of mangrove wood.** Test wood blocks of *Avicennia* and *Rhizophora* drastically lost weight after 12–18 months in mangroves of Southwest India, coinciding with peak fungal diversity on the wood (Maria et al. 2006). Significant wood borer activity accompanied the weight loss. Mouzouras (1989) using scanning electron microscopy demonstrated differential activity of fungi. The basidiomycetous fungus *Halocyphina villosa* caused significant weight loss of the wood of *Avicennia officinalis* in 24 weeks after inoculation. The anamorphic fungus *Trichocladium achrasporum* caused a rapid weight loss of wood of both *Avicennia officinalis* and *Xylocarpus granatum.* **Decomposition of the recalcitrant lignocellulosic substrate and fungal biomass build up transform mangrove wood into detritus with an enhanced nutritional value** (Sect. 2.4.2).

- Mangroves are tropical and subtropical marine vegetations found along marshy tidal areas of estuaries and coastal shorelines.
- Net primary production (NPP) results in the woody trunk, the dense foliage of leaves, underground roots, prop roots, pneumatophores, and viviparous seedlings of various mangrove tree species.
- Dead mangrove biomass is immediately available as substrates for colonization by bacteria and fungi.
- Fungi play an important role in decomposition of dead lignocellulosic material of mangroves.
- Mangrove wood is more difficult to degrade compared to leaves.
- "Manglicolous fungi" are indigenous, mangrove lignicolous fungi. About 300 species of fungi have been described in dead mangroves, of which nearly 80 species form the core group of manglicolous fungi.
- Fungi which were originally endophytes in living plants, as well as those that colonize from spores occurring in mangrove waters, grow as saprotrophs in dead mangrove wood.
- Diversity of mangrove fungi on wood depends on the host, location in the intertidal region, salinity and geographical location.
- A succession of species colonizes decomposing mangrove wood.
- Wood is initially colonized by a few fungi. Diversity increases with progressive decomposition of wood and declines in later stages. Some species colonize wood late during decomposition, while some are persistent throughout.
- Upon colonization of the substrate, mangrove fungi produce cellulases, hemicellulases, and lignin-modifying enzymes that break down the substrate and enable them to grow. Enzymatic degradation of wood by fungi results in weight loss of mangrove wood.
- Decomposition of the recalcitrant lignocellulosic substrate and fungal biomass build up transform mangrove wood into detritus with an enhanced nutritional value.

**Future Directions**

1. Fungal diversity of decomposing mangrove wood and woody parts has remained unexplored from many parts of the world. Some examples are Sunderbans in India and Bangladesh, which is the largest mangrove coverage on earth and mangroves along African coasts.
2. Many aspects of fungal degradation of mangrove wood and their contribution to outwelling in coastal waters in the form of fine, particulate organic material have not been quantified.

3. The diversity and role of facultative marine fungi in decomposing marine wood remains to be addressed.
4. The importance of wood-derived detritus in the food web of coastal waters has not been elucidated.

## 5.2 Fungi in Detritus of Mangrove Leaves and Seedlings

Mangrove leaves contribute approximately $67 \times 10^6$ metric tons of NPP per year at a global level. Decomposing mangrove leaves belonging to *Rhizophora mangle*, *R. apiculata*, and *Avicennia marina* and other mangrove trees have been studied intensely for saprotrophic fungi. Many experiments and observations have been carried out by placing the leaves in litter bags in the mangrove environment, recovering them periodically and analyzing for fungi and chemical changes (Fell and Master 1980; Benner and Hodgson 1985; Newell 1996a; Hyde and Jones 1988; Raghukumar et al. 1994b, 1995; Davis et al. 2003; Ananda and Sridhar 2004; Rajendran and Kathiresan 2007; Sridhar et al. 2012a).

Viviparous mangrove seedlings are yet another important nonwoody substrate for marine fungi. Members of the mangrove plant genus *Rhizophora* produce viviparous seedlings that germinate while still on the tree. They fall into the soil when ripe and grow into mature plants. Seedlings that do not establish themselves in the mangrove mud float in the water till they are deposited in a suitable area or decompose and become converted into detritus. Our understanding of fungi inhabiting viviparous mangrove seedlings come mostly from the studies of Steven Y. Newell in mangroves of southern Florida (Newell 1976). Newell studied fungal colonization in detached, uninjured seedlings, as well as those which were artificially injured.

### *5.2.1 Fungi and Decompositional Phases*

**Detrital formation from mangrove leaves follows the three typical phases of decomposition of large, particulate, marine detritus** (Sect. 2.4.2). Fungi play a key role in detritus formation. Duration and dynamics of the three processes varies according to geographic location and also the nature of the substrates. Thus, decomposition of mangrove leaves leading to fragmentation may take place in a few months, whereas those of the seedlings take over a year. **The important events during detrital formation are (1) mycosere, resulting in changes in species assemblages; (2) microbial biomass buildup; and (3) biochemical changes of detritus.**

**(1) The initial phase of leaching and early colonization phase in mangrove leaves lasts about a week.**

- Senescent mangrove leaves contain about 50% of lignocellulosic structural polymers and 40 to 60% of soluble, labile organics. Soluble organic compounds comprise sugars, cyclitols, amino acids, and phenolics. **Dead leaves of mangroves undergo a rapid abiotic leaching of soluble organic matter, as well as inorganic minerals such as K, Ca, Mg, and Mn as soon as they fall into the littoral and sublittoral zones of the mangroves.** Decomposition of detached viviparous mangrove seedlings starts if they fail to establish themselves in the soil and drift in the water or when they become senescent or injured and die. In the case of leaves, more than 50% of the phenolics or tannins are lost within 2–3 weeks.
- **Fungal decomposition is initiated during this phase by two different groups of fungi. Fungi from two sources, namely those which are already resident as endophytes and those which colonize from the surrounding waters are responsible for decomposition of mangrove leaves and viviparous seedlings** (Raghukumar et al. 1995).
- Mangrove leaves are inhabited by endophytic fungi even while they are alive and attached to the trees (Kumaresan and Suryanarayanan 2002; Sakayaroj et al. 2012; Li et al. 2016). Species of *Fusarium, Phyllosticta, Phoma, Colletotrichum, Sporormiella minima, Cladosporium, Glomerella cingulata*, and *Mycospharella* sp. are extremely frequent as mangrove endophytes. Some of these species may increase in numbers after the leaves fall and may play an important role in the initial stages of mangrove leaf degradation. Further studies are required to confirm whether these fungi were in an actively growing mycelial state and if they can be truly considered as endophytes. Seedlings too are inhabited by endophytic fungi even while they are alive and attached to the trees. Endophytic fungi persist during the initial stages of decomposition. Species of *Cladosporium, Alternaria alternata, Aureobasidium pullulans, Pestalotia* sp., and *Zygosporium masonii* are some of the fungi that have been found in such seedlings.
- **The initial leaching from leaves encourages colonization by r-strategy bacteria and fungi** (Sect. 2.4.2), **which utilize the soluble organics**. Abiotic removal of phenolics further encourages microbial growth. The straminipilan fungi, *Halophytophthora* spp. (Oomycetes) and the thraustochytrids (Labyrinthulomycetes), are efficient r-strategists in colonization of mangrove leaf detritus (Strongman et al. 1987). Zoospores of *Halophytophthora* and thraustochytrids show a chemotactic attraction to leaf leachates. They swim towards the fallen mangroves leaves which are the source of the leaching and settle on the surface (Leaño et al. 2000).
- **Colonization by the oomycete *Halophytophthora* and members of Labyrinthulomycetes is a significant event during the early decompostional phase.** A total of 13 species and 2 varieties of *Halophytophthora* are known (Fig. 5.4; Nakagiri 2000).

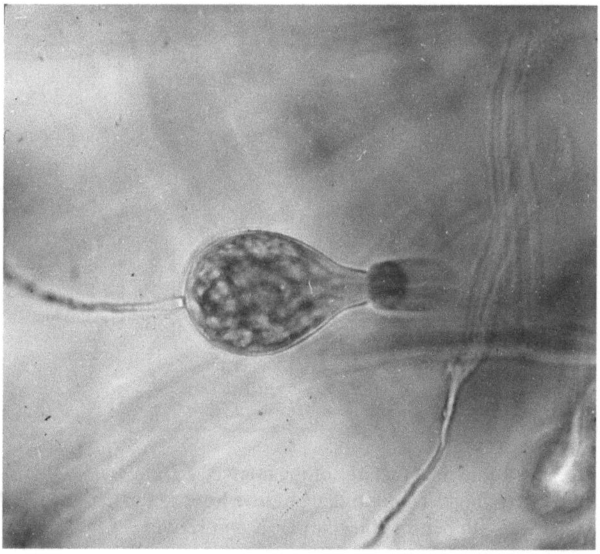

**Fig. 5.4** A zoosporangium of *Halophytophthora vesicula* containing zoospores (S. Raghukumar)

Species of *Halophytophthora* show a number of adaptive features that have made them highly successful in colonizing mangrove leaves (Fig. 5.5; Leaño et al. 2000).

Their zoospores can swim for 10 h or more at a rate of at least 100 microns per second. The process by which the zoospores locate mangrove leaves and attach to them takes place in less than 2 h of leaf fall. *Halophytophthora* can be isolated at a frequency of 100% from fallen mangrove leaves within 30 h (Newell et al. 1987). Culturing following surface sterilization has shown that the fungus penetrates the leaf surface and rapidly pervades the decaying leaf. The fungus is not only capable of tolerating the high levels of phenolics that are leached out, but is also capable of degrading them (Raghukumar et al. 1994b). *Halophytophthora* may rapidly utilize the labile organic matter that is leached out. It is further capable of utilizing the cellulose of the leaf cell walls. Different species of *Halophytophthora* show seasonal and salinity preferences. Many such as *H. avicenniae, H. masteri, H. operculata*, and *H. spinosa* var. *lobata* prefer leaves of particular species of mangroves (Nakagiri 2000). Of all the species, *Halophytophthora vesicula* (Anastasiou & Churchl.) Ho & Jong is the most cosmopolitan and versatile.

Although *Halophytophthora* is a very early colonizer in the case of mangrove leaves, it has not been observed to colonize freshly fallen mangrove seedlings. *Halophytophthora vesicula* colonized this substrate only after several months of decomposition.

- **Members of labyrinthulids and thraustochytrids, which are members of Labyrinthulomycetes, are common in mangrove habitats.** Zoospores of *Aurantiochytrium mangrovei* and *Ulkenia* sp. are chemotactically attracted to leaf extracts of many mangrove plants. Glutamic acid and pectin appear to be

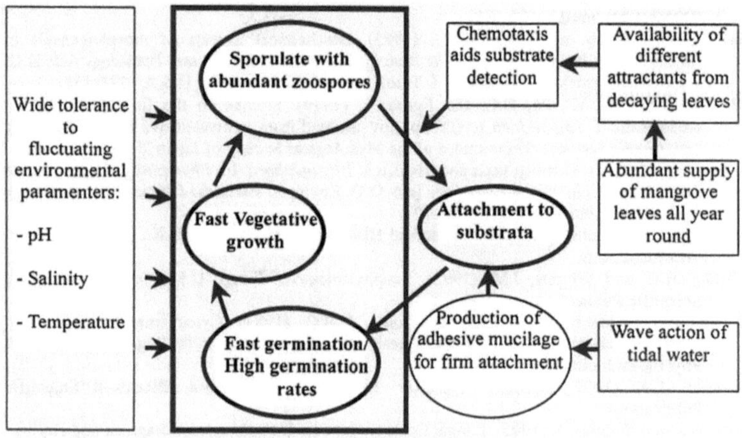


**Fig. 5.5** Biological features of the oomycete *Halophytophthora* species that aid them in successfully colonizing decomposing mangrove leaves (Source: Leaño, M. et al. 2000. Fungal Diversity 5: 131–151. With permission from Dr Kevin Hyde.)

some of the specific attractants (Fan et al. 2002). Thraustochytrids colonize mangrove leaves within 24 h of submergence (Bremer and Talbot 1995). Based on analyses of the signature polyunsaturated fatty acids of thraustochytrids, the omega-3 DHA or docosahexaenoic acid (22:6ω3) and the omega-6 docosapentaenoic acid (22:5ω6), Findlay et al. (1986) found that thraustochytrids colonized mangrove leaves immediately upon their fall. They are one of the initial and major colonizers of both mangrove leaf and seedling detritus and play an important role in their degradation. Thraustochytrids have been isolated both from the surface, as well as the interior of decaying leaves, suggesting their pervasion into the leaf. Species of *Thraustochytrium* and *Schizochytrium* have also been isolated from *Sonneratia* and *Rhizophora* leaf disks submerged in mangroves (Bremer and Talbot 1995). Culturable numbers of thraustochytids in leaf detritus of *Kandelia candel* reached values of $4.8 \times 10^3$ to $5.6 \times 10^5$ CFU $g^{-1}$(Wong et al. 2005). These species were isolated just after 24 h after immersion of leaf in mangrove waters. *Aurantiochytrium (Schiochytrium) mangrovei* (Raghuk.) R. Yokoy. & D. Honda as well as *A. limacinum* (D. Honda & Yokochi) R. Yokoyama & D. Honda are two of the commonest thraustochytrids in most mangroves. This species has been reported from many mangroves such as the west coast of India (Raghukumar et al. 1995), Panay island, Philippines (Leaño 2001), Hong Kong (Fan et al. 2002), and Viet Nam (Hong et al. 2011).

- **Rapid microbial colonization of dead mangrove leaves takes place following the initial leaching and colonization by *Halophytophthora* species and thraustochytrids.** The rapid leaching out of antimicrobial phenolics from leaves encourages colonization by bacteria and fungi. Following a substantial loss of phenolics, many facultative marine fungi such as species of *Acremonium*,

*Aspergillus, Cladosporium,* and *Phoma* colonize fallen mangrove leaves. Endophytic fungi that persist in early stages of mangrove leaf degradation, as well as other early colonizing facultative marine fungi, produce high amounts of various degradative enzymes, such as cellulases, lipases, pectinases, amylases, and proteases (Sakayaroj et al. 2012; Raghukumar et al. 1994b). Few obligate lignicolous mycetaen fungi seem to colonize leaf detritus at this stage. Fatty acid fungal markers increase rapidly from 0 to 2 weeks, indicating their colonization (Mfilinge et al. 2003).

Mangrove seedlings are also colonized by facultative marine fungi from the mangrove waters. *Septonema* sp., *Penicillium steckii,* and *Aspergillus repens* have been observed to colonize the seedlings in Newell's study in Florida mangroves.

**(2) The intermediate phase of biomass buildup and decomposition of leaves starts after the first week and lasts for several weeks after that. In the case of the seedlings, this phase starts after about 2 months and subsequently lasts for several months.** This stage is characterized by a low level of soluble organics, higher concentration of recalcitrant lignocelluloses, and mass loss of detritus. Little is known of such dynamics in mangrove seedlings.

- **Many fungi colonize mangrove leaf detritus in the intermediate phase than during the initial period of rapid leaching and microbial colonization. This includes terrestrial species, the stramenipilan fungus *Labyrinthula*, and a few obligately marine lignicolous fungi.** The latter belong to species of *Lulworthia*, such as *L. grandispora, Lanceispora amphibia, and Corollospora maritima*, and species of *Cirrenalia,* such as *C. basiminuta* and *Zalerion varium.* In a southwest Indian mangrove, the occurrence of marine lignicolous fungi increased after 2 weeks of submersion of fresh and dry leaf litter of *Rhizophora mucronata.* Several other anamorphic marine lignicolous fungi such as *Hydea pygmea, Periconia prolifica,* and *Trichocladium alopallonellum* have also been observed to colonize such leaves (Ananda et al. 2008; Sridhar 2009). The basidiomycete *Clathrus crispus* was reported to occur on the upper mangrove soil and decomposing leaves of *Rhizophora mangle* in Puerto Rico (Maldonado-Ramírez and Torres-Pratts 2005). *Halophytophthora vesicula* continues to inhabit leaf detritus of *Rhizophora apiculata* at this stage (Raghukumar et al. 1995).
- **Decomposing seedlings are also colonized during the intermediate phase by many obligate marine, lignicolous fungi.** These include *Lulworthia grandispora, Zalerion varium,* and *Keissleriella blepharospora*. The oomycetan stramenipile, *Halophytophthora vesicula*, colonized the seedlings only at this intermediate stage and not during the initial stages. *Thraustochytrium* sp. continued to occur in high numbers even during this phase.
- **Fungal abundance and biomass peak during the intermediate phase of decomposition of mangrove leaves.**
  - Peak densities of *Halophytophthora vesicula* in mangrove leaves occur by about 14 days of submergence (Newell et al. 1987).

- Fatty acid markers showed that thraustochytrid biomass peaked by 14 days in decaying leaves of *Kandelia candel* (Mfilinge et al. 2003).
- Fungi grow better in detritus of intermediate decomposition phase. In vitro studies have shown that their growth on early stages of detritus is poor because of the high amounts of tannins. On the other hand, they grow well on 21–28 days old, intermediate phase detritus (Raghukumar et al. 1994b).
- Peak culturable densities of mycetaen fungi in decomposing leaves of *Rhizophora apiculata* and *Avicennia marina* in Indian mangroves have been recorded by 20 and 30 days of decomposition, respectively (Rajendran and Kathiresan 2007).
- **Biomass of mycetaen fungi in leaf detritus varies from 0.05 up to 1.7% of detrital dry weight depending on different methods of estimation** (Table 5.5). Total microbial biomass in *Rhizophora mangle* leaves from Florida Bay, comprising both bacteria and fungi, was reported to be approximately 0.7% of detritus. An accurate estimate of fungal biomass in mangrove leaf detritus is still lacking. Maximum biomass of mycetaen fungi is generally achieved by 14- to 30-day-old detritus.
- Fungal biomass estimations would be more accurate if the biomass of straminipilan fungi, such as *Halophytophthora vesicula* and thraustochytrids are taken into account. For example, estimations based on fatty acid markers have shown that thraustochytrids can attain a biomass equivalent to 0.043% detrital dry weight (Findlay et al. 1986).

**(3) The late, fragmentation phase of mangrove leaf decomposition is characterized by continuing mass loss of detritus, a decrease in C/N ratio, a decline in recalcitrant lignocelluloses and a persistent microbial biomass. Biotic and abiotic processes result in shredding and fragmentation of the detritus into**

**Table 5.5** Fungal and bacterial biomass in decomposing mangrove leaves

| Host | Method | Maximum % biomass in dry wt detritus | | Reference |
|---|---|---|---|---|
| | | Fungi | Bacteria | |
| *Rhizophora mangle* leaves from Florida Bay | Direct detection | 0.08 (30 days) | 0.62 (30 days) | Blum et al. (1988) |
| *Rhizophora apiculata*, West coast of India | Direct detection | 0.05 (21 days) | 0.06 (35 days) | Raghukumar et al. (1995) |
| *Rhizophora mangle* leaves from Florida Bay | Ergosterol marker | 0.17 | – | Newell and Fell (1992) |
| *Rhizophora mangle* leaves from Florida Bay | Fatty acid marker | 0.05 | 0.005 | Findlay et al. (1986) |
| *Kandelia candel* | Fatty acid marker | 0–14 days | 84 days | Mflilinge et al. (2003) |
| *Bruguiera gymnorrhiza* | Fatty acid marker | 14–28 days | 28 and 84 days | |
| *Avicennia marina* | Extrapolated from numbers | – | 0.17 (40 days) | Robertson (1988) |

**finer particles** (Blum et al. 1988; Raghukumar et al. 1994b, 1995; Mfilinge et al. 2003).

- **Abundance and biomass of mycetaen fungi declines during the final phase.**
- **Field as well as experimental studies have shown that a few fungi, such as *Cirrenalia basiminuta*, appear to prefer detritus during the late decompositional phase**. Fungi grown on aged detritus have been shown to produce high levels of xylanase, in addition to pectinases, proteases, and amylases.
- **Thraustochytrid populations increase in the late stage detritus.** Although mycetaen fungi declined in diversity, abundance, and biomass in late stages of detritus of *Rhizophora apiculata* in mangroves of the west coast of India, thraustochytrids reached a peak by 60 days of decomposition, together with bacteria. High numbers of thraustochytrids have also been isolated in 40-day leaf detritus of *Avicennia marina* from Indian mangroves (Kathiresan et al. 2011). Fatty acid markers have indicated that thraustochytrid and bacterial biomass increase with age of detritus (Findlay et al. 1986). Thraustochytrid numbers were reported to decline only by 80 days in decomposing *Rhizophora mangle* leaves in Florida mangroves. Many thraustochytrids degrade cellulose, and this might help them to multiply during the late phase of detrital decomposition. *Halophytophthora vesicula* also may continue to inhabit leaf detritus at this stage.
- **Bacterial biomass may increase in late stages of decomposition**. Bacterial numbers may be high throughout decay. However, they are known to peak by 35 days on decomposing *R. apiculata* leaves and by 84 days on *Kandelia candel* and *Bruguiera gymnorrhiza.*

**Decompositional activities by fungi and bacteria cause significant biochemical changes in mangrove leaf detritus.** Initial abiotic leaching of soluble organics results in the enrichment of refractory lignocelluloses and other complex organic molecules in detritus (Raghukumar et al. 1994b, 1995; Bremer and Talbot 1995). The biochemical changes that are caused result from enzymatic degradation and utilization of various substrates, fungi-induced leaching of DOM from the detritus and fungal biomass buildup (Sect. 2.4.2).

- Fungi cultured from mangrove leaf detritus have been shown to elaborate a variety of enzymes such as cellulases, xylanase, pectinases, amylases, and proteases. Peak enzyme activity of individual enzymes depended upon the age of the detritus. For example, cellulases were most active during the initial phase of detrital degradation, while pectinases, amylases, and proteases were produced abundantly throughout decomposition. Both mycetaen fungi and the straminipilan fungi *Aurantiochytrium mangrovei* (*Schizochytrium mangrovei*) and *Halophytophthora vesicula* are capable of producing these enzymes. Another mangrove thraustochytrid, *Aurantiochytrium limacinum*, is also known to produce cellulases (Nagano et al. 2011). Total cellulolytic activity of *Kandelia obovata* leaf litter is known to peak within 6 weeks of decomposition

(Hodgkiss and Leung 1986). **Our knowledge of the enzymatic activities of fungi in detritus is still rudimentary.**

- **An important outcome of fungal degradation of complex organic molecules in detritus is release of DOM into surrounding waters and loss of weight of the particulate organic detritus.** Microbial degradation causes loss of carbohydrates, including cellulose and hemicellulose that had accumulated in the earlier phase. This process may last several weeks to months (Benner and Hodgson 1985). Among the structural carbohydrates, hemicellulose appears to be more labile, followed by cellulose, pectins, and gums. Lignin degradation may take place concomitantly with that of polysaccharides. Mycelial fungi as well as thraustochytrids inoculated onto detritus have been shown to cause rapid mass loss of mangrove leaf detritus under experimental conditions. Mass loss may depend on whether the leaves are below the low tide level and submerged or in the intertidal region. Submerged, decomposing leaves lose much of soluble organics and dry weight mass in 2–4 weeks. Leaves in the intertidal region are subject to a much slower loss.

  – Submerged leaves of *Rhizophora apiculata* in a tropical, Indian mangrove lost 30% of dry weight, as well as nearly 50% of reducing sugars and phenolics and nearly 25% of proteins within a week's time (Raghukumar et al. 1995).
  – Submerged decomposing leaves of *Rhizophora mangle* lost 40–60% of dry weight by about 3 weeks (Davis et al. 2003).
  – Submerged leaves of *Avicennia marina, Ceriops tagal,* and *Rhizophora stylosa* in an Australian mangrove lost half the initial Ash-free Dry Weight (AFDW) by 11, 27, and 39 days, respectively, while those in the intertidal region decomposed much more slowly. These showed a 50% loss of AFDW in 90, 128, and 226 days, respectively (Robertson 1988).

  These processes are significant to the food web. Leaching and release of DOM have profound implications in the "microbial loop" pathway of coastal and oceanic pelagic food web. Removal of structural polysaccharides makes the detritus more palatable to detritivores.

- **A large amount of phenolics is abiotically leached out during the initial stages.** Presence of phenolics inhibits microbial activities and deters detritivore feeding. However, some fungi that colonize early may also degrade phenolics. For example, *Halophytophthora vesicula* efficiently lowers phenolic contents in decaying mangrove leaves (Raghukumar et al. 1994b). More than 50% of the phenolics or tannins are lost within 2–3 weeks (Robertson 1988; Rajendran and Kathiresan 2007).
- **Microbial growth results in the lowering of C/N ratio.** Animals have a much higher C/N ratio compared to plant tissues. Hence, C/N ratio has generally been considered to be a good indicator of the nutritive value of detritus. Nitrogen may be lost during the early stages of decomposition, leading to a high and unsuitable C/N ratio. Subsequently, however, significant amount of nitrogen enrichment of

detritus leads to a substantial lowering of the C/N ratio. Likewise, the detritus also becomes highly enriched in P (Davis et al. 2003).

- C/N ratio of *Rhizophora mangle* leaves in subtropical mangroves declined from 90·6 to 40.7 by 70 days of decomposition (Cundell et al. 1979). C/N ratios in an Australian mangrove showed the lowest ratios by 80–160 days (Robertson 1988).
- Decomposing mangrove leaves of *Rhizophora apiculata* in a tropical Indian mangrove showed the lowest C/N ratio by 60 days of decomposition (Raghukumar et al. 1995). Despite increase in total nitrogen, total proteins in decomposed leaves may be negligible (Raghukumar et al. 1995). Total nitrogen in decomposing mangrove leaves of *Rhizophora apiculata* and *Avicennia marina* peaked by 40 days, coinciding with peak azotobacter concentrations that fix nitrogen (Rajendran and Kathiresan 2007).
- It is likely that most of the nitrogen is in the form of humic nitrogen, arising out of a complexing of microbial enzymes, carbohydrates, and phenolics with carbohydrates in the litter (Sect. 2.4.2).
- Decomposition of viviparous seedlings of *Rhizophora mangle* was accompanied by a steady increase in total nitrogen (Newell 1976).

**Unconditioned mangrove detritus contains a large amount of refractory compounds such as lignocelluloses** (Lee 1995). **Microbial enrichment and modification of detritus may be a prerequisite for many detritivores to feed on them**. Fungi and bacteria thus play a key role in the trophic web through the detrital pathway. A large variety of detritivorous animals such as crabs, crustaceans, gastropods, and fish consume mangrove leaf detritus.

- Sesarmid crabs drag enormous quantities, often up to 80% of the decomposing leaves into their burrows, where they are shred and consumed. The crab *Sesarma meinerti* is estimated to consume nearly 43% of leaf fall from *Avicennia marina* (Forsk.) Vierh. in a south African mangrove (Emmerson and McGynne 1992). Crabs assimilate around 10% of the consumed detritus and excrete most of it. The excreted, modified detritus may again serve as a source of microbial activity (Hutchinson et al. 2014). Thus, *Sesarma brockii* from an Indian mangrove preferred to consume 40-day decomposed *Avicennia marina* leaves. This aged detritus promoted growth, survival, and molting frequency of the crabs (Ravichandran et al. 2006).
- Detritivorous fish belonging to *Mugil* spp. (mullets) play an important in the trophic dynamics of the Sundarban mangrove ecosystem in India (Ray and Straškraba 2001). Other important leaf detritus consumers are the mangrove snails. Smaller invertebrates such as amphipods and isopods also shred and consume detritus.
- Mangrove detritus is abundant in mangrove tidal creeks and contributes to the nutrition of juvenile prawns belonging to *Penaeus merguiensis* de Man. They assimilate the detritus with an efficiency of about 13% (Newell et al. 1995).

- Fungi in mangrove detritus may provide detritivores with essential nutrients. Thraustochytrrids are rich in ω3 (omega-3) polyunsaturated fatty acids (PUFAs), which are essential in the nutrition of marine crustaceans and fish. These PUFAs reached a maximum by 2 weeks of decomposition in decaying mangrove leaves of *Kandelia candel* and in late stages of decaying *Bruguiera gymnorrhiza*, indicating their enrichment in terms of these essential nutrients (Mfilinge et al. 2003).

**Several fungi live in commensalistic association with mangrove animals or as saprotrophs in animal substrates.** Eighty four yeasts were isolated from three bivalve molluscs and four crab species in mangroves at Rio de Janeiro, Brazil, of which more than 44 were novel species. *Kluyveromyces aestuarii* predominated the yeast communities of 2 detritus feeding crabs, *Sesarma rectum* and *Uca* spp., (de Araujo et al. 1995; Kutty and Philip 2008). *Halophytophthora* has been found to grow in dead and live pupae of the biting midge *Culicoides subimmaculatus* in the upper intertidal region of Australian coastal waters (Stephen and Kurtböke 2011).

## 5.3 Fungi in Mangrove Sediments and Waters

Mangrove sediments are rich in organic matter. DOM and POM derived from mangrove detritus enrich mangrove soils and provide abundant nutrition for microorganisms residing therein. Mangrove sediments even just a few centimeters below the surface are anoxic. **Mycetaen fungi can be cultured from anaerobic mangrove sediments.** Most of these marine-derived fungi belong to terrestrial genera (Rai et al. 1969; Nayak et al. 2011).

**Chytrids and labyrinthulomycetes are common in mangrove sediments and waters** (Ulken 1983a, b). Brackish water habitats with low salinities have been found to harbor Chytridiomycetes and Blastocladiomycetes. *Phlycochytrium mangrovis* Ulken is a common inhabitant in the mangroves of Cananeia, Brazil; Hawaii, USA; and Veracruz, Mexico. Others include *Olpidium* sp., *Rhizophydium* sp., *Chytridium proliferum* Karling, *Chytriomyces multi-operculatus* Sparrow and Dogma, and *Catenaria anguillulae* Sorokin. Waters of higher salinities have been observed to contain more straminipilan fungi, comprising *Thraustochytrium multirudimentale*, *T. pachydermum*, *Schizochytrium* sp., and *Ulkenia visurgensis*.

**Dense populations of thraustochytrids and labyrinthulids, amounting up to 65700 cells per liter of sediments, have been reported from sediments of mangroves in the Red Sea and Malaysian coasts** (Ulken 1986). These numbers are apparently underestimates since the quantification was based on a culture method using the Most Probable Number technique and not direct cell counts. *Thraustochytrium aureum*, *T. aggregatum*, *T. kinnei*, *T. pachydermum*, *Schizochytrium aggregatum*, and *Labyrinthula* sp. have been found in these environments.

Anaerobic sulfate reduction is usually considered as the most important respiration process in these layers. Fungi are mostly aerobic organisms, and their activity in the anoxic sediments is expected to be poor. However, **fungal hyphae have been detected in anoxic sediments using Calcofluor staining and epifluorescence microscopy, suggesting that fungi may be active in anaerobic sediments**. A study on mycetaen fungi in anoxic mangrove sediments using 454 pyrosequencing of the nuclear ribosomal internal transcribed spacer 1 and 2 (ITS-1 and ITS-2) revealed a high diversity, Agaricomycetes of the Basidiomycota being dominant (Arfi et al. 2012). It is possible that some marine fungi are capable of tolerating low oxygen tensions and play an important role in turnover of the enormous amount of lignocellulosic organic matter in mangrove sediments compared to bacteria (Pointing and Hyde 2000).

**Yeasts too are abundant in mangrove sediments.** *Candida taylori, Kluyveromyces aestuarii, and Lachancea meyersii* have been reported from mangrove swamps in Florida, Bahamas, and Brazil. Some species such as *Kwoniella mangroviensis, Candida sharkiensis, C. rhizophoriensis, Cryptococcus mangaliensis,* and *Rhodosporidium paludigenum* appear to be specifically associated with mangrove habitats (Statzell-Tallman et al. 2008; Fell 2012; Jones and Fell 2012). Four marine basidiomycetous species, belonging to *Rhodosporidium fluviale*, *R. diobovatum*, *R. paludigenum,* and *R. sphaerocarpum*, have been isolated from sea water in mangrove habitats.

- Decaying leaves of *Rhizophora mangle*, *R. apiculata*, *Avicennia marina*, and other mangrove trees, as well as viviparous seedlings of *Rhizophora mangle*, have been extensively studied for fungi.
- Mangrove leaves follow the three typical, initial, intermediate, and final phases of decomposition.
- Mycosere, microbial biomass buildup, and biochemical changes are the important events of detrital formation.
- Fungi existing as endophytes and those that colonize fallen dead leaves and seedlings from surrounding waters carry out their decomposition.
- The initial leaching from detritus encourages colonization by r-strategy bacteria and fungi which utilize the soluble organics.
- The oomycete *Halophytophthora* is an extremely common fungus in mangrove leaves. This fungus shows a number of adaptive features that has made it highly successful in colonizing mangrove detritus.
- Members of labyrinthulids and thraustochytrids are abundant in mangrove detritus. *Aurantiochytrium mangrovei and A. limacinum* are two of the commonest thraustochytrids in most mangroves.
- Dead mangrove leaves are rapidly colonied by obligate and facultative marine fungi.

(continued)

- Fungal abundance and biomass peak during the intermediate phase of decomposition of mangrove leaves.
- Growth and enzymatic degradation by fungi alter the biochemistry of the intermediate phase detritus.
- Fungal and bacterial growth results in mass loss of detritus, biochemical changes, a decrease in C/N ratio, a decline in recalcitrant lignocelluloses, and a persistent microbial biomass. Biotic and abiotic processes result in shredding and fragmentation of the detritus into finer particles.
- Fungal biomass in leaf detritus reaches up to 1.7% of detrital dry weight.
- Microbial enrichment and modification of detritus may be a prerequisite for many detritivores to feed on them.
- Mycetaen fungi can be cultured from anaerobic mangrove sediments.
- Chytrids and labyrinthulomycetes are common in mangrove sediments and waters.
- Fungal hyphae have been detected in anoxic sediments.

## Future Directions

1. Diversity of fungi in mangrove leaf detritus in many parts of the world has been poorly studied.
2. Our knowledge of fungal biomass in mangrove leaf detritus is still rudimentary. Total fungal biomass, including mycetaen and straminipilan fungi, has not been elucidated so far.
3. A large amount of mangrove detritus is "outwelled" to nearby coastal waters. Microbial biomass thus transported out of the mangroves into surrounding coastal waters and buried in coastal sediments has not been quantified. Biochemical and molecular markers for fungi will help understand this.
4. Finer aspects on the role of fungi in detrital dynamics have not been addressed. For example, what is the role of fungi in DOM release and nitrogen enhancement in mangrove leaf detritus?
5. Diversity and activity of fungi in mangrove sediments and waters has not been studied in detail. Do anaerobic fungi inhabit mangrove soils?
6. Are fungi important in nutrition of mangrove detritivores? If so, in what manner?

# Chapter 6
# The Salt Marsh Ecosystem

*And what if behind me to westward the wall of the woods stands high?*
*The world lies east: how ample, the marsh and the sea and the sky!*
*A league and a league of marsh-grass, waist-high, broad in the blade,*
*Green, and all of a height, and unflecked with a light or a shade,*
*Stretch leisurely off, in a pleasant plain,*
*To the terminal blue of the main.*
*From* Sidney Lanier (1842–1881); *The Marshes of Glynn*

**Vast, extensive salt marsh vegetations stretch along many coasts of the world, particularly in temperate and high-latitude regions** (Fig. 6.1). These are the temperate equivalent of tropical mangrove ecosystems. The salt marsh ecosystem is driven by vegetations of halophytic, grass-like plants that thrive in low salinity or brackish water environments.

Some of the most extensive salt marsh vegetations in the world are those that occur along the Atlantic coast of North America. These are dominated by the smooth cordgrass, *Spartina alterniflora* Loisel., and the black needlerush *Juncus roemarianus* Scheele. The common reed, *Phragmites australis* (Cav.) Trin. ex Steud, is a cosmopolitan salt marsh plant. *Cyperus malaccensis* (Lam.) Palla. (Cyperaceae) is one of the dominant perennial sedges in the mangroves of the Southwest coast of India (Sridhar et al. 2012a).

Salt marsh plants are some of the most highly productive on earth. The estimated global coverage by salt marsh vegetations is $0.4 \times 10^6$ km$^2$ with an NPP of $0.44 \times 10^{15}$ g C per year, equivalent to $440 \times 10^6$ metric tonnes of carbon per year (Duarte and Cebrian 1996). Less than 30% of this is consumed by herbivores and nearly 230 million metric tonnes of C enter the decomposition cycle every year. Thus, detritus generated through microbial action on this enormous biomass from salt marshes assumes great importance in sustaining aquatic life in brackish water habitats.



S. Raghukumar, *Fungi in Coastal and Oceanic Marine Ecosystems*,
DOI 10.1007/978-3-319-54304-8_6

**Fig. 6.1** The salt marshes of Georgia, USA (Courtesy: The Georgia Sea Grant College Program. The University of Georgia)

Salt marshes are important habitats for crabs, flatfish, mullet, oysters, and shrimps, as well as birds. Microorganisms are responsible for decomposition of salt marsh grass, particulate detrital formation, and transport of an enormous amount of DOM that is released into coastal waters during their decompositional activities.

Salt marsh plants possess several unique features that make them ideal habitats for fungi when they senesce and die. A remarkable feature of salt marsh grass is that leaves and stems of the plant are not shed upon their death. Dead parts of the plant are colonized and decomposed by fungi while they are still attached to the living plants and not submerged, thus avoiding competition from bacteria that would predominate in the water. Dead plant parts are only partly immersed during the high tide. Upper parts of the plant are never exposed to seawater, while the lowermost parts are always submerged. Dead plant parts fall into marsh sediments only at a late stage of decomposition, following which further microbial decay predominated by bacteria and fragmentation take place.

Standing, decaying salt marsh grass offers ideal substrate for fungal growth. A number of detailed and exemplary studies on *Spartina alterniflora*, *Juncus roemerianus*, and *Phragmites australis* have brought to light the enormous diversity and biomass of fungi in decomposing parts of these salt marsh plants (Newell and Porter 2000; Buchan et al. 2002, 2003; Van Ryckegem and Verbeken 2005a, b; Van Ryckegem et al. 2007).

Earlier studies on marsh grass focused on decomposition of submerged detritus. Thus, they did not take into account the absence of abscission of shoot parts and the fact that most of the decomposition takes place in standing, decaying parts. As a result, prokaryotes were considered to have been the drivers of marsh grass-shoot decomposition in the 1950s (Newell 1993).

Starting in the 1960s and over a period of time, Jan Kohlmeyer, Roger D. Goos, Robert V. Gessner, and Steven Y. Newell carried out extensive studies on fungi in salt marshes along the Atlantic coasts of North America. This was followed by those of Kevin Hyde and others in salt marshes of Hong Kong and elsewhere. In addition to important, conventional studies on fungal diversity, these studies also used methodologies such as direct microscopy, index-biochemical measurements, quantification of sexual-reproductive structures, and transmission electron microscopy that have provided us a deep insight into the importance of fungi in coastal detrital formation (da Luz and Barata 2012). It is now known that fungi pervade and lyse standing shoots of smooth cordgrass and are major players in its decomposition (Newell and Porter 2000).

Fungi have been reported from about 50 different marsh plants. The most well-studied salt marsh plants with regard to fungal diversity and dynamics are *Spartina* spp., *Juncus roemerianus*, and *Phragmites australis*.

## 6.1 Fungi in Decomposing Salt Marsh Grass

Standing dead tissues of salt marsh grass are colonized by saprobic bacteria and fungi (Buchan et al. 2003).

Both obligate and facultative marine mycetaen fungi occur as saprotrophs on decaying marsh plants. Few studies have been carried out on straminipilan fungi in salt marshes.

### *6.1.1 Diversity of Fungi in Decomposing Salt Marsh Grass*

**Salt marsh grasses are one of the richest in terms of fungal diversity in an individual plant species.** Individual herbaceous plants on an average are considered to host 5–8 species of mycetaean fungi. Salt marsh plants, on the other hand, harbor a much greater diversity. Gessner and Kohlmeyer compiled a list of 101 species of mycetaen filamentous fungi from *Spartina* spp. in 1976. So far, more than

130 species fungi have been documented from this salt marsh grass. The needlerush *Juncus roemerianus* harbors about 120 species. *Phragmites australis*, a cosmopolitan grass that occurs in intertidal marshes and also in freshwater habitats, is host to nearly 80 species of fungi. Ascomycetes are the most prevalent in standing, decaying salt marsh grasses.

**Diversity of fungi in decomposing salt marsh grass is influenced by the varying cell wall chemistry, anatomy, and physiology of the plant substrate.** Thus, the composition of the fungal community is influenced by **individual species and different parts of salt marsh plants**. In addition, the geographical conditions where the marsh plants grow play an important role. Fungal assemblages also show a vertical distribution on the standing host (Gessner 1977; van Ryckegem et al. 2007; Al-Nasrawi and Hughes 2012).

- **Salt marsh fungi show a host preference** (Table 6.1). **Diversity studies based on the detection of sporulating structures and conventional morphological taxonomy as well as metagenomics based on automated ribosomal intergenic spacer analysis (ARISA) of fungi on the substrate have shown that different hosts may support distinct fungal communities** (Torzilli et al. 2006; da Luz and Barata 2012).

**Table 6.1** Common fungi found in different salt marsh plants

| *Spartina alterniflora/ S. maritima* | *Juncus roemerianus* | *Phragmites australis* |
|---|---|---|
| *Phaeosphaeria spartinicola*, *Mycosphaerella* sp., *Buergenerula spartinae* *Lachnum spartinae* *Halosarpheia retorquens* *Phialophorophoma littoralis* *Sphaerulina oraemaris* *Phoma* sp. *Dictyosporium pelagicum* *Passerinella obiones* *Lulworthia* sp. | *Loratospora aestuarii*, *Papulospora amerospora*, *Aropsiclus junci*, *Anrhostomela poesila*, *Physalospora citogerminans*, *Scirrhia annulata*, *Massarina ricifera*, *Tremataeia halophila* | ***Hong Kong Salt marshes (Poon and Hyde 2008):*** *Lignincola laevis*, *Trichoderma* sp., *Halosarpheia phragmiticola;* *Colletotrichum* sp. *Didymella* spp. *Phaeosphaeria* sp. *Aniptodera phragmiticola* *Cladosporium* sp. *Fusarium* sp. *Phomopsis* sp. ***Netherlands salt marshes*** **(Van Ryckegem and Verbeken** 2005a, b) *Septoriella* sp. *Phoma* sp. *Phomatospora berkeleyi* *Hendersonia* sp. *Puccinia phragmitis* *Stagonospora vexata* *Neottiosporina australiensis* *Pseudoseptoria donacis* *Deightoniella roumegueri* |

Fungal communities vary between the short and tall forms of *Spartina alterniflora*, *Juncus roemerianus*, *Distichlis spicata*, and *Sarcocornia perennis*. Eighty nine percent of fungi found in *Spartina* spp., *Juncus roemerianus*, and *Phragmites australis* are exclusively associated with one host. Only 9% of fungi are associated with two and 2% associated with all three host species.

**Many fungi that might grow as hyphae within the substrate may not sporulate.** For example, Walker and Campbell (2010), using morphological and molecular approaches to study fungi in *Spartina alterniflora* and *Juncus roemerianus*, found that more than 50% of the fungal signatures were found on both plants, but not their sporulating structures. Hence, a host plant may determine not only which fungi colonize them but also which of those are able to grow only vegetatively without sporulating on them.

- **Different parts of marsh plants may vary in fungal diversity**. Distinctly different ascomycetes are found in the leaf sheaths and true stems of decaying *S. alterniflora,* compared to the leaf blades. Thus, *Lachnum spartinae* is found in the leaf sheaths, but not in others (Newell and Porter 2000; Kis-Papo 2005).

The salt marsh plant *Spartina alterniflora* has been studied in greatest detail for fungi. *Phaeosphaeria spartinicola* Leuchtm. and *Mycosphaerella* sp. are the most frequent fungi on this plant (Fig. 6.2). Both these fungi possess bitunicate asci and belong to the Dothideomycetes. In addition, *Phaeosphaeria halima* (T.W. Johnson) Shoemaker & C.E. Babc. and *Buergenerula spartinae* Kohlm. & R.V. Gessner occur in about 2 to 40% of blades examined (Newell 2001a). *Buergenerula spartinae* is an ascomycete with unitunicate asci, whose position in the ascomycete classification is uncertain.

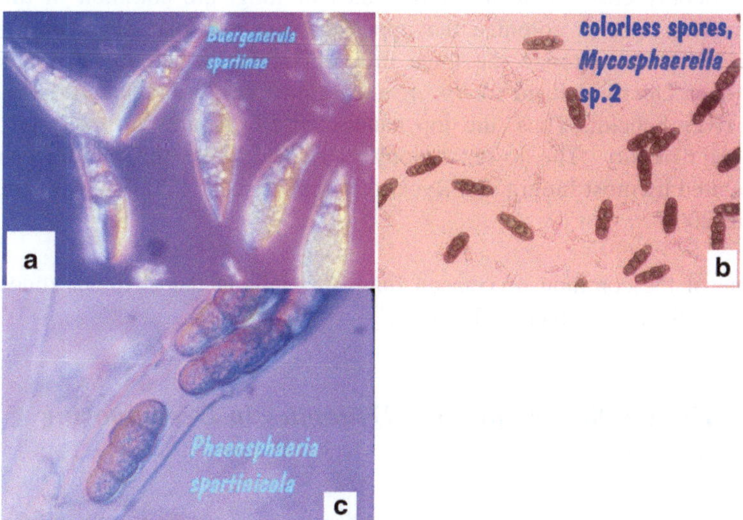


**Fig. 6.2** Three common ascomycetes found in *Spartina alterniflora* (Courtesy: Steve Newell. From: Georgia Coastal Ecosystems LTER)

- We have limited information on straminipilan fungi in the salt marsh ecosystem. Three species of a new oomycete genus *Salisapilia* were described from *Spartina alterniflora* detritus obtained from salt marshe sediments of southeastern Georgia (Hulvey et al. 2010). Another species *Halophytophthora masteri* Nakagiri & S.Y. Newell is also an inhabitant of salt marsh grass detritus (Nakagiri et al. 1994). However, species of *Halophytophthora* occur only in low frequencies in salt marsh grass detritus. On the other hand, another oomycete, *Pythium grandisporangium*, has been observed to occur in high frequencies of 40–45% on leaves of *Spartina alterniflora* (Newell 1992).
- **Fungal communities display a vertical zonation on salt marsh plants**. Standing, decaying salt marsh plant parts are exposed to three different environmental conditions that determine fungal diversity (Gessner 1977; Poon and Hyde 1998; Kohlmeyer and Volkmann-Kohlmeyer 2001; Barata 2002; Van Ryckegem and Verbeken 2005a, b; da Luz and Barata 2012).

  The subtidal is always immersed, and the upper parts are always exposed to air. Fungal diversity is determined by environmental factors that affect these. This submerged zone is inhabited by algae, barnacles, and obligate marine fungi, many of which are unitunicate ascomycetes.

  The intertidal zone is subject to extreme environmental conditions. They face varying moisture conditions of being periodically wet to different lengths of time. They are also exposed to salinity conditions ranging from freshwater to seawater. Temperature often reaches up to 40 °C. The intertidal zone is the most favorable place for colonization by a high diversity of typical salt marsh fungi as well as facultative marine fungi. Bitunicate ascomycetes belonging to Dothideomycetes and anamorphic states of fungi are common in this zone. Terrestrial species colonize the uppermost exposed part. Typical “indicator taxa” may characterize the three different zones in salt marsh plants. This has been clearly defined for *Phragmites australis* in a brackish tidal marsh of the river Scheldt. Thus, the top canopy of *Phragmites australis* is poor in fungal diversity. The lower canopy is richer, while the middle canopy is inhabited by most fungi (Van Ryckegem and Verbeken 2005b; Van Ryckegem et al. 2007).

Some of the common mycetaean fungi that inhabit the different zones of the salt marsh plants are given in Table 6.2.

### *6.1.2 Fungal Succession and Dynamics in Decomposing Salt Marsh Grass*

Salt marsh plants senesce and die from bottom towards top. These plants, as with most other grasses, do not abscise dead leaves or aboveground stems. **Most of our**

**Table 6.2** Vertical distribution of fungi in decomposing salt marsh grasses

| Vertical Zone | *Spartina alterniflora* | *Spartina maritima* | *Phragmites australis* |
|---|---|---|---|
| Intertidal | *Phaeosphaeria typharum; Drechslera halodes; Pleospora vagans, Buergenerula spartinae, Puccinia sparganioides; Phoma* sp. | *Phoma* sp., *Phaeosphaeria spartinicola; Stagonospora* sp.; *Buergenerula spartinae; Leptosphaeria pelagica; Ascochyta* sp.; *Dictyosporium pelagicum* | *Arthrinium* state of *Apiospora montagnei Camarosporium* sp.; *Colletotrichum* sp. *Didymella glacialis Lophodermium arundinaceum, Neottiosporina australiensis Phaeosphaeria* spp.; *Phialophoroma* sp.; *Phoma* sp. *Stictis* sp. *Trichoderma* sp. |
| Submerged | *Pleospora pelagica; Halosphaeria hamata; Leptosphaeria albopunctata; Passerniella obiones; Leptosphaeria pelagica, Lulworthia* sp. *Mycosphaerella* sp. | *Sphaerulina oraemaris; Phialophoroma litoralis; Passeriniella obiones; Halosarpheia retorquens; Sphaerulina albispiculata; Lulworthia* sp.; *Buergnerula spartinae* | *Halosarpheia phragmiticola; Lignincola laevis Colletotrichum* sp. *Trichoderma* sp. *Phomatospora berkeleyi; Phaeosphaeria* sp. |
| Fragmented litter | – | – | *Phomatospora berkeleyi; Phaeosphaeria pontiformis; Halosphaeria hamata; Massarina arundinacea* |

**knowledge on the role of fungi in decomposition of salt marsh grasses comes from studies on *Spartina alterniflora* in the southeastern US marshes by Jan Kohlmeyer, R.D. Goos, R.V. Gessner and Steve Newell.** Fungi play a key role in salt marsh grass detrital decomposition and transformation. Fungal dynamics over a period of time are more easily studied in litter bag experiments, whereby decomposing substrates are enclosed in litter bags with meshes and placed over the sediment of the habitat. However, natural conditions are not reproduced under such experimental conditions (Samiaji and Bärlocher 1996). Despite this, broad ideas about decomposition can be obtained through such experiments. Litterbag approach has been followed to study the decomposition of *Spartina maritima* in a salt marsh on the western coast of Portugal (Castro and Freitas 2000).

**Decomposition of dead salt marsh grass follows the three typical stages of macrophyte decomposition.**

1. **The initial phase of leaching and early colonization:** Salt marsh grass plants are inhabited by endophytic fungi, phylloplane fungi, and weak parasites when

they are healthy and senescent. These persist upon the death of the plant. *Spartina alterniflora* carries the parasitic fungus *Buergenerula spartinae* even when healthy and during senescence. The reed grass *Phragmites australis* is already colonized by fungi such as *Sporobolomyces* sp., *Alternaria alternata, Cladosporium* sp., *Septoriella* sp., *Phoma* sp. and *Phaeosphaeria* sp. when healthy. These exhibit a low tolerance to stress. These persist once the leaves become senescent. Immediately upon the death and during the initial stages of this phase of decomposition, salt marsh plants begin to lose soluble labile organic compounds through leaching.

2. **The intermediate phase of biomass buildup and decomposition: This second, slower decompositional phase in *Spartina* lasts up to a year. This is a highly fungal-dominated stage of decomposition.** During the early decay phase of salt marsh grass, the leaf blades are yellow or brown in color, remain attached to the stem, and do not collapse onto the sediment. Fungal colonization, decomposition, and biomass buildup peak during this period (Buchan et al. 2002, 2003). Microbial respiration rate, which is an indicator of microbial activity, peaks during the early stages of decay in blades of this salt marsh grass in a southeastern US salt marsh. This corresponds to peak fungal biomass. At the late decay phase, the blades become brown to black in color and collapse onto the sediment surface even while remaining attached to the stem.

   The ascomycetes *Phaeosphaeria spartinicola, P. halima*, and *Mycosphaerella* sp. dominate the decomposition of *Spartina alterniflora* from among various others that colonize and grow in it. Some of these appear during the early decay phases, while others colonize the late stages. A few occur during both stages of decomposition.

   *Phragmites australis* is characterized by colonization of fungi such as *Didymella glacialis, Microsphaeropsis* sp., *Neottiosporina australis, Cladosporium* sp. *Stictis* sp., *Lophodermium arundinaceum*, and *Stagonospora vexata*. These are more stress tolerant than the endophytic, parasitic, and phylloplane fungi that were present in the beginning (Van Ryckegem et al. 2007).

   *Juncus roemerianus* is colonized by the apothecial ascomycetes *Orbilia junci* and *Rivilata ius* at a stage when the tips of the leaves become brittle, the cuticle flakes off, and the surface turns rough.

   - **Fungal biomass that peaks during the intermediate phase of decomposition of salt marsh grass makes a significant contribution to the total organic matter** (Table 6.3).

     – **Biomass in decaying salt marsh grass is contributed by algae, bacteria, and fungi. Fungi are the most important decomposers of standing, dead *Spartina alterniflora*.** Individual biomass of these organisms has been estimated by Steven Y. Newell, who used a direct epifluorescence microscopy method for algae and fluorescent staining followed by

**Table 6.3** Biomass contribution of fungi to decomposing plant parts of various salt marsh grasses based on ergosterol estimation methods. *Ergosterol to biomass conversion is based on a conversion factor of 200*

| Host Plant | % Biomass in detritus | Remarks | Reference |
|---|---|---|---|
| *Spartina alterniflora*, Sapelo Island salt marshes, USA | 29 | Biomass during winter | Newell and Porter (2000) |
| *Spartina alterniflora* Sapelo Island Salt Marsh Ecosystem | 13.7 | 685 μg $g^{-1}$ ergosterol in early stage detritus; | Buchan et al. (2003) |
| *Spartina maritima* (Curtis) Fernald; salt marsh on the western coast of Portugal | 21.5 | 1077 μg $g^{-1}$ ergosterol | Castro and Freitas (2000) |
| *Phragmites australis*; tidal marsh of the Scheldt estuary, The Netherlands | 10.9 | 548 μg $g^{-1}$ ergosterol in ash-free dry weight, | Van Ryckegem et al. (2007) |
| *Juncus roemerianus* | 4.0 | – | Newell and Porter (2000) |
| *Spartina patens* | 5.3 | – | Newell and Fell (1992) |

epifluorescence microscopy for bacteria. Fungi were estimated by quantifying ergosterol, the signature compound of live fungi.

- Fungal biomass in decomposing *Spartina alterniflora* and *S. maritima* increases from less than 2% in senescent leaf to values of nearly 30% with progressive stages of decomposition. Similar high fungal biomass values have been reported for other salt marsh grasses.
- *Juncus roemerianus* and *Spartina patens* also support a substantial amount of fungal biomass, with ergosterol contents of hundreds of micrograms per gram of organic decaying material. This amounted to about 40 and 53 mg of biomass $g^{-1}$ detritus, respectively, in each of the hosts. Fungi thus contributed 4.0% of dry weight to *J. roemerianus* and 5.3% to *S. patens* detritus. However, the values were about half of the fungal biomass found in *S. alterniflora*.
- Fungi contributed to nearly 11% of detrital dry weight of *Phragmites australiensis* along the coast of the Netherlands (Van Ryckegem et al. 2007).
- Biomass buildup varies seasonally. Biomass and productivity of fungi increased in wet months, in naturally detached leaves, decreasing greatly in summer. Naturally detached leaves in *Spartina maritima* showed higher ergosterol concentrations of 939 and 1077 μg g−1 in March and December respectively, which amounts to 21.5% of detrital weight, and low levels during summer (<185 μg $g^{-1}$) (Castro and Freitas 2000). Fungal productivity in standing, decaying cordgrass has been found to be nearly 10 times greater in winter (3652 mg per $m^2$ per day) than in summer (369 mg per $m^2$ per day) (Newell and Porter 2000; Table 6.4).

**Table 6.4** Organic matter contribution by fungi and bacteria to the salt marsh grass ecosystem of *Spartina alterniflora*

| | mg production per day per $m^2$ salt marsh on dead plant parts | | g dry biomass per $m^2$ salt marsh | |
|---|---|---|---|---|
| Components | Summer | Winter | Summer | Winter |
| *Spartina alterniflora* | | | 588 | 183 |
| Fungi | 369 | 3652 | 19 | 52 |
| Bacteria | 10 | 35 | 0.02 | 0.07 |
| Fungi:Bacteria | 36.9 | 104 | 950 | 740 |
| % Fungi in Spartina | – | – | 3.2 | 28 |
| | mg production per day per $m^2$ salt marsh sediments | | g bacterial dry biomass per $m^2$ salt marsh sediments | |
| Bacteria | 750 | 265 | 44 | 44 |

3. **The late fragmentation phase:** The final phase of decomposition occurs when the leaves fall into the sediment and become fragmented. Smaller fragments that are generated during this decompositional phase possess a large combined surface and are more favorable for bacterial growth. **Both diversity and biomass of mycetaen fungi decline in this advanced stage of decomposition. Obligate marine fungi are better able to tolerate these constantly submerged, stress conditions** (Van Ryckegem et al. 2007; Kohlmeyer and Volkmann-Kohlmeyer 2001). The impoverished fungal community comprises many typical marine species. *Halosphaeria hamata* (Höhnk) Kohlm., *Massarina arundinacea* (Sowerby) Leuchtm., and *Phomatospora berkelyi* Sacc. colonize fragmenting litter of *Phragmites australis*. In *Juncus roemerianus*, obligate marine species such as *Lignincola laevis*, *Halosarpheia* spp., and a *Cirrenalia* sp. colonize detritus after the leaves and culms break off and fragment.

   Biomass of fungi in fragmented detritus of *Phragmites australis* declined to 3.2%, compared to 11% in standing, decaying detritus. Likewise, fungal contribution to total N dropped to about 10% compared to 34% in standing, decaying *P. australis* detritus. Fungi which contributed up to 34% of nitrogen in standing leaf detritus accounted for only up to 10% in fragmented detritus (Van Ryckegem et al. 2007).

   The complete decay and fragmentation may take over a year, much longer than for mangrove leaf, seagrass, and macroalgal detritus.

**Fungi cause several major biochemical alterations to salt marsh grass detritus. Enzymatic degradation, loss of weight, and enrichment with nitrogen are the hallmarks of fungal decomposition.**

- Lignocellulose is the major component of dead, standing *Spartina alterniflora* and constitutes 70–75% of the organic mass of the mature shoots of the smooth cordgrass (Hodson et al. 1984). **Marsh grass ascomycetes, such as *Phaeosphaeria spartinicola, P. obiones, Buergenerula spartinae,* and *Mycosphaerella* sp. 2, are efficient lignocellulose degraders.** Extracellular enzyme activity by fungi results in degradation of lignocellulose, leading to physical disruption of the structure (Newell and Porter 2000; Newell 2001a). Transmission Electron Microcopy (TEM) has revealed that cordgrass ascomycetes destroy the lignocellulosic secondary walls and the lignin-rich middle layer of fiber cells (Bergbauer and Newell 1992; Newell et al. 1996; Newell and Porter 2000). Decomposition follows the soft-rot pattern. Individual or mixed inocula of fungi brought about significant degradation of the lignocellulose fractions of *Spartina alterniflora*. Decomposition ranged approximately from 16–40% depending on the presence or absence of nitrogen enrichment or the species involved (Torzilli and Andrykovitch 1986). Microbial decay during the later part of this phase is very slow because only relatively refractory materials remain.

- **A significant amount of mass loss of detritus may be caused by fungal degradative activities during this fungus-dominated phase.** *Spartina* undergoes loss of up to 60% of the original organic mass (Newell and Porter 2000). By about 6 months, about 80% of the weight may be lost. Using a litterbag technique, it has been noticed that about 50% of the dry mass of decaying *Spartina maritima* leaves in a salt marsh on the western coast of Portugal declined rapidly after 1 month. After 13 months, the leaves had lost 88% of their original dry weight (Castro and Freitas 2000; Gessner and Goos 1973). Leaves of *Phragmites australis*, placed in litter bags on sediment surface, lost 50% of their initial ash-free dry weight within 7 months and nearly 80% by 11 months (van Ryckegem et al. 2007). It has been estimated that about 37% of this loss could be ascribed to fungal activities.
- **Following colonization of dead tissues of the salt marsh grass, fungi immobilize inorganic nitrogen from the environment in the form of fungal proteins.** Almost 100% of the protein content of the decomposing salt marsh grass is contributed purely by saprotrophic fungi within 1 month of decomposition (Newell et al. 1989). Fungal biomass contributed about 34% of the detrital nitrogen in the case of *Phragmites australis* (Van Ryckegem et al. 2007). Fungal biomass buildup in salt marsh grass may often be limited by inorganic nitrogen in the environment. An increase in ambient inorganic nitrogen content results in an increase of fungal productivity (Torzilli and Andrykovitch 1986).

**Fungal immobilization of nitrogen during the decompositional phase results in an increase in nitrogen and decrease in C/N ratios.** Leaf blades of *Spartina alterniflora* have a higher C/N ratio, higher organic matter content, and lower ash content during the early decay stages, compared to those during the late phases. The C/N ratio at this time ranges from 53.9 to 80.7. As decay progresses, C/N ratio drops to a minimum of 49.9 to 37.8 (Buchan et al. 2003). With further decomposition, the blades gradually collapse onto the marsh sediment. The C/N ratio of leaves of *Phragmites australis* placed on sediment surface decreased from 24.8% to 16.6% in 10 months (van Ryckegem et al. 2007).

- Fungal colonization of dead salt marsh plants results in an increase in lipids. Simultaneously, the fungi appear to metabolize the phenolics present, resulting in a decrease of total phenolics (Bergbauer and Newell 1992).

**Fungi make a substantial contribution to the organic matter in salt marsh ecosystems.** Organic matter production in the salt marsh ecosystem is the result of the combined production of the salt marsh plants, algae, bacteria, and fungi. Results based on dead, standing, and decomposing *Spartina alterniflora* are presented in Table 6.4.

- *Spartina alterniflora* has an annual production of 1313 g per $m^2$ per year. The standing crop of the plant is more in summer than in winter.
- Fungal as well as bacterial productivity are higher in winter than in summer.
- Fungi and bacteria both contribute to the detrital weight more in winter than in summer.
- Productivity and biomass of fungi are much higher than that of bacteria.
- Fungal biomass is 740 to 950 times more than that of bacteria in standing detritus of the smooth cordgrass.
- Bacterial productivity and biomass in detritus of the smooth cordgrass are much higher in sediments compared to fungi.

**The high fungal biomass available in *Spartina* detritus may be responsible for the high densities of detritivoral invertebrates in salt marshes** (Newell and Porter 2000). A number of invertebrate detritivores inhabit salt marshes. An important one in the Atlantic coast salt marshes is the shredder gastropod *Littoraria irrorata* (salt marsh periwinkle). In addition, a number of amphipods and gastropods live in these habitats.

- **Fungal colonization, decomposition, and biochemical alteration of detritus make salt marsh grass detritus much more palatable and nutritious than otherwise** (Newell and Bärlocher 1993). Lignocellulose of low digestibility and nutritional value is converted into fungal biomass. The detritus is enriched with fungal proteinaceous nitrogen. Almost the entire nitrogen in standing, decomposed salt marsh detritus may be present in the form of fungi (Newell et al. 1989). Such nitrogen is much more easily digestible by the periwinkle than proteins of the salt marsh grass (Bärlocher et al. 1989).
- **Fungi serve as nutrition to salt marsh detritivores.** Salt marsh periwinkles efficiently shred decaying leaves of *Spartina alterniflora*. They can ingest 7% of naturally decayed leaves per day and are capable of digesting 51% of the consumed detritus (Newell and Bärlocher 1993; Bärlocher and Newell 1994). The grass shrimp, *Palaemonetes pugio,* could assimilate carbon from detritus of *Spartina alterniflora* with an efficiency of 40% (Newell et al. 1995). Salt marsh periwinkles removed living-fungal mass more rapidly at a rate of 10% per day, than they removed leaf mass, and they can efficiently digest salt marsh-fungal mycelium. An experimental study on diet and growth of *Littoraria irrorata* using detritus showed that growth was highest on a diet of standing dead leaves of *Spartina alterniflora* (Bärlocher and Newell 1994). Growth was negatively related to phenolic content of the diet and positively to lipids. Detritus with specific fungi may be more palatable to detritivores than others. For example,

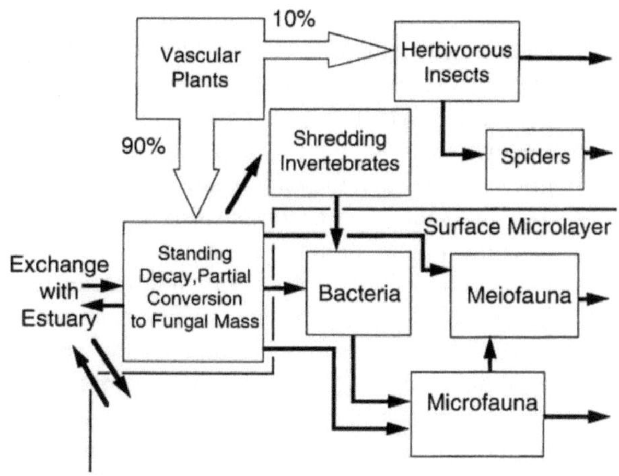


**Fig. 6.3** Flow of primary production from marsh grass shoots to the marsh food web (Source: Steven Y. Newell and David Porter. 2002. Microbial Secondary Production from Salt Marsh-Grass Shoots, and Its Known and Potential Fates. In: Concepts and Controversies in Tidal Marsh Ecology. Eds.: Weinstein, M.P., Kreeger, Daniel A. p. 159–185. Springer, Netherlands)

tissues with *Phaeosphaeria spartinicola* may be preferred for those colonized by *Buergenerula spartinae* (Newell 2001a).

- A diagrammatic model of the energy flow from salt marsh plants into decomposers and detritus feeders is presented in Fig. 6.3. When shoots of salt marsh plants decay, the energy mostly enters the fungal component that represents secondary production. Shredding invertebrates feed upon such detritus containing fungus. Once the leaves fall onto the sediment, the energy flows into the bacterial degradation pathway. The fragmented detritus may also enter the micro- and meiofaunal components of the food web.

- Salt marsh ecosystems are some of the most productive on earth. Vast stretches of their vegetations are found particularly in temperate and high-latitude regions.
- The salt marsh plant *Spartina alterniflora* has been studied most extensively for fungi, followed by *Juncus roemerianus* and *Phragmites australis*.
- Dead parts of salt marsh plants are extensively colonized and decomposed by fungi while they are still attached to the living plants.
- Salt marsh plants harbor a large diversity of fungi. Many obligate, mycetaen, marine fungi are found exclusively in decomposing salt marsh plants. *Buergenerula spartinae, Phaeosphaeria spartinicola*, and *Mycosphaerella* sp. are three dominant ascomycetes in standing *Spartina alterniflora* detritus.
- The composition of fungal communities in decomposing salt marsh grass is influenced by the host species, plant part, and the stage of decomposition. They also display a vertical zonation on salt marsh plants. Most of our

(continued)

knowledge on the role of fungi in decomposing salt marsh grasses comes from studies on the smooth cordgrass *Spartina alterniflora* in the southeastern US marshes.

- Endophytic fungi, phylloplane fungi, and weak parasites inhabit salt marsh grass plants when they are healthy and during senescence.
- The first phase of decomposition commences with death and a rapid leaching out of organic compounds. The second phase corresponds to the decompositional phase. Organic matter production in the salt marsh ecosystem is the result of the combined production of the salt marsh plants, algae, bacteria, and fungi. Fungi are the major decomposers of standing, dead marsh plants. Fungal biomass comprises a significant portion of the dry weight of standing, decomposing salt marsh grass. They often contribute more than 20% of the total dry weight mass of *Spartina alterniflora* detritus, much more than bacteria and algae. Decomposition of *Spartina alterniflora* is characterized by colonization and growth of various fungi, dominated by ascomycetes. The final phase of decomposition occurs when the leaves fall into the sediment and become fragmented. Both diversity and biomass of mycetaen fungi decline at this stage and bacteria prevail. Aquatic, zoosporic fungi belonging to Straminipila may colonize fragmented detritus in salt marshes. Species of the oomycete *Salisapilia, Halophytophthora* spp., and *Pythium grandisporangium* colonize *Spartina altemiflora* detritus. The complete decay and fragmentation may take over a year, much longer than for mangrove leaf, seagrass, and macroalgal detritus.
- Efficient lignocellulose degradation by salt marsh grass ascomycetes, such as *Phaeosphaeria spartinicola, P. obiones, Buergenerula spartinae,* and *Mycosphaerella* sp., results in physical disruption, release of DOM, and loss of weight. Fungal immobilization of nitrogen during the decompositional phase may result in an increase in nitrogen and decrease in C/N ratios. Microbial decay during the later part of this phase is very slow because only relatively refractory materials remain.
- Fungi are responsible in making salt marsh grass detritus highly palatable and nutritious to detritivores. The high fungal biomass available in *Spartina* detritus may be responsible for the high densities of detritivoral invertebrates in salt marshes. Some of the major detritivores in salt marsh ecosystem that benefit from fungi are the shredder gastropod *Littoraria irrorata* (salt marsh periwinkle), amphipods, and gastropods.

## Future Directions

1. North American salt marshes and a few others have been well researched. Diversity and dynamics of fungi in many salt marshes in many other geographical parts of the world have not been studied. This is particularly true of tropical salt marshes.
2. Most studies have focused on mycetaen fungi. Little is known of the ubiquitous straminipilan fungi, the Labyrinthulomycetes in salt marsh ecosystems.
3. A few studies have shown that yeasts are a component of the salt marsh ecosystem. However, they have been poorly studied so far.
4. One study has shown that salt marsh ascomycetes may be involved in the release of dimethylsulfide to the atmosphere. DMS is an "anti-greenhouse" gas. It will be interesting to study the role of fungi in this further.
5. Based on the enormous biomass and nitrogen contribution of fungi to salt marsh grass detritus, it has been surmised that they are of tremendous importance to detritivores. A few experimental studies have confirmed this. More studies on the contribution of nutrients by fungi to detritivores will help to provide a more complete picture.
6. As with mangroves, the role of fungi in outwelled detritus to coastal waters remains unknown. Biochemical and molecular markers for fungi will help understand this.

# Chapter 7
# The Seagrass Ecosystem

> *There's nothing wrong with enjoying looking at the surface of the ocean itself, except that when you finally see what goes on underwater, you realize that you've been missing the whole point of the ocean. Staying on the surface all the time is like going to the circus and staring at the outside of the tent.*
>
> Dave Barry

Heaps of detritus washed ashore are often a strong pointer to the vast, unseen, underwater meadows of **seagrass that thrive in shallow, protected, quiet, well-lit waters in both temperate and tropical coastal environments** (Fig. 7.1). Seagrasses are often found on the seaward side of mangroves or landward side of coral reefs. They are also found in clean estuaries and lagoons. **Seagrass vegetations rival salt marsh grasses in their productivity and support a complex ecosystem** (Duffy 2006).

About 50 species of the marine angiosperms, the seagrasses, are known. Seagrass vegetations serve several important functions. Their occurrence along with coral reefs or mangroves is characterized by an enormous flux of organic and inorganic material between the two systems, as well as migration of animals. They are important nurseries for many marine animals and support a number of commercially useful fish and shellfish. Seagrass meadows trap and bind sediments, thus preventing erosion. Upon death, seagrass detritus supplies substantial amounts of nutrition to organisms living within the seagrass ecosystem, as well as to those in adjacent waters and sediments.

Seagrass meadows have a global cover of about 60,000 $km^2$. Their NPP amounts approximately to $0.49 \times 10^9$ metric tons of C $yr^{-1}$. Only about 18% of this is grazed directly by herbivores, sea turtles, dugongs, or sea cows including manatees and waterfowl. **More than 50% of seagrass production enters the detrital food web** (Duarte and Cebrian 1996). Sea urchins (echinoderms), sea cucumbers, and crustaceans are voracious feeders of seagrass detritus.

**Fungi play a crucial role in functioning of the seagrass ecosystem.** Fungal parasites and pathogens can negatively affect their productivity. Mutualistic fungi may be essential for the health of seagrasses. Finally, fungi transform dead



S. Raghukumar, *Fungi in Coastal and Oceanic Marine Ecosystems*,
DOI 10.1007/978-3-319-54304-8_7

**Fig. 7.1** Underwater meadow of the seagrass *Thalassia testudinum* (turtle grass) in the South Pigeon Creek estuary, southeastern San Salvador Island, eastern Bahamas (Courtesy: James St. John, Ohio University Newark Campus, US)

seagrasses into palatable, nutritive detritus, through their saprotrophic, decompositional activities.

**Mycetaen as well as straminipilan fungi inhabit seagrasses, both as symbionts and saprotrophs. Among Mycetae, both obligate as well as facultative marine fungi occur in seagrasses.**

## 7.1 Symbiotic Fungi in Living Seagrasses

**Seagrasses defend themselves from surface, epiphytic growth, as well as internal microbial invasions by production of phenolic compounds.** Such compounds are known to have antifouling properties. For example, leaf extracts of the seagrass *Thalassia testudinum* inhibited surface settlement of zoospores from the thraustochytrid *Schizochytrium aggregatum*. Few thraustochytrids inhabit surfaces of the seagrass, probably because of the inhibitory property of this compound. A sulfated glycoside purified from the seagrass has been shown to be responsible for this antifouling property (Jensen et al. 1998).

**Despite the chemical defense of seagrasses, many fungi inhabit them as endophytes and parasites.**

### 7.1.1 Endophytic Fungi

Fungi colonize internal tissues of seagrasses in a symptomless manner. Their presence has been discovered by culturing and metagenomics, as well as direct microscopy. All parts of seagrasses contain fungi. Most of these are facultative marine fungi and hence belong to terrestrial species.

**Direct microscopic observations have shown that fungi inhabit healthy leaves of *Thalassia* and *Zostera* as endophytes** (Alva et al. 2002). Various patterns of colonization of internal tissues have been noticed in healthy leaves of *Thalassia testudinum*. These include colonization by intercellular hyphae in the mesophyll underlying the aerenchyma tissue of the seagrass, intracellular localization within a single aerenchyma cell, localization within single subepidermal cells, and colonization of the abaxial surface.

**Endophyte frequency may depend on the host plant and the plant part. Rhizomes harbor more endophytes than leaves.**

**Culture methods that employ surface sterilization have shown that facultative marine fungi are the most common as seagrass endophytes** (Alva et al. 2002; Venkatachalam et al. 2015).

- Terrestrial species belonging to *Acremonium* sp., *Gamsia sp., Arthrinium arundinis, Fusarium* sp., *Phoma* sp., *Trichoderma* sp., and *Penicillium chrysogenum* have been commonly found in *Thalassia testudinum* along coastal waters of Hong Kong and Philippines.
- The seagrass *Zostera marina* harbored high frequencies *of Trichoderma* sp., *Arthrinium arundinis, Amphobotrys* sp., *Cladosporium* sp., and *Paecilomyces* sp., while *Z. japonica* was inhabited by *Alternaria* sp., *Cladosporium* sp., and *Trichoderma* sp. Ten different seagrasses belonging to Hydrocharitaceae and Cymodoceaceae from the coastal waters of Tamil Nadu, India, were seen to harbor mostly species of *Aspergillus, Paecilomyces*, and *Penicillium*. Despite the fact that endophytes colonize seagrasses, their frequencies are lower compared to terrestrial plants.
- **Metagenomic studies based on LSU, ITS1, ITS2, and 5.8S rDNA sequence analyses have detected a high diversity of endophytes belonging to terrestrial species of fungi in seagrasses.** Majority of endophytes detected in the seagrass *Enhalus acoroides* in Thailand belonged to the Ascomycota (98%), while the Basidiomycota accounted for only 2%. The predominant ascomycetous orders were Hypocreales, followed by Eurotiales and Capnodiales, respectively. Species belonging to *Aspergillus, Penicillium*, and *Cladosporium* occurred in high frequencies. An unknown diversity of fungi belonging to Hypocreales has been found in this habitat (Sakayaroj et al. 2010, 2012).

**Roots of the Mediterranean seagrass, *Posidonia oceanica,* are colonized by special mutualistic fungi termed DSEs (dark septate endophyte)** (Fig. 7.2; Torta et al. 2015; Vohnik et al. 2015a, b). These fungi resemble DSEs found in roots of

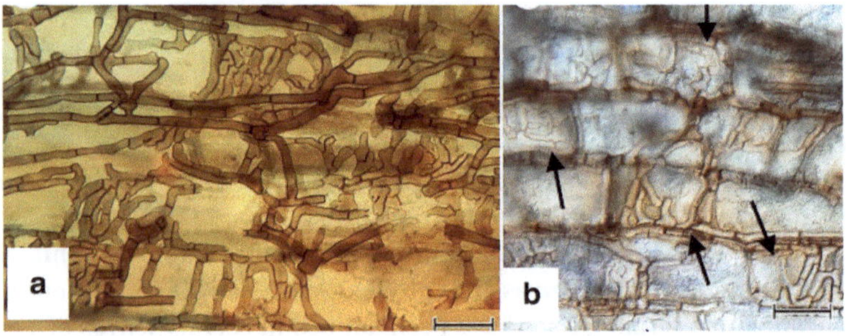


**Fig. 7.2** Morphology of the dark septate endophyte colonization in *Posidonia oceanica* roots (Source: Martin Vohník, Ondřej Borovec, Ivan Župan, David Vondrášek, Miloslav Petrtýl & Radka Sudová. 2015. Anatomically and morphologically unique dark septate endophytic association in the roots of the Mediterranean endemic seagrass *Posidonia oceanica*. Mycorrhiza 25: 663–672. Springer)

many terrestrial plants. As the name suggests, they produce melanized, septate, inter-, and intracellular hyphae and microsclerotia. Roots anchored on rock seem to harbor more of these than those on sand. The most dominant of these appears to represent a new monotypic lineage within the Pleosporales, Ascomycota. One of these has been described as a species of the ascomycete fungus *Lulwoana*. The consistent occurrence of DSEs within roots of this seagrass suggests a mutualistic relationship as with mycorrhizal fungi in terrestrial plants.

## *7.1.2 Fungal Diseases of Seagrass*

**A disease termed the "Wasting Disease" caused extensive, catastrophic losses of the seagrass *Zostera marina* (eelgrass) along the Atlantic coasts of North America and Europe in the 1930s.** The malady was estimated to have wiped out nearly 90% of the seagrass population. It briefly surfaced again in the 1980s, along the coasts of New Hampshire and Maine of USA. The disease had been reported much earlier when it caused a serious devastation of the seagrass population in Chesapeake Bay of the USA in 1889.

**The causal agent of the "wasting disease" is the straminipilan fungus *Labyrinthula zosterae* Porter & Muehlstein** (Fig. 7.3a; Tutin 1938; Short et al. 1986; den Hartog 1987; Muehlstein et al. 1988; Sullivan et al. 2013). This fungus has been regularly cultured from affected lesions of the plants. However, its role as the causal agent of the "wasting disease" was unclear for many years. It was only in the 1980s that studies based on **Koch's postulates confirmed this fungus to be the causal agent of the disease**.

**The disease initially appears as small brown dots on the leaf blades of the seagrass. Subsequently, the spots coalesce and became larger, black, and brown streaks** (Fig. 7.3b). Entire shoots may then be lost, often resulting in

**Fig. 7.3** The wasting disease of *Zostera marina* (eel grass). (**a**) Spindle-shaped cells of *Labyrinthula zosterae* in culture. (**b**) Leaves of *Zostera marina* in different stages of the disease. (**a** and **b** Courtesy of Daniel Martin, University of South Alabama USA). (**c**) *Labyrinthula* cells spreading between cells of *Thalassia hemprichii* leaf tissue (Source: Raghukumar, S. Ecology of the marine protists, the Labyrinthulomycetes (Thraustochytrids and Labyrinthulids). Europ. J. Protistol. 38: 127–145 (2002) © Urban & Fischer Verlag. With permission from Elsevier)

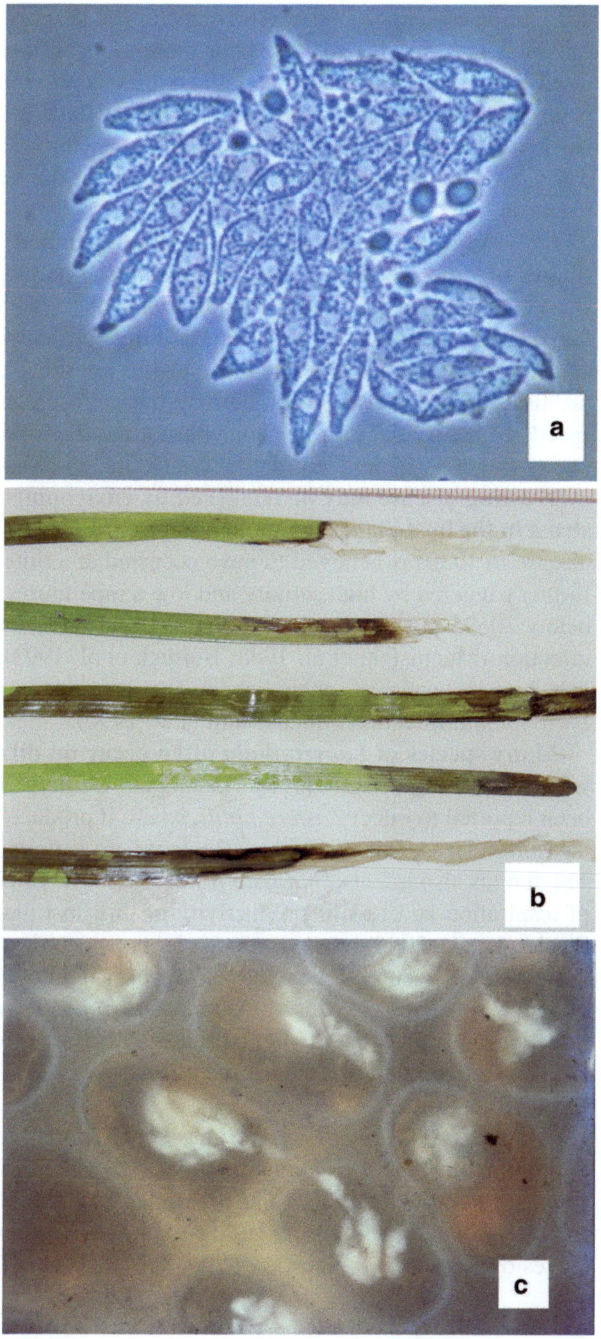

decimation of seagrass beds. Rhizomes of the seagrass may also be affected. The rhizomes may survive for 1 year or more after death of the stem and leaves. The disease spreads through contact with an infected growing plant or drifting detached plant parts. Although the disease results in dark lesions, the pathogen can be isolated more frequently from green and greenish-brown tissue. Hence, the hub of pathogen activity seems to lie in green tissue of the leaf, rather than in the darker patches.

***Labyrinthula zosterae* appears as spindle-shaped motile cells within the leaf tissues of *Zostera marina*.** An idea of this can be gleaned from Fig. 7.3c, which shows a similar invasion of the leaf tissue of the seagrass *Thalassia hemprichii* by *Labyrinthula* cells in coral reef islands of the Lakshadweep in the Arabian Sea. The cells are motile, move at a rate of up to 175 μm $min^{-1}$, penetrate cells, and invade the tissue. They subsequently destroy the cell contents. Lesions may expand at speeds of nearly 0.8 mm per hour during the day. Infection results in the loss of photosynthetic efficiency of the leaves (Ralph and Short 2002).

**Wasting disease may be triggered by environmental conditions that lead to stress in the host plant** (Hemminga and Duarte 2000). The outbreak of the wasting disease in 1930s is believed to have occurred at a time of high water turbidity and further triggered by high salinity and low temperatures. It is believed that salinities below 20–25 ppt reduce disease activity and allow eelgrass to recover from the infection (Muehlstein et al. 1988; Burdick et al. 1993). It has also been suggested that higher levels of phenolics in the host plant make it more resistant to *Labyrinthula* infection (Buchsbaum et al. 1990).

**Many species of *Labyrinthula* often occur on different seagrasses, however without causing serious losses** (Leaño and Damare 2012). *Labyrinthula* has also been reported to infect *Zostera noltii,* where it produces similar symptoms as in the "wasting disease" of *Z. marina*. However, no serious losses have been observed. The fungus also infects *Thalassia testudinum*. The infection results in a higher rate of respiration and loss of productivity of this host plant (Durako and Kuss 1994; Steele et al. 2005). *Labyrinthula* is considered obligately marine, as with labyrinthulomycetes in general. However, an interesting disease on turf grass used for golf courses, caused by *Labyrinthula terrestris*, has been reported. The infection causes the "rapid blight disease" (Bigelow et al. 2005; Craven et al. 2005). Perhaps true to its marine nature, the infection is caused when saline water is used to sprinkle the grass.

The seagrass *Halodule wrightii* has been reported to be infected by yet another labyrinthulomycete belonging to aplanochytrids. This fungus, *Aplanochytrium schizochytrops* (J.A. Quick) C.A. Leander & D. Porter, was recovered at a high frequency of nearly 80% from yellow to brown patches on the basal and outer living leaves of this seagrass (Quick 1974).

## 7.2 Saprotrophic Fungi in Seagrass Detritus

**Dead and decomposing seagrass is a rich substrate for fungi, which are found in high diversity, attain substantial biomass, and bring about biochemical changes in the detritus.** However, the role of fungi in seagrass decompositional dynamics has not been studied in as much in detail as with mangrove or salt marsh grass detritus.

**Endophytic fungi present in living seagrass may persist in the tissues after death of the plant and participate in its decomposition.** Most of the seagrass endophytic fungi are facultative marine fungi, belonging to terrestrial species.

**Facultative marine fungi from ambient water freshly colonize detritus of seagrass, thus joining the endophytic fungi that had persisted.**

- Species of *Acremonium, Chaetomium, Graphium, Humicola*, and *Penicillium* have been isolated from surface sterilized, decomposing leaves of *Thalassia hemprichii* from the coral reef islands of the Lakshadweep in the Arabian Sea, where this seagrass forms extensive beds (Sathe and Raghukumar 1991). Fine fragments of decaying *T. hemprichii* harbored a higher diversity of fungi, compared to green, light brown and large fragments of dark brown decomposing leaves. This is in interesting contrast to the salt marsh grasses, where diversity and biomass of fungi declines when detritus becomes fragmented.
- Terrestrial species of fungi have also been detected in the seagrass *Posidonia oceanica* (L.) Delile, an endemic and seriously threatened seagrass of the Mediterranean Sea, based on metagenomic methods. The most frequent such marine-derived fungi belonged to the genera *Penicillium, Cladosporium*, and *Acremonium* (Panno et al. 2011). Their active role as facultative marine in detritus of *P. oceanica* requires further confirmation.

**Seagrass detritus becomes colonized by a number of obligate, marine, mycetaen fungi.** (Table 7.1; Kohlmeyer and Kohlmeyer 1979; Cuomo et al.

**Table 7.1** Some obligate marine fungi in four species of seagrasses

| *Zostera marina* | *Thalassia testudinum* | *Posidonia oceanica* | *Cymodocea manatorum* |
|---|---|---|---|
| ***Lulworthia* sp.,** | ***Lulworthia* sp.,** | ***Lulworthia* sp.,** | ***Lulworthia* sp.,** |
| ***Corollospora maritima*** | ***Corollospora maritima*** | ***Corollospora maritima*** | |
| ***Varicosporina ramulosa*** | ***Varicosporina ramulosa*** | *Halotthia posidoniae* | |
| | *Lindra marinera* | *C. intermedia* | |
| | *L. thalassiae* | *Pontoporeia biturbinata* | |
| | *Corollospora lacera* | *Papulospora halima* | |
| | *Dendryphiella salina* | | |

1985). Many of these are lignicolous fungi and are regularly found also on allochthonous wood and other lignocellulosic detritus.

- The genus *Lulworthia* appears to be the most common obligate mycetaen fungi in decomposing seagrasses, followed by *Corollospora maritima* Werderm. The fungus *Halotthia posidoniae* Kohlmeyer may be specific to the seagrass *Posidonia oceanica*.
- Different fungi may be prevalent on certain plant parts. In the case of *Posidonia oceanica,* it was found that *C. maritima* occurred in all parts of the plant, except the rhizome. *Halotthia posidoniae* and *Pontoporeia biturbinata* (Durieu & Mont.) Kohlm. were confined to the rhizome, while *Papulaspora halima* Anastasiou, *Lulworthia* sp., and *Phoma* sp. were confined to the old leaf sheaths and the stems. In addition, different fungi may occur in detritus at different depths. For example, *H. posidoniae* was found at a depth of 30 m, while *Corollospora intermedia* I. Schmidt was found only between 3 and 6 m.

**Straminipilan fungi are frequently found in decomposing seagrass, in addition to facultative and obligate marine mycetaen fungi**. The aplanochytrid *Aplanochytrium minutum* (S. W. Watson & Raper) C. A. Leander & D. Porter and the thraustochytrid *Thraustochytrium motivum* Goldstein have been reported in decomposing *Thalassia hemprichii* from the Lakshadweep Islands of the Arabian Sea (Sathe and Raghukumar 1991).

**Fungi may contribute substantially to seagrass detrital carbon.**

- Blum et al. (1988) concluded that microbial biomass in seagrass detritus belonging to *Halodule wrightii, H. decipiens, Thalassia testudinum*, and *Syringodium filiformis* was poor. According to them, the total microbial biomass varied from 0.12 to 0.7% of detrital dry weight in these seagrasses.
- **Sathe and Raghukumar** (1991) **estimated a high total fungal biomass of 3.4% in detritus of *Thalassia hemprichii* in the Lakshadweep islands of the Arabian Sea**. The discrepancy could be based on the techniques used. While Sathe and Raghukumar, detected fungal hyphae in unfragmented leaf detritus and estimated biomass, Blum et al. homogenized the detritus, which might have resulted in loss of fungal details. Detritus of *Thalassia hemprichii* is densely pervaded by fungal hyphae, suggesting a large fungal biomass in detritus of this seagrass.

**Several fungi in seagrass detritus may be capable of removing phenolics and tannins, thus making the substrate more available for microbial colonization.** Many seagrass fungi produce oxidoreductase and tannase enzymes (Panno et al. 2011). Saprotrophic fungi from seagrasses produce a variety of degradative enzymes that help in breaking down detritus. Out of a number of seagrass-derived fungi tested in the laboratory, over 75% have been found to degrade plant structural polysaccharides, while only one produced lignin-modifying enzymes (Alva et al. 2002). A facultative marine fungus, *Flavodon flavus*, isolated from *Thalassia hemprichii* from Lakshadweep islands was shown to be a remarkable producer of lignin-degrading enzymes (Raghukumar, 2008).

## 7.3 Fungi in Sediments of Seagrass Vegetation

Soil in the immediate vicinity of plant roots is termed the rhizosphere. Rhizosphere fungi in terrestrial ecosystems are known to confer benefits to the plant by providing them specific nutrients. **Seagrass sediments might harbor rhizosphere fungi.** Rhizosphere of the seagrass *Posidonia australis* from shallow waters near Perth, Western Australia, have been shown to harbor fungi (Kuo et al. 1981). These penetrate the epidermal cells and lyse the walls of the hypodermal cells. They then inhabit the lumen of the hypodermal and cortical cells of the root. It is believed that rhizosphere fungi in seagrasses might perform a function similar to those in land plants.

**Thraustochytrids are found to be particularly abundant in marine sediments underlying seagrass (*Posidonia oceanica*) beds** (Bongiorni et al. 2005a, b). Thraustochytrids largely dominated over heterotrophic nanoflagellates biomass in *P. oceanic*a beds characterized by the presence of highly refractory organic matter dominated by structural carbohydrates, up to 70% of the biopolymeric C fraction. Thraustochytrids contributed 0.124 μgC $g^{-1}$ sediment, compared to 0.044 μgC $g^{-1}$ contributed by heterotrophic nanoflagellates. Thraustochytrids reached up to 50% of total protist abundance (as sum of thraustochytrids, nanoflagellates, and ciliates). A rough estimate of thraustochytrid contribution to the total extracellular enzymatic activities in two types of sediments showed that these organisms may contribute up to 4% of the total β-D-glucosidase, 1% of aminopeptidase, and 1.4% of alkaline phosphatase activities. However, these data could be underestimates since they were based only on culturable thraustochytrids.

## 7.4 Fungi as Food for Seagrass Detritivores

**Many animals consume both living and detrital seagrass containing microbial biomass.** The purple sea urchin, *Paracentrotus lividus* is the most dominant herbivore of the seagrass *Posidonia oceanica* in the Mediterranean waters. The sea urchins prefer brown leaves to green leaves as food, apparently because the former are colonized by fungi and bacteria. It is possible that sea urchins are unable to digest the cellulose present in green leaves of *P. oceanica* because they lack appropriate enzymes. Hence, they may prefer to graze on brown leaves that contain cellulolytic marine fungi which break down cellulose and make the detritus more palatable. Two fungi that inhabit the seagrass *P. oceanica*, namely, *Corollospora maritima* and *Lulworthia* sp., are known to produce cellulases that can break down complex ligno-cellulose compounds. Partially digested seagrass detritus of *P. oceanica* has been detected in the gut contents of the sea urchin *P. lividus* (Traer 1979). This material is excreted by the sea urchin, which thus is responsible for the initial step of the seagrass detritus formation. The feces becomes overgrown by bacteria and fungi. Pieces of brown leaf material of *P. oceanica* present in the fecal pellets of sea urchins have been shown to produce ascocarps of the obligate marine fungus *Corollospora maritima* upon incubation (Cuomo et al. 1985). In addition to *Paracentrotus lividus*, crustaceans and the sea cucumbers or holothurians feed upon detritus in seagrass beds. These may utilize fecal and detrital material arising from seagrasses as a food resource.

- Seagrasses produce chemical defenses against microbial colonization. Yet, many mycetaen fungi inhabit them as endophytes. The straminipilan fungus *Labyrinthula* is also found in living seagrasses.
- Endophyte frequency may depend on the host plant and the plant part. Rhizomes harbor more endophytes than leaves.
- Facultative marine fungi are the most common as seagrass endophytes.
- Roots of the Mediterranean seagrass, *Posidonia oceanica*, may harbor special mutualistic fungi termed DSEs.
- The straminipilan fungus *Labyrinthula zosterae* causes the "wasting disease" of the seagrass *Zostera marina* (eelgrass). The fungus caused a serious devastation of the population of the seagrass in the 1930s. *Labyrinthula zosterae* appears as spindle-shaped motile cells within the leaf tissues of *Zostera marina*.
- Endopytic fungi that are present in the living seagrass may persist in the tissues after death of the plant and actively participate in its decomposition. Facultative and obligate marine, mycetaen, as well as straminipilan fungi are active decomposers of seagrass detritus.
- Seagrass may harbor mutualistic rhizosphere fungi. Thraustochytrids occur in high numbers in seagrass sediments.
- Seagrass detritus constitutes the food for many detritivores such as sea urchins and sea cucumbers in the seagrass ecosystem.

## Future Directions

1. Seagrasses are autochthonous marine vascular plants. How unique is the fungal diversity in these in different geographical areas, seasons, and hosts?
2. Can metagenomics unravel the diversity of fungi in living and decaying seagrasses?
3. Do seagrass endophytes produce anti-grazing compounds like those produced by endophytes of terrestrial plants?
4. What is the combined biomass of straminipilan and mycetaen fungi in seagrasses? How important is their biomass to detritivores?
5. How do fungi enzymatically decompose seagrass detritus and help in nutrient recycling?

6. How prevalent are "DSEs" in rhizomes of seagrasses and what is their importance as mutualistic fungi?
7. Estimates of the transport of detritus with fungi through outwelling to coastal and oceanic waters will help in understanding their role in such ecosystems.
8. What is the balance between the endophytic phase of *Labyrinthula* versus its transformation into a parasite?

# Chapter 8
# The Macroalgal Ecosystem

*Obtaining all of medicinal plants and, indeed, all of jewels,*
*O gods, churn the ocean, then you will gain the nectar.*
*Hindu Purana: "Samudramanthana" of Mahabharata,*
*Adi Parvan*

**A striking feature of rocks and solid substrata in well-lit surfaces in coastal environments is the dense vegetation of diverse filamentous and thalloid macroalgae or seaweeds that often covers them** (Fig. 8.1). Green algae grow abundantly in the intertidal zone. Red algae mostly prefer subtidal conditions. The green algae (Phylum Chlorophyta) and the red algae (Phylum Rhodophyta) are true plants, belonging to the Kingdom Viridiplantae. Brown algae, which belong to the Kingdom Straminipila, prefer the subtidal regions or grow under constantly wetted littoral regions. Several brown algae, the kelps, form magnificent underwater forests. Kelp forests are abundant in temperate and polar waters and support an entire ecosystem. *Laminaria* and *Fucus* are well-known kelps. A vast diversity of macroalgae inhabit coral reefs.

**Macroalgae are some of the most important primary producers of the coastal oceans in tropical and temperate waters and support a number of herbivores and grazers.**

Algal beds and vegetations cover an area of about $7.4 \times 10^6$ km$^2$ and have a global NPP of nearly $3.15 \times 10^9$ metric tons of carbon (Fig. 2.3). Macroalgal vegetations are highly productive ecosystems. About 30% of the NPP of macroalgae is consumed by herbivores. Nearly 40% of their NPP and more than 75% of that of coral reef algae enter the decomposition pathway. **Macroalgae, therefore, contribute enormously to marine detritus and nutrition of coastal detritivores.**

Fungi are frequently associated with macroalgae. The first algae-inhabiting fungi were described in the late nineteenth century, more than a hundred years ago. The first systematic studies on fungal parasites in marine algae were carried out by A.D. Cotton and D.K. Sutherland at the beginning of the twentieth century. After a long gap following this, several authors, namely Irene Wilson, Jan Kohlmeyer, and many others during the 1960s, contributed tremendously to our knowledge on fungi in marine algae (Chap. 3).



S. Raghukumar, *Fungi in Coastal and Oceanic Marine Ecosystems*,
DOI 10.1007/978-3-319-54304-8_8

**Fig. 8.1** Macroalgal vegetations. (**a**) A sublittoral kelp forest of *Fucus* sp. (**b**) An intertidal kelp vegetation of the "sea palm" *Postelsia palmaeformis*. *(Photos by S. Raghukumar)*

**Fungi are associated with marine macroalgae in various ways similar to their occurrence in land plants and play important ecological roles in the macroalgal ecosystem.**

**Mutualistic fungi aid in sustaining the productivity of macroalgae. Pathogens have the potential of affecting the macroalgal population and negatively affecting productivity and diversity. Commensalistic fungi are casual inhabitants that do not cause harm or confer benefits to the algae. Saprobic detrital fungi could be the key to making decomposing macroalgae more palatable to detritivores.**

**Obligate and facultative marine fungi inhabit macroalgae. Obligate marine fungi that are found exclusively in living and decomposing algae are termed "algicolous fungi." Mycetaen as well as straminipilan fungi occur as algicolous fungi. About 80 species of obligate marine mycetaen fungi have been described in algae** (Kohlmeyer and Volkmann-Kohlmeyer 2003). Most of these belong to Ascomycota.

Marine algae produce a variety of polyphenolic compounds and various metabolites that deter fouling of their external surfaces (Jensen et al. 1998; Bhadury and Wright 2004). These have antibacterial, antialgal, antimacrofouling, and antifungal properties. Antimicrobial metabolites produced by the algae play an important role in preventing fungal invasion and colonization. However, **several fungi seem to have developed adaptations to tolerate or overcome the toxic effects of antimicrobial compounds produced by algae. Multicellular, perennial algae with structural complexity may be more favorable for fungal colonization than delicate, short-lived ones.** Brown and red algae harbor fungi much more commonly than green algae. Some of the well-studied algal genera are *Ascophyllum*, *Fucus*, *Laminaria,* and *Sargassum* among brown algae and *Ballia*, *Chondrus, and Dilsea* among red algae. Fungi may live as symbionts in living algae or as saprotrophs in decomposing ones.

## 8.1 Symbiotic Fungi in Macroalgae

**Symbiotic fungi in marine algae may occur as endophytes, mutualists, or as parasites** (Table 8.1).

### *8.1.1 Endophytic Fungi in Marine Algae*

**Many fungi grow asymptomatically in the internal tissues of algae.** This is confirmed by surface sterilization and plating on culture media of living algal thalli. Mycetaen as well as straminipilan endophytic fungi can be easily cultured from their internal tissues (Sathe-Pathak et al. 1993; Zuccaro et al. 2008; Suryanarayanan et al. 2010; Suryanarayanan 2012).

**Table 8.1** Examples of various forms of symbiotic association of fungi with marine macroalgae

| Substrate specificity | Symptoms | Ecological role | | |
|---|---|---|---|---|
| | | Commensals | Mutualists | Parasites |
| Found only in algae (Algicolous) | Endophytes and epiphytes (non-symptomatic) | – | Mycophycobionts (*Mycophycias ascophylli*, *Blodgettia confervoides*) | *Chadefaudia*, *Thalassoascus*, *Pontogeneia*, *Retrostium*, *Hispidocarpomyces*, *Thalassochytrium gracilariopsis*, *Chytridium polysiphoniae* |
| | Non-endophytic (symptomatic) | – | Lichens (*Verrucaria* spp., *Pyrenocollema* spp.) | *Phycomelaina laminariae*, *Haloguignardia cystoseirae*, *Spathulospora*, *Pythium porphyrae*, *Eurychasma dicksonii*, *Petersenia pollgaster* |
| Broad range of substrates | Endophytes (non-symptomatic) | Facultative marine fungi; Obligate marine fungi such as *Acremonium fuci* and *Lulworthia*; *Aplanochytrium*; *Labyrinthula* | | |
| | Non-endophytic (symptomatic) | | | *Lindra* spp.; *Labyrinthula* sp. |

- **Most algal endophytic fungi are facultatively marine and belong to anamorphic species of terrestrial ascomycetes.**
  - Endophytic fungal assemblage in marine algae may be different from that in terrestrial plants. The common endophytes in terrestrial plants are genera such as *Phyllosticta*, *Colletotrichum*, *Phoma*, and *Pestalotiopsis*. These seem to be absent in marine algae. **Species of *Aspergillus*, *Cladosporium,* and *Penicillium* appear to have adapted well to marine environmental conditions, resulting in colonization of marine algal hosts in the form of endophytes.** Terrestrial fungi belonging to *Acremonium*, *Phoma,* and yeasts inhabit healthy *Sargassum cinereum.* The straminipilan fungus *Aplanochytrium minutum (Labyrinthuloides minuta)* occurs as an endophyte in *Sargassum cinereum*, while *Ulkenia visurgensis* (Ulken) Gaertner occurs as an epibiont.
  - Loque et al. (2010) used sequence data to identify marine-derived fungi associated with the marine algae *Adenocystis utricularis*, *Desmarestia anceps,* and *Palmaria decipiens.* Of these, 27 belonged to filamentous mycetaen fungi and 48 to yeasts. The most frequently detected were the terrestrial fungi *Geomyces pannorum* and the yeast *Metschnikowia australis.* These have also been isolated earlier from other marine habitats (Zuccaro et al. 2008). Most of the fungi detected were anamorphic taxa and yeasts. It is not clear if these marine-derived, terrestrial species of fungi are facultatively marine or not.

- **Diversity and density of colonization of endophytic fungi vary with the host species of algae.** Some species of fungi have a broad host specificity, while others seem restricted to a few hosts. *Aspergillus terreus* was the most frequently isolated endosymbiont in *Caulerpa scalpelliformis*, *Halimeda macroloba*, *Ulva lactuca*, *Ulva fasciata*, *Lobophora variegata*, *Padina gymnospora*, *Stoechospermum marginatum*, *Sargassum ilicifolium*, *Portieria hornemannii*, and *Gracilaria edulis* in coastal waters of Tamil Nadu. Species of *Cladosporium* have been isolated from a number of marine algae such as *Caulerpa racemosa*, *C. sertularioides*, *Ulva lactuca*, *Padina gymnospora*, *Sargassum wightii*, *Turbinaria conoides*, *Turbinaria.* sp., *Portieria hornemannii*, *Grateloupia lithophila*, *Halymenia* spp., and *Fucus serratus*. Green algae are densely colonized by endophytes, but have a low diversity. Thus, the green alga *Ulva lactuca* from coastal waters of Tamil Nadu, India, showed 100 % frequency in endophyte occurrence. **Brown algae have the highest diversity of fungal endophytes.** Twenty-five species were isolated from a single algal host, *Turbinaria* sp. The brown alga *Fucus serratus* from a rocky-shore site on the northeastern side of Helgoland Island, Germany, harbors a high frequency of terrestrial species of fungi such as species of *Cladosporium*, *Engyodontium,* and *Fusarium*.
- **Obligate marine, mycetaen fungi are common endophytes of marine algae. The brown alga *Fucus serratus* harbors a very high density of *Halosigmoidea (Sigmoidea) marina* (Haythorn & E.B.G. Jones) Nakagiri, K.L. Pang et E.B.G. Jones, which is an anamorph of the obligate marine**

**fungus, *Corollospora* sp.** This species grows systemically inside algal thalli. Other obligate marine species in this brown alga are *Acremonium fuci* and *Dendryphiella salina*. Spores of *A. fuci* can germinate only in the presence of *Fucus serratus* tissues and not in seawater, indicating its host specificity. Molecular techniques such as 28S rRNA gene PCR-denaturing gradient gel electrophoresis (DGGE) and phylogenetic and real-time PCR analyses have revealed the presence of many more obligate marine species such as *Lindra* and *Lulworthia* in *Fucus serratus* (Zuccaro et al. 2008).

- **Fungi belonging to Chytridiomycota are also known as algal endophytes**, in addition to facultative and obligate marine members of Ascomycota. Thus, the chytridiomycete *Coenomyces* inhabits the green algae *Cladophora* and *Rhizoclonium* without causing any external symptoms. They can be isolated by seawater incubation of the algal thalli (Raghukumar 2006).
- **Stramenipilan fungi occur frequently as algal endophytes**. *Labyrinthula* sp. can be observed directly within thalli of several green algae such as *Cladophora*, with no visible symptoms of its presence (Raghukumar 1987). Aplanochytrids can be regularly cultured from marine algae, where they might live as endophytes. *Aplanochytrium minutum* has been isolated from green and red algae and also consistently cultured from the brown alga *Sargassum cinereum* and *Padina tetrastomatica* following surface sterilization and culturing, where they do not cause any external symptoms. Another species, *Aplanochytrium kerguelense* Bahnweg et Sparrow, has also been described from the brown alga *Sargassum* (Sathe-Pathak et al. 1993; Raghukumar 2002).

### *8.1.2 Mutualistic Associations Between Fungi and Algae*

**Fungi form important mutualistic associations with a number of algae in the form of mycophycobionts and lichens.** Each organism helps in the survival of the other in such associations. (1) Mycophycobiosis is an obligate algal–fungal association in which the fungal partner is immersed within the photosynthetic partner.

(2) Lichens are obligate fungal–algal, mutualistic associations in which the fungi form the external structure surrounding an algal core. Fungi involved in these associations belong to the Ascomycota.

The photosynthetic partner is called the photobiont.

**Mycophycobionts are those in which the alga predominates and no prominent external symptoms are seen** (Table 8.2; Kohlmeyer and Kohlmeyer 1979; Kohlmeyer and Volkmann-Kohlmeyer 2003; Hawksworth 2000). Hence, mycophycobiosis is essentially an endophytic mode of fungal life. This relationship is believed to be mutualistic.

- ***Mycophycias (Mycosphaerella) ascophylli* (Cotton) Kohlm. & Volkm.-Kohlm. is an ascomycete belong to Eurotiomycetes and lives in the brown**

**Table 8.2** Examples of mycophycobiontic association between fungi and marine algae

| | Causal Organism | Host | References |
|---|---|---|---|
| Ascomycota, anamorph | *Blodgettia confervoides* Harvey | Green alga *Cladophora* species | Hawksworth (1988) |
| Ascomycota, Dothideomycetes | *Mycophycias ascophylli*; Anamorph: *Septoria ascophylli* | Brown algae *Ascophyllum nodosum* and *Pelvetia canaliculata* | Kohlmeyer and Volkmann-Kohlmeyer (1998) |
| | *Mycophycias apophlaeae* | Red alga *Apophlaea lyallii* | |
| Ascomycota, Sordariomycetes | *Mastodia tessellata (syn. Turgidosculum complicatulum)* | Green algae *Prasiola borealis* and *P. crispa* ssp. *antarctica* | Kohlmeyer and Kohlmeyer (1979) and Hyde et al. (1998) |

**algae *Ascophyllum nodosum* (Knotted Wrack) and *Pelvetia canaliculata* (Channeled Wrack).** This association was first discovered in 1913 by A.D. Cotton and later by G.K. Sutherland along the coast of England. Mycelium of the fungus pervades the interior of the algae without causing any visible symptoms (Hawksworth 1988). Maximal hyphal density occurs in the apex of the alga, followed by the middle thallus and receptacles. Old thalli contain few hyphae. The fungus forms large hyphal nodes in intercellular spaces. These are suspected to be dispersal points for nutrients from the algal host to the fungus. The fungus becomes visible only when it produces ascocarps on the surface of the algae, which appear as tiny dots on the receptacles of the algae, which are the fertile portions. The life cycles of the fungus and the algae appear to be intimately related. Ascospore release coincides with the release of the algal oospores. **It has been surmised that the fungus helps the alga in overcoming desiccation when the alga is exposed to air during the intertidal period.** However, this mutualistic relationship is still debated (Jones et al. 2012). Hyphae of *Mycophycias ascophylli* can also penetrate the red alga *Polysiphonia lanosa*, which grows epiphytically on the brown algae. Filaments of the infected alga are longer and have a greater apical diameter. The fungus grows in culture. The anamorph of this fungus is *Septoria ascophylli*.

- **Another example of mycophycobiosis is that of the fungus *Mastodia tessellata* (Hook. f. & Harv.) Hook. f. & Harv., which is found in the Antarctica and subantarctic waters of the Southern Ocean.** The fungus is an ascomycete with unitunicate asci and lives in association **with the green algae *Prasiola borealis* and *P. crispa* ssp. *antarctica*. The fungal association is believed to confer the benefit of resistance to desiccation at low tides to its algal partners.**

**Table 8.3** Some common marine lichens

| Supralittoral | Littoral | Sublittoral |
|---|---|---|
| • *Caloplaca marina* (Wedd.) Zahlbr. ex Du Rietz; grows on rocks<br>• *Lichina confinis* (O.F. Müll.) C. Agardh: grows on rocks | • *Verrucaria ditmarsica* Erichs., *V. maura* Wahlenb., *V. mucosa* Wahlenb., *V. striatula* Wahlenb.: On rock, wood, *Littorina*, barnacles, and algae. Photobionts include green algae. *Lichina pygmaea (*Lightf.) C. Agardh; grows on rocks.<br>• *Collemopsidium halodytes* (Nyl.) Grube & B.D. Ryan; grows on barnacle shells. | • *Collemopsidium sublitorale* (Leight.) Grube & B.D. Ryan; grows on limpets and barnacles causing pitting to the shell.<br>• *Pharcidia balani* (Winter) Bouch, *P. rhachians* Kohlm., *P. laminariicola*: Grows on shells of molluscs and cirripedes; epiphytic algae on *Laminaria* sp.<br>• *Leioplea pelvetiae* (Sutherland) Kohlm. and Kohlm.; grows on the brown alga *Pelvetia canaliculata*.<br>• *Halographis runica* Kohlm. & Volkm.-Kohlm.; grows on coralslabs and snail shells in Belize and Australia. |

**Marine lichens are found in the supralittoral, littoral, and immediate sublittoral of many oceans** (Table 8.3; Fig. 8.2; Kohlmeyer and Kohlmeyer 1979; Kohlmeyer and Volkmann-Kohlmeyer 2003; Hawksworth 2000; Jones et al. 2009). Lichens are well-known examples of obligately mutualistic partners. The fungal partners of marine lichens belong to Ascomycota. Green algae and cyanobacteria are the most common algal photosynthetic partners or photobionts of marine lichens. "Primtiive lichens" are non-obligate associations in which the algal or fungal partner can also live independently of each other. The mutualistic association, thus, is not obligate and is loose. The fungal partner does not develop a well-differentiated cortical layer as in true lichen associations. About 700 lichens are known from marine or maritime habitats (Hawksworth 2000). Marine lichens are cosmopolitan but are particularly abundant in temperate and polar regions of the world. Lichen diversity depends on their exposure to seawater (Table 8.3).

Their vertical distribution corresponds to three zones.

- The supralittoral zone harbors lichens that are immersed or sprayed with seawater only at high tide.
- The intertidal or littoral zone is inhabited by diverse lichens that grow on rock surfaces. Species belonging to *Verrucaria, Collemopsidium, and Lichina* are the most common lichens in this intertidal zone. All three are ascomycetes belonging to the Eurotiomycetes.
- Only a few lichens have been reported from below the low tide level, under completely submerged conditions. Common among these is *Pharcidia, which occurs* on brown algae, *Laminaria digitata,* or shells of marine animals, e.g., molluscs. The fungus belongs to the Dothideomycetes.

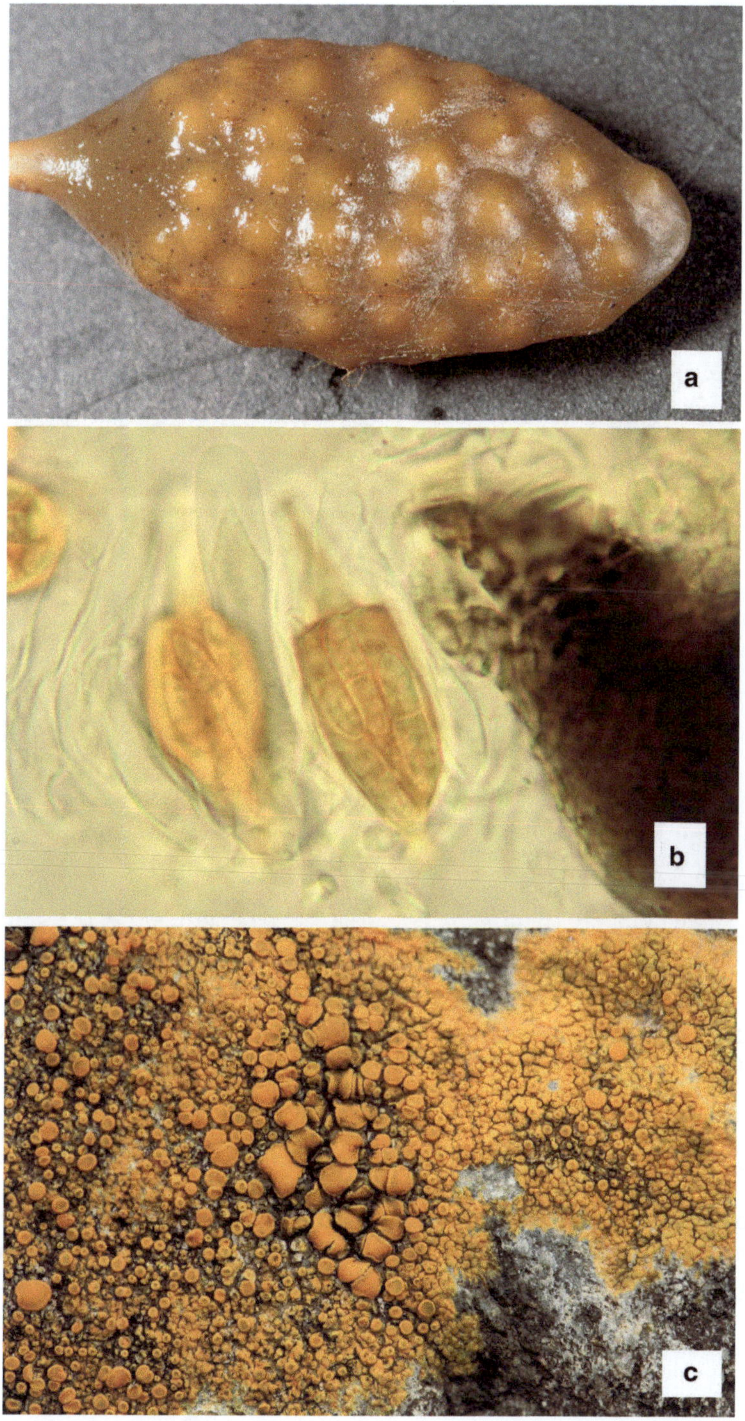


**Fig. 8.2** Mutualistic associations of fungi with macroalgae. (**a** and **b**) *Mycophycias ascophylli* in the brown alga *Ascophyllum nodosum*. (*Credit: Malcolm Storey. http://www.bioimages.org.uk/*)

**Fig. 8.2** (continued) *html/r164289.htm*) (**a**) Ascocarps appear as *minute black dots* on the receptacle. (**b**) Asci within ascocarps of the fungus embedded within the alga. (**c–e**) Marine lichens (*Credit:David Fenwick (www.aphotomarine.com)*. (**c**) *Caloplaca marina*. (**d**) *Lichina pygmaea*. (**e**) *Verrucaria mucosa*

Photobionts of marine lichens are mostly cyanobacteria such as *Hyella* or various green algae. The lichen *Verrucaria tavaresiae* is unique in having a brown alga, *Petroderma maculiforme*, as a photobiont.

### 8.1.3 Parasites of Macroalgae

**Fungal diseases of marine algae are quite prevalent and can be detected through a diligent search of algal hosts** (Porter 1986; Raghukumar 2006; Tables 8.1 and 8.4). **Some of these fungi are obligate parasites (biotrophs), while others are facultative or opportunistic parasites.** All groups of algae, namely, Chlorophyta (green algae), Rhodophyta (red algae), and Phaeophyta (brown algae), are prone to fungal diseases.

**Fungi belonging to Mycetae are common on large, filamentous, and thalloid macroalgae. Particular parasites are restricted to brown, red, or green algae, suggesting a coevolution of the fungal parasites and their hosts.** For example, members of the ascomycete genus *Haloguignardia* are found on various brown algae, while *Spathulospora* spp. infect red algae. Some are restricted to particular organs. Thus, *Lindra thalassiae* inhabits air vesicles of the brown alga *Sargassum*, where it causes lesions. These fungi have different effects on their algal hosts (Kohlmeyer and Kohlmeyer 1979).

**Several fungi parasitize algae without causing any external symptoms.** These are parasitic endophytes. The reason to call them parasites is that they live within their hosts and derive their nutrition from them. Besides, they are not known to confer any advantages to their hosts as mutualists would do. Many of these are ascomycetes. The presence of such parasites is very difficult to discern by examining the morphology of their hosts. One may detect them by culturing or by metagenomics. Their presence becomes apparent on their hosts only during those times when they begin to reproduce by producing sporulating structures on their algal hosts. These too may often escape notice.

- **Species of *Pontogeneia* Kohlm. parasitize a variety of green and brown algae** (Table 8.4) without producing external symptoms. This is a unitunicate ascomycete belonging to Sordariomycetes. **Another example is that of the ascomycete *Chadefaudia*,** which belongs to the most diverse of obligately marine mycetaen fungi, the Halosphaeriales. **All six species of this ascomycete live as parasites in various red algae.**
- **A few chytrids live as parasites without causing distinct symptoms** (Gleason et al. 2012a). **The chytridiomycete *Thalassochytrium gracilariopsis* is a biotrophic, endosymbiotic, polycentric species that specifically infects the red alga *Thalassochytrium gracilariopsis*.** Multinucleate, septate hyphae of the fungus penetrate between the cells, and their haustoria penetrate the medullary cells. Large, endobiontic zoosporangia are produced within the algal tissues. Tips of their discharge tubes protrude out. The infection does not kill the host although thylakoid membranes become disorganized and starch reserves are degraded during sporulation of the fungus. Numerous oval, orange zoospores are released during sporulation. The fungus has not been cultured so far.

**Table 8.4** Examples of some fungal parasites of marine algae

| Taxonomic group | Causal organism | Algal host | Remarks |
|---|---|---|---|
| Kingdom Mycetae: Chytridiomycota | *Chytridium polysiphonae* E. Cohn | Red algae *Polysiphonia; Ceramium*, *Centroceros*, and brown alga *Pylaeiella* | Raghukumar (2006), Amon (1984), Gleason et al. (2012a) |
| | *Rhizophydium littoreum* | Green algae, *Bryopsis* and *Codium* | |
| | *Thalassochytrium gracilariopsis* | Red alga *Gracilariopsis* sp. | |
| | *Olpidium rostiferum* | Green alga *Cladophora* spp. | |
| Kingdom Mycetae: Ascomycota | *Chadefaudia* Feldm.; Halosphaeriales | Various red algae | |
| | *Pontogeneia* Kohlm. | Green algae *Codium*, *Valoniopsis pachynema*, *Castagnea chordariiformis*, *Microdioctyon* and *Egregia* and brown algae *Padina* and *Halopteris* | |
| | *Haloguignardia* Cribb & J.W. Cribb | Brown algae *Cystoseira*, *Halidrys,* and *Sargassum* | Apt (1988) |
| | *Spathulospora* Cavaliere & Johnson | Several red algae | Jones et al. (2009) |
| | *Lindra thalassiae* | Brown alga *Sargassum* sp. | |
| Kingdom Mycetae: Basidiomycota | *Mycaureola dilseae* | Red alga *Dilsea carnosa* | |
| Kingdom Straminipila: Oomycota | *Pythium porphyrae* | Red alga *Porphyra yezoensis* | Raghukumar (2006), Hyde et al. (1998) |
| | *Pontisma lagenidiodes* | Green alga *Chaetomorpha media* | |
| | *Sirolpidium bryopsidis* | Green alga *Cladophora* spp. | |
| | *Petersenia palmaria* | Red alga *Palmaria mollis* | |
| | *Petersenia pollgaster* | Red alga *Chondrus crispus* | |
| | *Olpidiopsis porphyrae*, *Olpidipsis* spp. | Several species of green, red, and brown algae | Marano et al. (2012), Fletcher et al. (2015) |
| | *Eurychasma dicksonii* | Filamentous brown algae *Pylaiella littoralis* and *Ectocarpus* sp. | |
| Kingdom Straminipila: Hyphochytriomycota | *Anisolpidium rosenvingei* | Brown algae *Ectocarpus*, *Hincksia*, *Sphacelaria*, *Pylaiella littoralis* | |
| Kingdom Straminipila: Labyrinthulomycetes | *Labyrinthula* | Green algae *Chaetomorpha*, *Cladophora,* and *Rhizoclonium*; Cyanobacterium *Lyngbya* | Raghukumar (2006) |

- **Another biotrophic chytrid parasite is *Chytridium polysiphoniae*, a monocentric species that infects *Pylaiella littoralis* and many other species of brown algae** (Phaeophyta). The sporangia grow on the surface of the host cells and the rhizoids penetrate into the cytoplasm of host cells. Zoospores of *C. polysiphoniae* attack the distal uniseriate parts of brown algae. In culture, individual infected host cells become depleted and eventually die, whereas the host alga will continue to grow. This fungus can infect a large number of species from many parts of the world. **Massive epidemics have been observed along the European coast.** The prevalence can often exceed 10% of *Pylaiella* thalli at sites in Shetland (Küpper and Müller 1999). This chytrid is also a parasite of many red algae, such as *Polysiphonia* sp., *Callithamnion* sp., *Ceramium* sp., and *Centroceras* sp.

**Some fungal parasites cause light to dark discolorations of their hosts or various distinct symptoms** (Tables 8.1 and 8.4; Fig. 8.3).

- *Lautitia danica* causes a blackening of reproductive host tissue on the red alga *Chondrus crispus* by production of fungal stromata (Stanley 1992). The fungus is tissue specific and restricted to the reproductive cystocarpic and tetrasporangial regions. Few sporocarps are present in the cystocarp during the early stages of growth, while the entire cystocarp becomes blackened at later infection stages. The sporocarps protrude through the epidermis of the host.
- ***Phycomelaina laminariae*** (Rostr.) Kohlm. causes the "stipe blotch of kelp" on brown algae belonging to *Laminaria* species. This is an ascomycete belonging to the Phyllachorales of the Sordariomycetes. Hyphae pervade the stipe of the algae, causing a blackening of the surface and form superficial ascocarps. The fungus causes damage to the tissues which slough off, exposing the alga to damage by other organisms.
- ***Haloguignardia cystoseirae*** Cribb & J.W. Cribb is an obligate marine fungus that parasitizes the brown algal genera *Cystoseira*, *Halidrys*, and *Sargassum* (Apt 1988). It is a unitunicate ascomycete of the family Lulworthiaceae (Order Lulworthiales, Class Sordariomycetes) which typically contains obligate marine fungi. It has been reported from the Sargasso Sea, California, Australia, and New Zealand. It is probably a weak parasite and causes galls in the form of subglobose, ellipsoidal, or elongated protuberances on the algae (Kohlmeyer and Kohlmeyer 1979). The galls contain a mixture of algal tissues and fungal hyphae. They also bear spermogonia and ascocarps, the sexual structures of the fungus. Another ascomycete *Massarina cystophorae* forms galls on the brown algal genus *Cystophora*.
- ***Spathulospora*** Cavaliere et Johnson is a biotrophic parasite. All five species are parasitic on red algae, particularly species belonging to *Ballia*. This obligately marine ascomycete belongs to the Lulworthiales of Sordariomycetes. *Spathulospora* has been reported from southern temperate waters of the Pacific, Atlantic, and Indian Oceans (Kohlmeyer and Volkmann-Kohlmeyer 1975; Kohlmeyer and Kohlmeyer 1979). The fungus appears to penetrate just a single cell of the host, but probably derives its nutrients from the adjacent cells as well. It develops a stroma on the host surface. Ascocarps are found in the stroma.

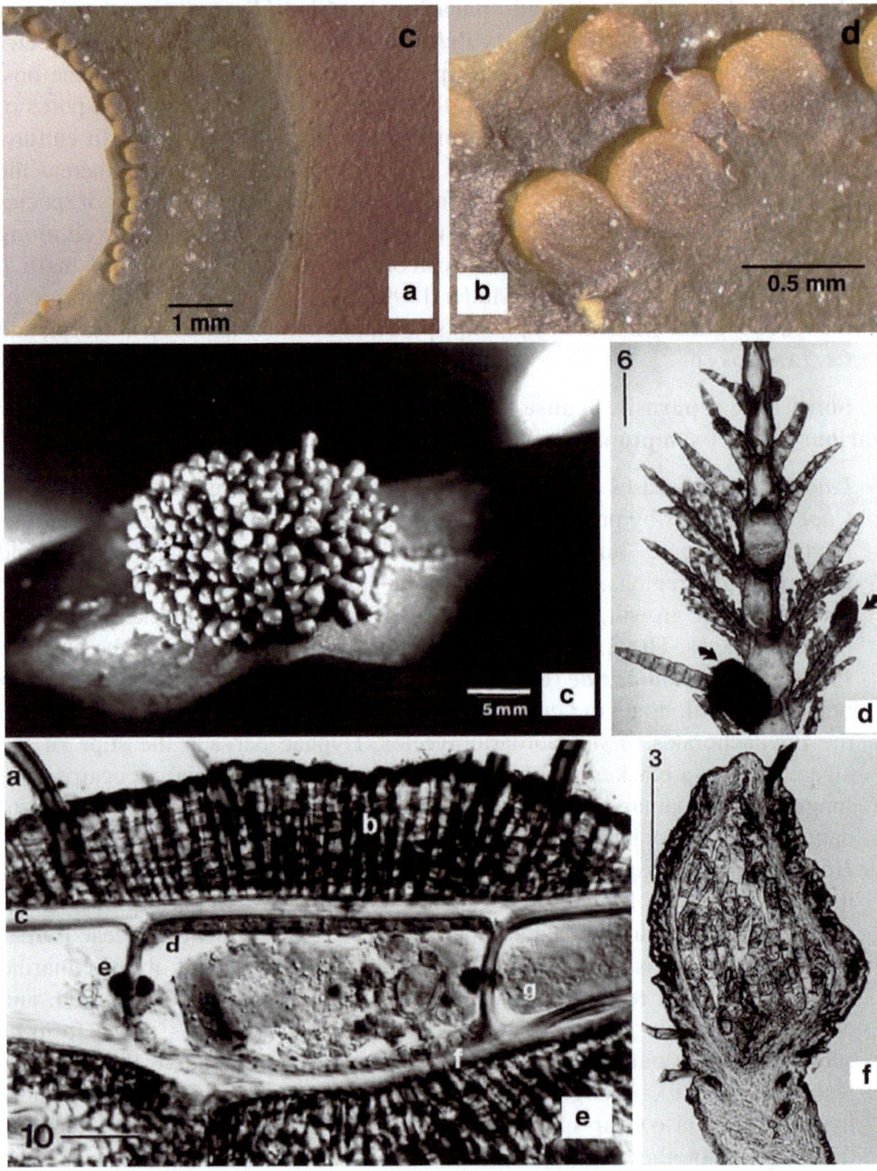


**Fig. 8.3** Fungal diseases of marine macroalgae. (**a**, **b**) Basidiocarps of the fungus *Mycaureola dilseae* on the red alga *Dilsea carnosa* around a necrotic lesion. (*Source: Binder M. et al. (2006). Evolutionary relationships of Mycaureola dilseae (Agaricales), a basidiomycete pathogen of a subtidal rhodophyte. American Journal of Botany 93: 547–556. With permission from the Botanical Society of America.* (**c**) Galls induced by the ascomycete *Haloguignardia* on frond of the alga *Cystoseira osmundacea*. (*Source: Kirk E. Apt* 1988. *Galls and tumor-like growth on marine algae. Diseases of Aquatic Organisms 4: 211-217. With permission from Inter-Research*). (**d–f**) *Spathulospora* spp., a fungal parasite of red algae. (**d**) Young thalli (*arrows*) of *S. antarctica* on branches of the red alga *Ballia callitricha*. (**e**) A longitudinal section of young thalli (*arrows*) of

- ***Mycaureola dilseae*** Maire & Chemin is the only marine basidiomycete parasitic on an alga. It occurs exclusively on the red alga *Dilsea carnosa*. The fungus belongs to the family Physalacriaceae of the Order Agaricales under the Ustilaginomycetes. It occurs in temperate climates and sporulates during the colder months when the temperature is below 15 °C. The fungus causes 1 to 2 mm necrotic lesions in the blades of the red alga, the lesions being distinctly green in color. However, no abnormal growth such as hyperplasia and hypertrophy occur. The algal tissue becomes damaged in these discolored zones. The lesions may fuse. A distinct border is found in the discolored zones between infected and uninfected algal tissues. Algal tissue in the central part of the lesion breaks down, leaving a hole surrounded by a ring of basidiocarps (Fig. 8.3a, b). Basidiocarps appear as small white protuberances in clusters, circles, or arcs (Stanley 1992).

**Straminipilan fungi, particularly those belonging to the Oomycetes, "exhibit the greatest diversity of parasitic species" by virtue of their evolution and adaptation to an aquatic habitat through production of motile zoospores** (Porter 1986; Marano et al. 2012). Most oomycete parasites of marine algae belong to the basal clade in the evolution of Oomycetes. It is believed that the Oomycetes originated in the sea, and this is probably a major reason for their ability to infect marine algae (Beakes et al. 2014). Diseases caused by these organisms appear to be host-specific and show characteristic symptoms such as changes in color, rot lesions, and abnormal growth in the host (Li et al. 2010). Oomycetes that parasitize red algae may obtain cues for infection from specific carbohydrates present in their hosts.

- **The Red Rot disease of *Porphyra*** (Fig. 8.4)*: Porphyra* species, called Nori in Japanese, is a highly popular edible red alga cultivated extensively in Japan, Korea, and other oriental countries. The Nori cultivation industry earns more than 1 billion US dollars per year. The red rot disease, called Akagusare-byo in Japanese, causes serious losses to *Porphyra* farming, resulting in great economic losses, amounting to 40 to 60 million US dollars annually (Blouin et al. 2011; Kim et al. 2014). The disease was first reported by Arasaki in 1947 and the disease agent, the oomycete *Pythium porphyrae* Takahashi et Sasaki was discovered by 1968.

The disease appears as small red spots, which gradually change color to amaranth and green (Fig. 8.4a). The spots coalesce later into holes. Spots develop mostly at the end of the frond. The holes are about 4–6 mm in diameter. Favorable disease conditions can result in coalescence of the spots and rot of the fronds. Red

**Fig. 8.3** (continued) Spathulospora lanata on Ballia hirsuta. (**f**) A longitudinal section of the ascocarp of *S. adelpha* on its red algal host. *(Reprinted from Inderbitzin P. et al. 2004. The phylogenetic position of Spathulospora based on DNA sequences from dried herbarium material. Mycological Research. 108: 737–748, with permission from Elsevier)*

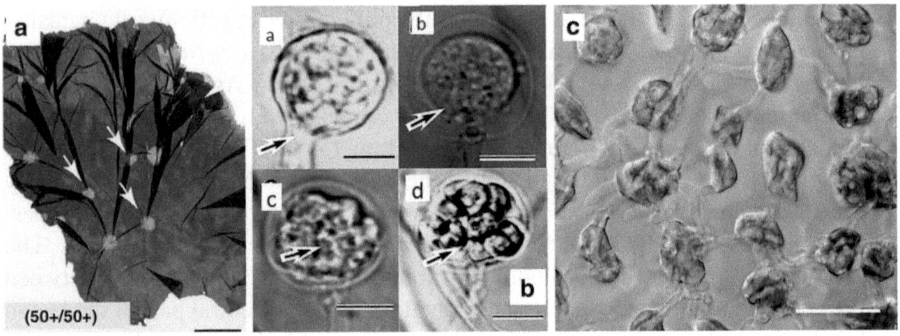


**Fig. 8.4** The Red Rot Disease of the red alga *Porphyra* caused by *Pythium porphyrae*. (**a**) Symptoms of the Red Rot disease on cultivated *Porphyra*, "nori." *(Reprinted from Gachon, C.M.M. et al., 2010. Algal diseases: spotlight on a black box. Trends in Plant Science 15: 633–640, with permission from Elsevier).* (**b**) Different stages of the development of zoosporangia in *Pythium porphyrae*. *(Source: Addepalli, M.K. and Fujita, Y.* Regulatory role of external calcium on *Pythium porphyrae* (Oomycota) zoospore release, development and infection in causing red rot disease of *Porphyra yezoensis* (Rhodophyta) *FEMS Microbiol. Lett., 2002,* 211. 253-257, *by permission of Oxford University Press.).* (**c**) Mycelium of *Pythium porphyrae* pervading the thallus of Korean *Pyropia (Porphyra)*. (*Source: Gwang Hoon Kim et al. 2014. A revaluation of algal diseases in Korean Pyropia (Porphyra) sea farms and their economic impact. Algae 29: 249-265. Kind permission of the Korean Phycological Society.)*

rot reduces the quality and market value of *Porphyra* products even if they narrowly escape critical damage before harvest, because the sheet products obtained from infected thalli appear to be less lustrous, uneven, and discolored (Addepalli and Fumita 2002).

The zoosporangium (Fig. 8.4b) produces a long tube with a vesicle at the tip into which zoospores move out before being liberated. The fungus spreads by means of heterokont biflagellate zoospores, typical of the Straminipila. The infection process proceeds along standard fungal infection steps, namely: (1) adhesion of the zoospore and encystment, (2) germination, (3) appressorium formation and penetration of the cell, and (4) growth of the fungus as hyphae within the algal host. *Pythium porphyrae* requires calcium for zoosporangia formation, release of zoospores, and their germination. The inter- and intracellular mycelium invades the host cells in the early stage of the development of infection. Some of the hyphae can penetrate through the host cells (Fig. 8.4c).

The rapid spread of zoospores often results in serious damage to the alga. The fungus survives in seafloor sediments from which they may infect during favorable conditions (Kawamura et al. 2005). Countermeasures for the disease include immersing the *Porphyra* cultivating nets into an organic acid, early harvesting, exposing the *Porphyra* cultivating nets to the air, and short-term freezing of the nets at around -20 °C for several days. However, these treatments are effective only when they are implemented at the early stages of infection. Farmers have also made a constant effort to obtain *Porphyra* strains with beneficial properties by means of natural mutation and artificial selection.

Another serious pathogen of the related red alga *Porphyra yezoensis* is yet another stramenipile, *Olpidiopsis porphyrae* Sekimoto, Yokoo, Kawamura et Honda (Sekimoto et al. 2007), which causes the "Tsubozyo-kin-byo" disease. This fungus has been reported to occur together with *Pythium porphyrae* in China. This single-celled endobiotic fungus is holocarpic and reproduces by biflagellate zoospores.

- **Many straminipilan fungi frequently occur as parasites in brown algae. *Eurychasma dicksonii* (E.P. Wright) Magnus is a common parasite in cold and temperate waters** (Fig. 8.5a, b). The fungus has a broad host range and infects a number of brown algae, as also a few members of red algae, often causing epidemic outbreaks in the algae (Strittmatter et al. 2009). At least 45 species of filamentous as well as thallic brown algae are infected by *E. dicksonii*. This parasite significantly affects populations of the filamentous brown alga *Pylaiella littoralis*, causing "brown tides" (Marano et al. 2012). The fungus also infects *Ectocarpus siliculosus* (Dillwyn). *Eurychasma dicksonii* is an endobiotic, single-celled, holocarpic fungus, where the entire thallus is converted into a zoosporangium. It reproduces by producing biflagellate zoospores. The fungus is amenable to culturing. *Ectocarpus siliculosus* is a brown algal genome model and a study on the proteomics of this alga has shown that 21 algal proteins accumulated differentially in response to infection by the *E. dicksonii*. These include classical algal stress response proteins such as a manganese superoxide dismutase, heat shock proteins 70, and a vanadium bromoperoxidase. The response of the alga to its fungal parasite overlaps with its response to herbivore grazing and abiotic stresses (Strittmatter et al. 2015).
- **Parasitic fungi of the family Anisolpidiace belonging to hyphochytriomycetes are parasitic in a number of brown algae such as *Ectocarpus*, *Cladostephus*, *Sphacelaria*, and *Pylaiella*** (Fig. 8.5c). *Anisolpidium rosenvingei* is a monocentric parasite of the cosmopolitan filamentous alga *Pylaiella littoralis*. T.W. Johnson, a pioneering marine mycologist, studied *Anisolpidium ectocarpii* infection of *Ectocarpus siliculosus* on the east coast of North America. The fungus infects only unilocular and pleurilocular sporangia of the algae and can cause epidemics in populations of this host. Up to 70% of the population of *P. littoralis* at one site along the European coast was infected by *A. rosenvingei at* the end of the growing season. This parasite probably disrupts the reproductive processes of the host. Field data collected over 3 consecutive years suggested a correlation between the prevalence of the parasite and the decline in host populations. Both vegetative cells and pleurilocular sporangia in the host are infected by this parasite. *Anisolpidium ectocarpii* produces thick-walled zygotes during sexual reproduction. These zygotes become resting spores that can possibly survive a wide range of temperatures. Both *A. ectocarpii* and *A. rosenvingei* are found on hosts in the intertidal zone where temperatures and salinities vary considerably. Resistance to environmental extremes is likely to be a characteristic of thick-walled resting spores (Marano et al. 2012).

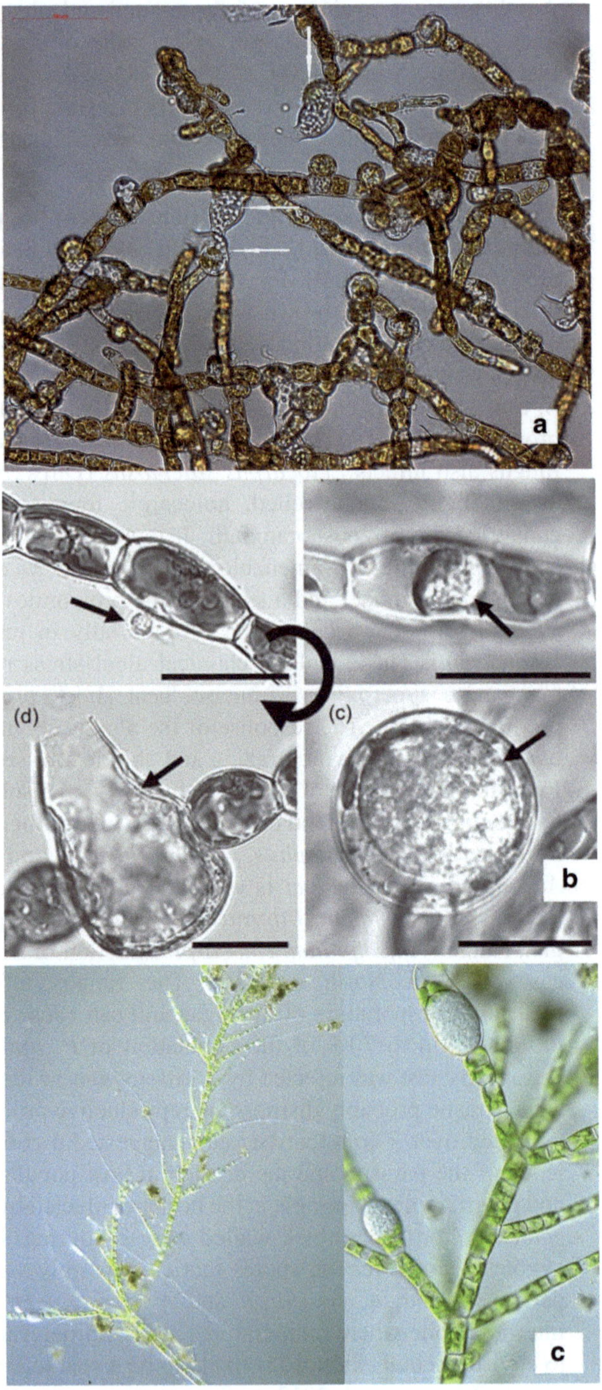


**Fig. 8.5** (**a**, **b**) The oomycete parasite *Eurychasma dicksonii* in the brown alga *Ectocarpus*. (**a**) Note the fungal zoosporangia (*arrows*) in the algal filaments. *(Courtesy: Dr Claire Gachon, Scottish Marine Institute, UK.)* (**b**(a)) Encysted zoospore on the algal filament (*arrow*). (b) Young fungal thallus inside the alga. (c) Conversion of the holocarpic thallus into zoosporangium with zoospores (*arrow*). (d) Empty zoosporangium after zoospore discharge. *(Courtesy: Martina Strittmatter et al., 2016. Infection of the brown alga Ectocarpus siliculosus by the oomycete Eurychasma dicksonii induces oxidative stress and halogen metabolism. Plant, Cell & Environment. 39: 259–271. With permission from John Wiley & Sons).* (**c**) The hyphochytriomycete fungal parasite *Anisolpidium* (*arrows*) infecting cells of the green alga *Stigeoclonium (Courtesy: Dr Yuuji Tsukii, Dr. Sci., Hosei University, Tokyo, Japan. From: Protist Images Website: http://protist.i.hosei.ac.jp/pdb/images/EuMycetae/Anisolpidium/sp_2.jpg)*

- **The commercially important red alga *Chondrus crispus Stackh.* is parasitized by *Petersenia pollagaster* (Petersen) Sparrow**. *Chondrus crispus* is cultivated in mariculture facilities for k-carrageenan. The first epiphytotic outbreak of the disease was reported in 1980 from a mariculture facility in Canada which was growing the red alga. The fungus is an endobiotic, holocarpic, obligate biotroph. It has not been cultured so far. The fungus causes necrotic lesions which attain a size of up to 1 cm in diameter. Laboratory studies have shown that the optimal temperature for infection is 20°C, being significantly less at lower and higher temperatures. The entire life cycle of the parasite, from infection, growth, and release of zoospores, is completed in 48–72 h at a temperature of 15–20°C.
- **A number of green algae become infected by mycetaen and straminipilan fungal parasites.** The genus *Cladophora* is host to the oomycetes *Sirolpidium bryopsidis* and *Pontisma lagenidioides*, the chytridiomycete *Olpidium rostriferum,* and the labyrinthulomycete *Labyrinthula* species. The oomycete *Pontisma lagenidioides* is the cause of an extremely prevalent infection of the green alga *Chaetomorpha media* along the west coast of India. The alga grows on intertidal rocks. Infected cells appear brownish and the infection spreads from the apex of the filaments downwards. Cells are colonized by the unicellular, holocarpic thallus of the fungus. The fungus is a biotroph and has not been cultured. Chlorophyll contents decline, the chloroplasts are destroyed, and phaeopigments accumulate in the cell. The infection is a regular annual feature during the period following the southwest monsoon period from October to January (Raghukumar 2006).
- ***Labyrinthula* spp., the straminipilan fungus belonging to Labyrinthulomycetes, infects a wide range of algal hosts.** The fungus parasitizes green algae such as *Chaetomorpha*, *Cladophora,* and *Rhizoclonium* and cyanobacteria such as *Lyngbya* (Raghukumar 2006). It is also known as a hyperparasite, growing upon on the fungal sporangia of *Pontisma lagenidioides*, which infects other algae.

**Algae may respond to fungal infection in various ways.**

- Several seaweeds have been shown to respond to fungal attack, by the production of antifungal compounds. The brown *alga Lobophora variegata* produces the compounds lobophorolide and tolytoxin, which are toxic towards the parasitic ascomycetes *Lindra thalassiae* and *Dendryphiella salina* (Kubanek et al. 2003).
- Spatially resolved (imaging) microscopic measurements of chlorophyll-fluorescence kinetics have shown details of the inhibition of photosynthesis of *Pylaiella littoralis* infected by *Chytridium polysiphoniae* (Gachon et al. 2006; Gleason et al. 2012a).

- Macroalgae or seaweeds are some of the most important primary producers of the coastal oceans where they often form dense vegetations.
- Obligate marine fungi that are found exclusively in living and decomposing algae are termed "algicolous fungi." About 80 species are known.
- Fungi are associated with algae as mutualists, commensals, parasites, and saprotrophs.
- Facultative marine fungi belonging to anamorphic species of terrestrial ascomycetes as well as obligate marine mycetaen and straminipilan fungi grow as endophytes in algae. Most algal endophytic fungi are facultatively marine. Species of *Aspergillus*, *Cladosporium,* and *Penicillium* are common macroalgal endophytes. *Halosigmoidea marina* and *Acremonium fuci* are obligately marine macroalgal endophytes. Diversity and density of colonization of endophytic fungi vary with the host species of algae. Brown algae have the highest diversity of fungal endophytes.
- Many fungi form an obligate mutualistic association with algae in the form of mycophycobiont or lichens. Examples of mycophycobionts are *Mycophycias ascophylli* in the brown algae *Ascophyllum nodosum* and *Pelvetia canaliculata* and *Mastodia tessellata* in the green algae *Prasiola borealis* and *P. crispa* ssp.
- Macroalgae are infected by a wide variety of fungi. Fungi parasitize all groups of algae, namely, Chlorophyta (green algae), Rhodophyta (red algae), and Phaeophyta (brown algae). Some of these fungi are obligate parasites (biotrophs), while others are facultative or opportunistic parasites. Specific parasites are restricted to brown, red, or green algae, suggesting a coevolution of the fungal parasites and their hosts.
- Several fungi, such as the ascomycetes *Pontogeneia* and *Chadefaudia*, and the chytridiomycetes *Thalassochytrium gracilariopsis and Chytridium polysiphoniae*, parasitize algae without causing any external symptoms. Some fungal parasites cause light to dark discolorations of their hosts. Many fungal parasites such as *Haloguignardia cystoseirae* in the brown algal genera *Cystoseira, Halidrys,* and *Sargassum* and *Spathulospora* in red algae and *Mycaureola dilseae* cause malformations of their host algae.
- Fungi belonging to the Oomycetes of the Kingdom Straminipila are the most significant marine algal pathogens. The Red Rot disease of *Porphyra*, an economically important infection, is caused by *Pythium porphyrae*. Brown algal parasites belong to *Eurychasma dicksonii* and species of *Anisolpidium* of Hyphochytriomycetes. The commercially important red alga *Chondrus crispus* is parasitized by *Petersenia pollagaster*. A number of green algae become infected by mycetaen and straminipilan fungal parasites. *Labyrinthula* infects a wide range of algal hosts.

**Future Directions**

1. What is the role of endophytic fungi in macroalgae?
2. Fungal diseases of algae are expected to be highly prevalent. Diligent search will reveal many more diseases than known presently.
3. What are the interactions between parasitic fungi and their algal hosts? What are the mechanisms of algal defense and fungal infection strategies?
4. The dynamics of fungal diseases in regulating algal populations deserves attention.

## 8.2 Saprobic Fungi in Marine Algae

An estimated $950 \times 10^6$ metric tons of macroalgal NPP enters the decompositional phase (Fig. 2.3). Thus, an enormous amount of detritus is available for growth of fungi and bacteria.

### *8.2.1 Diversity of Fungi in Macroalgal Detritus*

**A number of fungi colonize and decompose dead macroalgae**. Large, thalloid brown algae, such as *Fucus vesiculosus*, *F. serratus*, *Macrocystis* spp., *Laminaria* spp., *Sargassum* spp., and *Ascophyllum nodosum,* are host to a high diversity of fungi, many of which are also found on decomposing wood.

**Many obligately marine ascomycetes which are found in decomposing wood also grow in decomposing algae.**

- The brown alga *Fucus vesiculosus* harbors *Corollospora intermedia* I. Schmidt, *C. pulchella* Kohlm., Schmidt and Nair, *Crinigera maritima* I. Schmidt and *Lulworthia salina* (Linder) Cribb and Cribb, as well as the anamorphic fungi, *Acremonium fuci and Phoma* spp. (Kohlmeyer and Kohlmeyer 1979; Zuccaro et al. 2008). Other fungi associated with macroalgae are *Asteromyces cruciatus*, *Corollospora intermedia*, *Dendryphiella arenaria*, *Dendryphiella salina*, and *Varicosporina ramulosa* (Genilloud et al. 1994).
- Many saprophytic, obligate marine species grow upon large macroalgae that are washed up and cast on the beach. Many such fungi grow out from the decomposing algal tissues and produce ascocarps or conidia firmly attached to sand grains. These are called the "arenicolous fungi." Some of the common arenicolous fungi are the anamorphic fungi *Asteromyces* and *Varicosporina* and the ascomycetes *Arenariomyces trifurcatus* and *Corollospora* species such as *C. maritima*, *C. pulchella*, and *C. angusta* (Sridhar et al. 2012b).

**Marine-derived fungi from macroalgal detritus include terrestrial mycetaen species.** Many have been detected in the brown alga *Fucus serratus* from the North Sea coast using molecular techniques. Several were also isolated in culture from the same macroalga in British coasts (Zuccaro et al. 2008; Miller and Jones 1983). Likewise, many terrestrial species of fungi have been recorded from algae in the Antarctic using plating techniques, without surface sterilization (Loque et al. 2010). However, only a few of these have been isolated from surface-sterilized material, suggesting that just a small portion of the detected fungi were facultatively marine and active in algal detritus (Zuccaro et al. 2008). Further studies are required to confirm the diversity of facultative marine species in decaying algae.

**Yeasts are important components of the detrital mycoflora on decomposing algae** (Kutty and Philip 2008). ***Metschnikowia zobellii* is a dominant yeast on surfaces of decomposing macroalgae** from the kelp *Macrocystis pyrife* (Van Uden and Branco 1963). High numbers of this yeast have been cultured from *Fucus vesiculosus*, *Ascophyllum nodosum*, *Odonthalia dentate*, and *Saccharina lattissima* at White Sea Biological Station in White Sea, a marginal sea of the Arctic (Kachalkin 2014). Yeasts have been found to be abundant in decomposing giant kelp at Macquarie island in Antarctica. Other predominant yeasts from marine algae are *Torulopsis* sp., *Candida albicans*, *C. natalensis*, *Trichosporon cutaneum*, *Endomycopsis chodatii*, *Metschnikowia reukaufii*, *Pichia farinose*, *Kluyveromyces aestuarii*, *Candida marina*, *Torulopsis torresii*, and *Torulopsis maris*. Seaweeds have been reported to be a reservoir of *Candida* yeasts in inshore waters.

**Labyrinthulomycetes occur in high numbers in decomposing macroalgae.** The thraustochytrid *Ulkenia visurgensis* (Ulken) Gaertner and the aplanochytrid *Aplanochytrium minutum (Labyrinthuloides minuta)* are found in high frequencies in the decomposing brown alga *Sargassum cinereum* (Sathe-Pathak et al. 1993).

### *8.2.2 Dynamics of Fungi in Macroalgal Detritus*

**Macroalgae are more labile to microbial decomposition compared to lignocellulosic detritus derived from vascular plants** such as those from mangroves, salt marsh plants, and seagrasses. Therefore, their detrital dynamics are different.

**Macroalgae that are submerged in the sea may differ in decomposition processes compared to those cast ashore in the intertidal region. Straminipilan fungi such as thraustochytrids and r-strategy fungi are more likely to play a key role in submerged macroalgal degradation than k-strategy lignicolous fungi**. The rapidly degradable submerged algal detritus is less likely to be decomposed by lignicolous fungi that adapt a k-strategy and which are regularly encountered in decomposing vascular plant detritus. **On the contrary, dead macroalgae that are cast ashore in the intertidal region are likely to harbor lignicolous marine fungi more than straminipilan fungi**. Bacteria are likely to play a more important role in submerged macroalgal degradation than fungi.

**Important contributions to our knowledge on the decomposition of macroalgae come from the studies on kelps *Laminaria pallida* and *Ecklonia maxima*** in coastal waters of South Africa by R.C. Newell and colleagues and laboratory studies on decomposition of green, red, and brown algae along the North Sea coast (Newell et al. 1980, 1982; Lucas et al. 1981; Stuart et al. 1982; Rieper-Kirchner 1989, 1990). However, these studies did not take fungi into account. There are actually few studies that consider fungi. The limited picture of fungal dynamics of submerged macroalgal detrital formation has come from field observations combined with laboratory studies on decomposition of the brown alga *Sargassum cinereum* from the west coast of India and *Fucus serratus* from the British coast (Miller and Jones 1983; Sathe-Pathak et al. 1993; Sharma et al. 1994). A broad picture of microbial dynamics of macroalgal decomposition has emerged from these studies.

**Macroalgal decomposition dynamics by fungi comprises the early leaching phase, microbial biomass buildup phase, and fragmentation phase.** Thus, they follow the same broad steps as known for mangrove leaves and salt marsh grass detritus (Sect. 2.4.2).

1. **DOM is constantly released even by healthy algae and forms an important source of nutrition for saprotrophic microorganisms.** Nearly 25% of the net particulate primary production of live kelps, *Laminaria pallida* and *Ecklonia maxima*, has been estimated to be released as dissolved organics in the form of mucilage from broken tips of the kelps. These contain sugars, proteins, polyols, laminaran, and alginate. Most of these are utilized by bacteria within 48 h and the rest by 6–10 days. The role of fungi in this process is not known.
2. **The initial stage of algal decomposition is characterized by abiotic leaching.** Endophytic fungi that were present in healthy algae (Sect. 8.1.1) may persist during the early stages of algal decomposition. For example, the population of *Acremonium fuci*, which is present in healthy tissues of *Fucus serratus,* may increase in dead algae (Zuccaro et al. 2008). The straminipilan fungus *Aplanochytrium minutum*, which occurs as endophyte, persists during the decomposition phase of *Sargassum cinereum.*

   Death of the alga immediately triggers the abiotic leaching phase, which lasts for about 7 days as judged by mass loss. This is characterized by a rapid loss of soluble organics and phenolics, which contribute to the DOM pool available for microbial growth. The leaching at once stimulates colonization of the detritus by fungi and bacteria present in the ambient waters. **Straminipilan thraustochytrids with an r-strategy may further colonize macroalgae immediately upon their death.** For example, *Ulkenia visurgensis* colonizes *Sargassum cinereum* only upon the death of the alga. However, populations of initial colonizers may remain low because of the presence of phenolics. Yeast populations have been observed to be low in living algae compared to counts in the surrounding seawater, but to increase when decomposition starts (Kutty and Philip 2008).

The colonization further accelerates the leaching process and release of DOM.

3. The decompositional phase, starting from about a week, is characterized by further colonization of detritus by fungi and their growth. Thraustochytrids, which had colonized the detritus earlier, increased in numbers and attained their greatest population densities by day 16 in the case of *Fucus serratus*. Fungi and bacteria peaked by 21 days in the case of *Sargassum cinereum*. Several marine-derived fungi have been reported from *Fucus serratus*. *Alternaria*, *Nigrospora oryzae*, *Aspergillus*, and *Paecilomyces* occur as facultative marine fungi in *Sargassum cinereum* during this stage. Fungal biomass increases slowly till about 15 days. Detritus of red algae such as species of *Ceramium* and *Chondrus crispus* are also inhabited by saprophytic fungi.

   Experimental evidence, based on inoculation of various fungi on *Sargassum cinereum* detritus of different ages, has shown that aged detritus of about 1–2 weeks supports optimal fungal growth. Thus, the mycetaen fungi *Acremonium* sp. and *Lindra thalassiae* and the straminipilan fungi *Pythium* sp., *Ulkenia visurgensis, and Aplanochytrium (Labyrinthuloides) minutum* grew best on 7- and 14-day-old detritus.

   *Ecklonia maxima* stranded on beaches has been reported to be decomposed by bacteria rather than fungi.
4. **The recalcitrant detrital phase may set in early during decomposition of macroalgae compared to vascular plants by about 21 days.** Fungal populations decline during this period. Bacterial populations, however, remain high, probably owing to the fact that breakdown of detritus into finer particles increases the surface area available for bacterial colonization.
5. **Feces of algal herbivores is an important source of nutrients for microorganisms.** A large portion of macroalgae is directly grazed by herbivores. In the case of *Laminaria pallida* and *Ecklonia maxima*, feces of these animals containing partly digested algae are an important source of microbial growth and decomposition.

**Biochemical changes and fungal biomass buildup occur during macroalgal decomposition.**

- Dead macroalgae have been reported to rapidly lose 60–90% of their weight within 2–3 weeks (Rieper-Kirchner 1990). Nearly 50% of mass loss occurs in the case of *Sargassum cinereum* by one week. Reducing sugars, proteins, and alginate show a gradual decrease, presumably caused by microbial growth and decomposition (Sathe-Pathak et al. 1993). More than 75% of the carbohydrates and 25% of the phenolics in *Fucus serratus* are known to be leached out within a week (Miller and Jones 1983).

  Bacteria in kelp beds utilize the DOM much better than the particulate organic matter of the detritus. These dynamics are not understood for fungi. Experimental evidence suggests that fungi are capable of bringing about a number of changes in detritus, the sum total of which would define the biochemistry of the detritus.

- As decomposition progresses beyond the leaching phase in *Sargassum cinereum*, fungal growth resulted in a decrease of carbohydrates and increase of protein values in detritus, whereas free amino acid contents always decreased substantially.
- Fungi utilize a wide variety of carbon sources from the alga such as laminaran, cellulose, and alginate. They produce a number of degradative enzymes such as amylase, protease, and cellulase. *Dendryphiella salina*, *D. arenaria*, *Varicosporina ramulosa*, *Lindra thalassiae,* and *Lulworthia* spp. produce laminarinase and utilize the polysaccharide for growth. Laminarinase activity in the fungus *Dendryphiella salina is tightly* associated with fungal mycelia and not leached out extracellularly (Kis-Papo 2005). *Acremonium fuci* can utilize fucose as efficiently as glucose. Thraustochytrids produce proteases while growing in detritus.
- Individual fungi, such as *Ulkenia visurgensis*, *Aplanochytrium minutum*, *Acremonium* sp., *and Lindra thalassiae,* play different and various roles in altering the biochemical composition of *Sargassium cinereum* detritus.
- Microbial activities lower the C/N ratio of macroalgal detritus during decomposition.
- **Despite experimental evidence for the capability of fungi to grow and decompose macroalgal detritus, the actual biomass of fungi and bacteria in detritus appears to be low.** Biomass of bacteria in detritus of *Sargassum cinereum* was estimated to be only 1.1% of detrital C. That of thraustochytrids, estimated by an immunofluorescence technique, was nearly 15 times less and amounted to only 0.07% C. This might be because of the limitation of the technique in detecting cells that were embedded in detritus. The low microbial biomass may be disproportionate to their role in biochemical transformation. By virtue of enzymatic degradation, release of nutrients and making them available to others in the trophic web and enrichment of detritus in terms of essential nutrients, microorganisms, including fungi, may play a major, as yet undetermined role.

**Live and decomposing macroalgae are an important source of food for filter feeding and detrital feeding organisms.** Diversity of the fauna in an algal ecosystem is an indicator of the utilization of macroalgal detritus by them. For example, filter feeders have been observed to be the most important fauna in kelp beds of *Ecklonia maxima* and *Laminaria pallida* in South African coastal waters. Hence, suspended matter, comprising fine particulate macroalgal detritus, algal mucus, as well as fecal matter, colonized by microorganisms and fragments later would be an important source of nutrition for them (Lucas et al. 1981).

The importance of microorganisms as feed is more likely to be protein enrichment of detritus than in terms of biomass. For example, only 10% of macroalgal production is converted into bacterial biomass in *L. pallida* and *E. maxima* kelp ecosystem. The trophic significance of bacteria as a carbon resource is likely to be small in coastal waters, but large in terms of the kinds of nutrients that they supply. The same may be true of fungi.

- **A large part of macroalgal NPP enters the decompositional phase.**
- **Decomposing algae are colonized and degraded by obligate marine mycetaen and straminipilan fungi as well as facultative marine fungi.**
- **"Arenicolous fungi" are obligate marine fungi that utilize dead algal POM in intertidal beaches and produce ascocarps attached to sand grains.**
- **Among many yeasts growing on surfaces of decomposing macroalgae, *Metschnikowia zobellii* is the most dominant.**
- **Macroalgae that are submerged in the sea differ in decomposition processes compared to those cast ashore in the intertidal region. Fungi with r-strategy are more likely to play a key role in submerged macroalgal degradation than k-strategy lignicolous fungi. Macroalgae cast ashore in the intertidal region are likely to harbor k-strategy lignicolous marine fungi.**
- **DOM released by healthy algae is an important source of nutrition for saprotrophic microorganisms.**
- **The initial stage of algal decomposition is characterized by abiotic leaching. Endophytic fungi that were present in healthy algae persist during the early stages of algal decomposition.**
- **Fungi and bacteria present in the ambient waters are earlier colonizers of dead macroalga. Straminipilan thraustochytrids with an r-strategy may further colonize macroalgae immediately upon their death. More fungi and bacteria colonize macroalgae with progressive decomposition.**
- **The recalcitrant detrital phase may set in early during decomposition of macroalgae compared to vascular plants by about 21 days.**
- **Feces of algal herbivores is an important source of nutrients for microorganisms.**
- **Biochemical changes and fungal biomass buildup occur during macroalgal decomposition.**
- **Dead macroalgae rapidly lose labile organics and phenolics, leading to a substantial loss in weight.**
- **Fungi produce a number of degradative enzymes and utilize a wide variety of carbon sources from the alga such as laminarin, cellulose, and alginate.**
- **Microbial activities lower the C/N ratio of macroalgal detritus.**
- **The actual biomass of fungi and bacteria in macroalgal detritus appears to be low.**
- **Live and decomposing macroalgae are an important source of food for filter feeding and detrital feeding organisms. The importance of microorganisms as feed is more likely to be nutrient enrichment of detritus than in terms of biomass.**

### Future Directions

All aspects of saprotrophic fungi in marine algae have been very poorly studied. Some of these are:

1. Diversity of obligate and facultative marine fungi.
2. The abundance and diversity of straminipilan fungi and yeasts that reside on algal surfaces and utilize algal DOM and POM.
3. Role of endophytic fungi during the initial phases of decomposition.
4. Biomass of fungi in particulate algal detritus.
5. Enzymes produced by fungi in decaying algae.
6. Fungi in fecal pellets of algal herbivores and filter- and deposit feeders that feed on algal detritus.
7. Importance of fungi in algal detritus in the nutrition of algal detritivores.

# Chapter 9
# The Coral Reef Ecosystem

*We would shout and swim about, The coral that lies beneath the waves Oh what joy for every girl and boy, Knowing they're happy and they're safe*
Ringo Starr—*Beatles: Octopus' Garden*

The sheer beauty and diversity of coral reefs that lie beneath turquoise waters are sufficient incentives to pursue the study of fungi in this ecosystem. Coral reefs occur along shallow, tropical coastlines where the marine waters are clean, clear, and warm (Fig. 9.1). **Coral reefs are one of the most productive ecosystems in the world. They equal tropical rainforests in terms of biological diversity.** The importance of coral reefs lies in their high productivity, sustenance of economically useful plant and animal life, prevention of land erosion, and tourism. They contribute to fisheries in many ways (Sheppard et al. 2009). The reefs alone sustain abundant fishing. Approximately, one-third of the world's fish species are said to live on coral reefs. Coral reefs also support the productivity of coastal waters and fisheries. The tremendous productivity of reefs also influences adjacent offshore waters and contribute to fish productivity.

Coral reef ecosystems are surrounded by waters with limited nutrients, particularly N and P. Despite this, they maintain high productivity and biological diversity. They are able to achieve this because energy is highly conserved within the ecosystem through a tight recycling and an efficient transfer between trophic levels. Ecosystems with low energy transfer efficiency between trophic levels will result in a larger availability of energy to the pool of decomposers. On the contrary, those with high energy transfer between trophic levels will make less energy channeled into the decomposer pool.

The keystone organisms of coral reef ecosystems are the photosynthetic, dinoflagellate algae called zooxanthellae which live in symbiotic, mutualistic association with scleratinian or hard (stony) corals that belong to Cnidaria. Each coral is made up of a number of individual polyps that live together in association with zooxanthellae. Scleractinian corals secrete a spectacular variety of calcium carbonate skeletons around them and are responsible for the building of coral reefs. Soft corals or gorgonians that are devoid of a calcium carbonate skeleton occur at lower



S. Raghukumar, *Fungi in Coastal and Oceanic Marine Ecosystems*,
DOI 10.1007/978-3-319-54304-8_9

**Fig. 9.1** Coral reefs. (**a**) Branched hermatypic corals with a swarm of fish. (**b**) A stony coral in the foreground with a pink soft coral in the background. *(Credit: Venkat Charloo, Barracuda Diving, Goa, India)*

depths in the reefs. The enormous diversity of scleractinian corals supports an equally diverse community of algae, fish, and a myriad other plant and animal life. Although coral reefs occupy less than 0.1% of the world's ocean surface, they are home to about 25% of all marine species, These include fish, molluscs, worms, crustaceans, echinoderms, sponges, tunicates, and other cnidarians. Coral reefs are found between 30° N and 30° S of the equator and are at their most luxurious in tropical waters.

Convincing evidence for presence of fungi in reefs came only in the 1980s. Kendrick et al. reported filamentous microboring fungi in coral skeletons in 1982, and Kohlmeyer and Kohlmeyer described novel ascomycetes from coral slabs between 1987 and 1990. Raghukumar and Balasubramanian reported thraustochytids in coral waters in 1991.

We have begun to appreciate the presence and importance of fungi in coral reefs even more in recent years (Golubic et al. 2005; Raghukumar and Ravindran 2012; Yarden 2014). However, our present information on fungi in coral reefs is rudimentary. It nevertheless points out to some exciting discoveries on the biology of fungi in coral reefs. Fungi in coral reefs are found in living corals, various other sedentary and motile coral invertebrates, coral sediments and waters, the algal beds associated with coral reefs, and the seagrass vegetations. Fungi associated with algae and sea grass vegetations have been discussed earlier.

## 9.1 Fungi in Scleractinian Corals

A coral, together with protists, microbes, and other associated animals, constitutes the coral holobiont. **Fungi are an important component of the coral holobiont and grow both on the surface (epibiontic) and interior (endobiontic) of scleractinian corals.** According to a metagenomic study, 38% of microbial rDNA sequences obtained from *Porites astreoides* at Bocal del Toro, Panama, corresponded to mycetaen fungi, indicating their predominance in corals (Fig. 9.2; Wegley et al. 2007). Endobiontic fungi that live within mineral surfaces, such as the calcium carbonate skeletons of corals, are endolithic fungi.

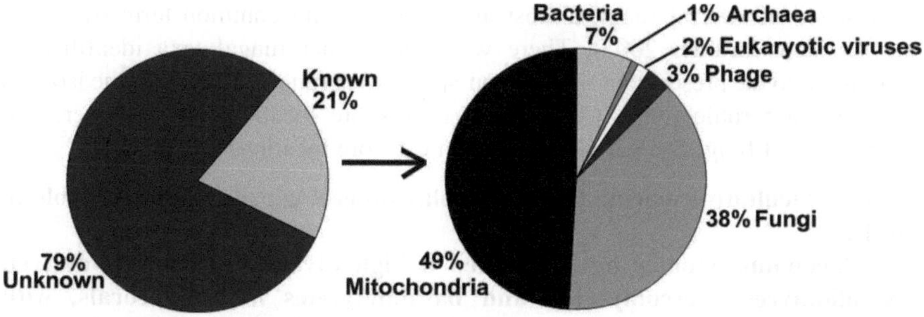


**Fig. 9.2** Microbial metagenome associated with *Porites astreoides*. Note that only 21% of the sequences belong to known organisms, out of which fungi comprise the majority of 39% *(Source: Wegley, L. et al. 2007. Metagenomic analysis of the microbial community associated with the coral Porites astreoides. Environmental Microbiology 10: 2707–2719. With permission from John Wiley & Sons)*

### 9.1.1 Diversity of Fungi in Scleractinian Corals

**A large diversity of endolithic fungi inhabit calcium carbonate skeletons of living as well as dead coral. They may occur in close association with the polyps in living corals.** They often intermingle with endolithic algae, frequently parasitizing the latter.

**Most of the endolithic coral fungi are facultative marine fungi, belonging to terrestrial species** (Kendrick et al. 1982; Ravindran et al. 2001; Yarden et al. 2007). Terrestrial species of fungi have been regularly isolated from corals in many parts of the world, including the Caribbean, South Pacific, Great Barrier Reef, Andaman, and Lakshadweep coral islands of the Bay of Bengal, Arabian Sea.

- Careful culturing from coral tissues has yielded terrestrial species of fungi. Species of *Alternaria, Aspergillus, Aureobasidium pullulans, Fusarium, Penicillium, Cladosporium, Fusarium, Hormonema dematioides, Phialophora bubakii*, and *Phoma* occur in a variety of healthy scleractinian corals. Facultative marine fungi are present in healthy, bleached, and diseased specimens of *Porites lutea* from Arabian Sea (Ravindran et al. 2001).
- *Scolecobasidium* sp. occurs commonly in dead patches of various corals in the Andaman and Lakshadweep coral islands of the Bay of Bengal and Arabian Sea, respectively (Raghukumar and Raghukumar 1991). The fungus was isolated after partial dissolution of the coral skeleton with EDTA followed by culturing.
- A fungal species morphologically resembling *Aspergillus* was detected in the coral skeleton of *Porites lobata* on Moorea island near Tahiti, French Polynesia (Priess et al. 2000).
- The terrestrial species *Curvularia lunata* as well as a non-sporulating, unidentified fungi were detected in *Porites lutea* from the Lakshadweep Island in the Arabian Sea using polyclonal immunofluorescence probes raised against the two cultures isolated from the same coral. *Conidia*-like structures were observed within the coral skeleton by the authors (Ravindran et al. 2001).
- Report on fungi isolated from Australian Great Barrier Reef corals indicated presence of several fungi, almost all of them being common terrestrial ones (Morrison-Gardiner 2002). There were 54 distinct fungal taxa identified in addition to the presence of several non-sporulating fungi. Majority of the isolates were anamorphic fungi. Corals from near shore locations showed increased presence of fungi compared to that from offshore locations.

**Thus, facultative marine fungi may play an ecologically important role in corals.**

**Metagenomic studies have revealed a high diversity of marine-derived chytridiomycetes, ascomycetes, and basidiomycetes in hard corals, with many novel lineages** (Wegley et al. 2007; Amend et al. 2012). A study based on D1 and D2 variable regions of the large-subunit (28S) ribosomal DNA as well as mitochondrial ribosomal nucleic acid (mRNA) amplicons of mycetaen fungi in the coral *Acropora hyacinthus* has detected the presence of typical "indicator species" belonging to the terrestrial genera *Stagonospora*, *Alternaria*, *Edenia*, *Verticillium*, *Phlebia*, *Phaeosphaeria musae*, *Westerdykella*, *Cerrena*, *Exidia*, *Exidiopsis*, *Dendryphiella*, and *Malassezia globosa*. *Malassezia* is a yeast belonging to the Ustilaginomycotina (Basidiomycota). A concordance between the rDNA and mRNA taxonomic diversity suggests that the marine-detected fungi from this coral may truly represent facultative marine fungi which are metabolically active (Amend et al. 2012).

**Certain obligately marine, mycetaen fungi such as species of the genus *Koralionastes* Kohlm. *and Corallicola* Volkm.-Kohlm. & Kohlm., belonging to the family Koralionastaceae of the Ascomycota, are corallicolous and are found exclusively in coral reefs** (Kohlmeyer and Volkmann-Kohlmeyer 1992). Five species of this genus are known to inhabit dead coral skeletons and coral slabs in shallow water of the Atlantic Ocean, Belize, Central America and Pacific Ocean, Queensland, Australia, together in association with crustaceous sponges (Jones et al. 2009).

**Thraustochytrids densely colonize coral surfaces and mucus.**

- White mucus coatings are widely prevalent on surfaces of scleractinian corals. Labyrinthulomycetes-like organisms are present on the surface and in the tissue of the solitary coral *Fungia granulosa* at the Gulf of Eilat (Harel et al. 2008). Molecular investigation identified them as *Labyrinthula* and *Oblongichytrium multirudimentale*. These species were also found in the coral gastrodermal layer in areas devoid of zooxanthellae. Mucus on this coral as well as on the coral *Favia* sp. contain 5–30 micron-sized aggregates of microbes. The aplanochytrid *Aplanochytrium* sp. and species of *Thraustochytrium*, besides a number of diatoms, occur in such corals (Arotsker et al. 2011).
- A novel thraustochytrid has been identified in the mucus of *Fungia granulosa* (Harel et al. 2008).
- *Thraustochytrium motivum*, *Aplanochytrium (Labyrinthuloides) minutum*, *A. kerguelense*, and *Ulkenia visurgensis* have been isolated from polyps and fresh mucus of the corals *Pocillopora* sp. and *Acropora* sp. as well as from floating and attached mucus detritus in coral islands of Lakshadweep in the Arabian Sea (Raghukumar and Balasubramanian 1991). Fresh mucus collected from corals yielded only 20 thraustochytrids $ml^{-1}$ mucus, while mucus detritus attached to corals yielded a large abundance of $1.9 \times 10^6$ thraustochytrid $g^{-1}$ mucus.

It has been surmised that when present in the corals, Labyrinthulomycetes provide them with nutritional sources that aid survival during bleaching events (Harel et al. 2008).

## *9.1.2 Endolithic Mycelial Fungi in Corals and Bioerosion*

**Endolithic mycelial fungi are present in both live and dead corals. Their presence within corals can often be discerned by dark bands caused by them** (Fig. 9.3a).

- Alternating black and white bands caused by mycelial fungi were reported to occur within corals belonging to *Porites* sp. from Indonesian coral reefs (Bak and Laane 1987) and in *Porites lobata* from Mayotte Island in the Mozambique Channel and Moorea Island in French Polynesia (Priess et al. 2000). Fungi are present throughout the coral interior. Another endolith, the green alga *Ostreobium queketti*, grows as bands within *P. lobata*. The bands are caused when fungi grow and result in destruction of the alga. The black color may be attributed to the dematiaceous nature of the fungal hyphae, or to an organic pigment excreted by the fungus as in *P. lobata*. Alternating black and white

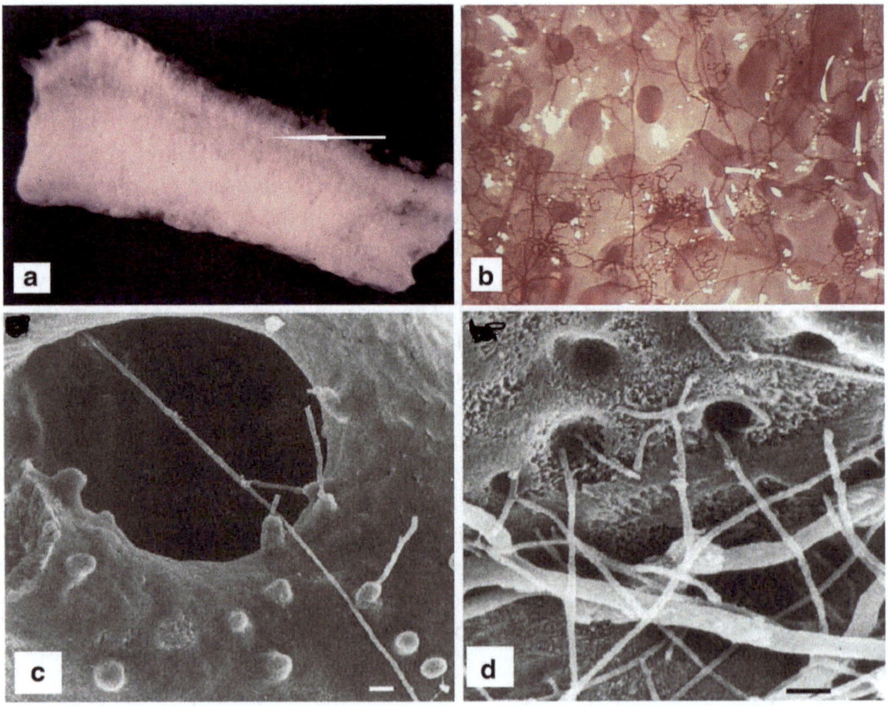


**Fig. 9.3** Fungi in coral skeletons. (**a**) Section through the coral *Montipora tuberculosa* showing a dark subsurface band of fungal mycelium (*arrow*) (S. Raghukumar). (**b**) Photomicrograph of fungal mycelium within the coral *Montipora tuberculosa*. (**c**, **d**) Resin casts of fungal hyphae in corals *(Source: T. Le Campion-Alsumard et al. 1995. Fungi in corals: symbiosis or disease? Interaction between polyps and fungi causes pearl-like skeleton biomineralization. Mar. Ecol. Prog. Ser. 117: 137–147. With permission from Inter-Research Science Center)*

bands may be related to the organic matter accumulation depending upon the season.

- Yet another terrestrial fungus, *Scolecobasidium* sp., forms a distinct dense brown to black zone of 0.5 to 1.5 cm width immediately below the surface layer of scleractinian corals (Fig. 9.3a; Raghukumar and Raghukumar 1991).

**Hyphae of endolithic fungi densely pervade coral skeleton and cause bioerosion** (Fig. 9.3b–d; Golubic et al. 2005). Bioerosion is a biological process by which organisms grow within mineral substrata and cause their removal. Presence of fungi in coral skeletons is believed to date back even to the lower Paleozoic period (see Kendrick et al. 1982).

Endolithic fungi appear to infect the coral soon after its larva settles and can be detected even in young, growing corals. Fungal hyphae often occur even at the tips of coral structural spines (pali), just after a few hours after their deposition by the polyp. They later thoroughly colonize the coral skeleton. Thus, fungi may associate with the coral early in life and grow with the carbonate skeleton to maintain their position just under the coral tissue.

A section through a coral invariably reveals a dense colonization of the calcium carbonate skeleton by bioeroding fungal hyphae. Fungal hyphae pervade the skeleton of *Porites lobata* up to 5 mm depth and extend up to the very tip of skeletal spines on the coral surface (Le Campion-Alsumard et al. 1995). Septate dark brown fungal mycelia have been found in the subsurface of dead patches in sections of five species of massive corals from the Andaman Islands in the Bay of Bengal (Fig. 9.3a, b; Raghukumar and Raghukumar 1991). Fungi pervade up to 3 cm depth from the surface in *Porites lutea* from Lakshadweep islands in the Arabian Sea (Ravindran et al. 2001). They may occasionally be seen as deep as 7–8 cm below surface. Fungal biomass in this coral has been estimated to constitute up to 0.05% of coral wet weight. The fungus *Scolecobasidium* contributed 3–5 mg of biomass per $cm^3$ of a coral in the Andaman Islands of the Bay of Bengal.

**Bioeroding organisms grow under low light intensities and low oxygen concentrations.** Therefore, their biological mechanisms are of great interest. Bioerosion of corals by fungal hyphae has been described in detail by S. Golubic, Le Campion-Alsumard, Bryce Kendrick, C. Raghukumar, and others (Raghukumar and Ravindran 2012).

The mechanism of bioerosion has been experimentally elucidated by inoculating *Aspergillus versicolor* and *Penicillium stoloniferum* in the coral *Siderastrea sidereal* and incubating the corals for 28 days in seawater. The process of bioerosion was studied using the resin-cast method in which dehydrated coral specimens are infiltrated with resin. The resin is allowed to polymerize under heat. The carbonate is then removed with acid, and the resin-cast borings are studied by scanning electron microscopy. Septate fungal mycelium has been clearly demonstrated in skeletons of hard corals using this method (Fig. 9.3c, d; Le Campion-Alsumard et al. 1995; Bentis et al. 2000; Golubic et al. 2005).

Fungal hyphae within corals of *Porites lobata* reside in the surface 4–5 mm of the coral skeleton in the same layer as the polyps and coexist with them

(Le Campion-Alsumard et al. 1995). Fungal hyphae generally do not enter the spaces occupied by coral polyps. Whenever the hyphae exit from the coral skeleton and begin to grow within the space occupied by polyps, the polyps respond by deposition of a dense, hemispherical to conical, aragonite skeletal material ahead of the fungal hypha (Fig. 9.3c). The protrusions in turn become penetrated by the hyphae, inducing the deposition of new layers of aragonite. Once the polyps grow, evacuate the earlier space and move upward, hyphae grow into the open spaces.

The terrestrial fungus *Aspergillus* appears to reside in the coral *P. lobata*. Fungal hyphae permeated the coral skeleton, entered the pore spaces, and produced the typical *Aspergillus*-type sporulating structures (Golubic et al. 2005).

### *9.1.3 Symbiotic Relationships Between Endolithic Fungi and Corals*

**Endolithic fungi derive their nutrition from the organic matrix of carbonate skeletons. Under normal conditions of symbiosis, fungi and corals live in a balanced equilibrium of commensalism or mutualism. However, a disturbance of this equilibrium may turn this relationship into that of a host and parasite** (Le Campion-Alsumard 1995; Amend et al. 2012; Raghukumar and Ravindran 2012).

**Coral-associated fungi may be mutualists with corals, deriving some or most of the nutrients from the corals and in turn contributing essential nutrients to them.**

- **Fungi may be involved in the cycling of recalcitrant nitrogen metabolism for uptake by zooxanthellae in corals** (Fig. 9.4). Endolithic fungi play a role in the nitrogen cycle within the coral holobiont. For example, many endolithic fungi have been detected in the coral *Porites astreoides* using metagenomic techniques. Several fungal functional genes are also expressed within the coral, apparently from these fungi. Some of these relate to denitrification, including reduction of nitrate/nitrite to ammonia and assimilating ammonia for use in biosynthesis (Wegley et al. 2007). Such an association will be extremely useful to corals because they often reside in nitrogen-impoverished oligotrophic waters.
- The basidiomycetous yeast *Cryptococcus* sp. may produce a transient cryoprotective effect in *Pocillopora damicornis*, selectively enhancing the survival of skeletogenic cell types (Domart-Coulon et al. 2004). This was demonstrated through experiments on interactions between coral cells and the yeast strain F19-3-1. Coral cells established in primary culture obtained a short-term extension of their survival in the presence of the yeast.
- Corals in the tropical oceans protect their tissues from damaging doses of ultraviolet radiation. This protection is considered to be offered by the symbiotic

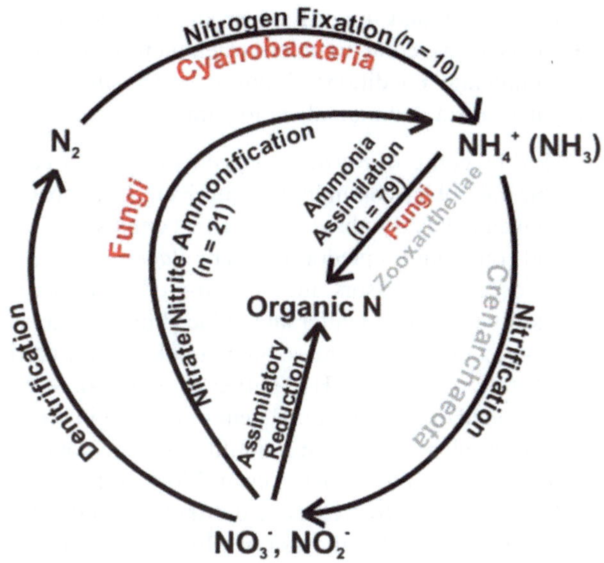


**Fig. 9.4** A proposed model for nitrogen cycling in the *Porites astreoides*-associated microbial community showing the role of fungi in ammonification and ammonium assimilation processes *(Source: Wegley, L. et al. 2007. Metagenomic analysis of the microbial community associated with the coral Porites astreoides. Environmental Microbiology 10: 2707–2719. With permission from John Wiley & Sons)*

zooxanthellae living in their tissue through mycosporine-like amino acids (MAA). Some members of MAAs are found in fungi (Dunlap and Shick 1998). It will be worthwhile to study how much of MAAs are contributed by the coral-associated fungi in corals and what role they play in absorbing and dissipating UV energy.

**A number of possible fungal diseases affecting scleractinian corals have been reported in literature.**

- A "lower marine fungus," presumably an oomycetan fungus, has been reported to be associated with black line disease of the star corals, *Montastrea annularis* (E&S), in Caribbean reefs (Ramos-Flores 1983).
- *Scolecobasidium* sp., a terrestrial fungus, was associated with necrotic patches of scleractinian corals, *Porites lutea*, *Goniopora* sp., and *Montipora tuberculosa* in the Arabian Sea (Raghukumar and Raghukumar 1991).
- Frequency of corals yielding culturable fungi increased four times among corals affected by the brown band syndrome and skeletal eroding band syndrome than in healthy ones, suggesting their role in causing the disease (Yarden et al. 2007).
- A dark mycelial fungus with perithecia, probably belonging to Ascomycetes, has been observed in black bands of *Porites* species in the eastern part of the Indonesian Archipelago. Partial dissolution of such black bands with EDTA (ethylenediamine tetracetic acid) revealed the presence of branched filaments (Bak and Laane 1987).

**Adverse environmental conditions may trigger endolithic fungi to become parasitic in corals** (Thurber et al. 2009). The coral holobiont is made of metazoans, protists, and microbes. The microbiome of corals is made of bacteria, archaea, protists, algae, and fungi. Each group of organisms contributes to the health and resilience of the coral holobiont when their populations are in balance as in normal environmental conditions. Environmental stresses may shift the balance from commensalism and mutualism to parasitism.

- Healthy corals of *Porites compressa* in Hawaii harbor a predominantly autotrophic microbial community comprising Cyanobacteria, Proteobacteria, and the zooxanthellae Symbiodinium. This composition shifted to heterotrophic microbial communities predominated by *Bacteriodetes*, *Fusobacteria*, and fungi when exposed to stressors in the form of increased temperature, nutrients, and dissolved organic matter and reduced pH. Fungi, as judged by DNA sequences, showed the most increase among eukaryotes, particularly with increased temperature or lowered pH (Thurber et al. 2009). Stress also resulted in increase in the abundance of microbial genes involved in virulence, stress resistance, sulfur and nitrogen metabolism, motility and chemotaxis, fatty acid and lipid utilization, and secondary metabolism. Most of these fungi belonged to Ascomycota. Corals stressed by excessive nutrients in ambient water showed a nearly threefold increase in fungi, most of which (~75%) resembled *Harpochytrium*, which is a parasitic Chytridiomycota belonging to *Monoblepharidales*.
- Remarkably similar results were obtained in the case of the coral *Acropora millepora*, which was stressed through a natural thermal bleaching event at Magnetic Island in the Great Barrier Reef (Littman et al. 2011). This metagenomic study showed that the microbial community shifted from a predominantly autotrophic to heterotrohic composition and increases in genes associated with the metabolism of fatty acids, proteins, simple carbohydrates, phosphorus, and sulfur. The proportion of virulence genes were also higher in bleached corals indicating an increase in microorganisms capable of causing disease.
- Coral polyps secrete an aragonitic calcium carbonate deposit ahead of fungal hyphae growing towards the space occupied by them. This is a defense reaction similar to the response of molluscs to foreign bodies. Such a relationship suggests an inherently parasitic relationship between fungi and scleractinian corals. This deposition is similar to the response of molluscs to foreign bodies.

**The relation of fungi to corals may span a continuum from mutualist to commensalist to parasite depending on environmental context and overall coral health.**

## 9.2 Fungi in Gorgonians

The occurrence of a widespread disease in 1990s of the sea fan *Gorgonia* in Caribbean reefs brought coral reef fungi to limelight. The disease, "aspergillosis," is caused by the fungus *Aspergillus sydowii*.

Gorgonians (sea fans, or soft corals) are members of Cnidaria like the scleractinian corals. Gorgonians are also colonial organisms consisting of polyps embedded in a matrix. Many gorgonians live in a symbiotic association with zooxanthellae, as with scleractinian corals. Unlike scleractinian corals, however, they do not secrete a calcium carbonate skeleton.

**Healthy and diseased gorgonians harbor fungi.**

Marine-derived fungi belonging to Ascomycota, comprising many dematiaceous anamorphs such as *Alternaria*, *Aspergillus*, *Cladosporium*, *Curvularia*, *Fusarium*, *Nigrospora*, *Scolecobasidium*, *Penicillium*, *Gloeotinia*, *Rhodotorula*, *Stachybotrys*, and *Xylaria* and also pleosporaceous taxa, as well as yeasts have been isolated using plating methods or detected using molecular sequences from *Gorgonia ventalina*, *Annella* sp., and other gorgonians from Puerto Rico, Singapore, and Mu Ko Similan National Park, Andaman Sea, Thailand (Toledo-Hernandez et al. 2008; see Sakayaroj et al. 2012).

### *9.2.1 The "Sea Fan Disease" or "Aspergillosis" of Gorgonian Soft Corals*

**The "aspergillosis" disease of the sea fan *Gorgonia*, caused by the fungus *Aspergillus sydowii*, has assumed alarming proportions and has kindled world-wide interest because of its potential harm to the coral reef ecosystem** (Fig. 9.5; Raghukumar and Ravindran 2012).

Mass mortalities of sea fans belonging to *Gorgonia* species were first reported in the 1980s from reefs of Costa Rica and Santa Marta (Colombia). **The disease assumed epizootic proportions in *Gorgonia ventalina* and *G. flabellum* in the Caribbean during 1995 and 1996**. More than 90% of gorgonians were affected by the disease.

**The disease syndrome of aspergillosis sea fan disease is now well known.** Infection of the corals commences with an erosion of the outer rind tissue, thus exposing the central core (Smith et al. 1998)**. Both tissue and skeleton disappear in the diseased sea fans.** The necrotic tissue along the edges of the lesions is either lighter or dark purple in color, in contrast to the coloration of the healthy tissue. The eroding tissues contain fungal hyphae. The purple color is the result of sclerites produced by the coral in response to the disease. Formation of sclerites results in galls that contain the fungus. It appears that corals that show less defense in the form of sclerite formation are liable to mortality.

Immune response of the host against the pathogen included production of antifungal lipid metabolites (Kim et al. 2000) and a physical barrier through melanin deposition to prevent fungal expansion (Petes et al. 2003), production of peroxidases that show antifungal activity (Mydlarz and Harvell 2007) and production of exochitinase against the fungus *Aspergillus sydowii* (Douglas et al. 2007).

**Fig. 9.5** Aspergillosis on the sea fan (*Gorgonia ventalina*). Necrotic patches appear colored *(Source: Ein-Gil, N. et al. 2009. Presence of Aspergillus sydowii, a pathogen of gorgonian sea fans in the marine sponge Spongia obscura. The ISME Journal 3: 752–755)*

**The fungus *Aspergillus sydowii* was regularly isolated from diseased portions of the gorgonian during the outbreak of the disease in the Caribbean.** *Aspergillus sydowii* is a terrestrial anamorphic species belonging to Ascomycota. Many other terrestrial species of fungi, such as *Aspergillus*, *Cladosporium*, *Fusarium*, and *Nigrospora*, have also been commonly associated with diseased corals from the Caribbean, Australia, Singapore, and the Andaman Seas (Geiser et al. 1998; Koh et al. 2000; Yarden et al. 2007). However, *A. sydowii* is the most frequent, consistent, and abundant fungus in diseased gorgonians. *Aspergillus* species also caused a disease in the sea fan, *Annella* sp., in Southeast Asia after the 2004 tsunami (Phongpaichit et al. 2006). **Experimental evidence, including Koch's postulates, have established *Aspergillus sydowii* to be the causal agent of the disease.** The disease could be established in healthy colonies by inoculating them with pure cultures of *A. sydowii* or by grafting infected tissue onto healthy tissue (Smith et al. 1996).

*Aspergillus sydowii* is now accepted to be the causative agent of the aspergillosis sea fan disease.

**What is the source of the fungus? Several possibilities have been suggested.**

- African dust may be one source of the pathogen in Caribbean gorgonian corals (Shinn et al. 2000). The presence of spores of *A. sydowii* in the African dust was suggested as a proof of it being the primary source of the coral pathogen (Weir-Brush et al. 2004). However, *A. sydowii* could not be isolated from dust collected from Africa and the Cape Verde Islands in the Atlantic. Hence, African dust may not be the real or the only source of this coral pathogen (Rypien 2008).
- Other organisms in the coral reef may be vectors of the pathogen. For example, *Aspergillus sydowii* has been isolated from the sponge *Spongia obscura* in coral reefs of Bahamas (Ein-Gil et al. 2009). The sponge could be a passive carrier of the disease. In another instance, the snail, *Cyphoma gibbosum*, was demonstrated to transmit the pathogen through feeding on the gorgonians (Rypien and Baker 2009).

**The triggering mechanism for the disease is not clear.** Aspergillosis is probably the result of two distinct factors.

1. **The disease may be caused by distinct, pathogenic strains of the fungus**, compared to others of the same species. Genetic structure of global samples of *A. sydowii*, including isolates from diseased corals, diseased humans, and environmental sources, revealed that disease-causing isolates are not genetically distinct from environmental isolates (Rypien 2008). These authors further suggested that no specific virulence factor in the pathogen was responsible and that any isolate of *A. sydowii* could cause aspergillosis. The opposite that the pathogenic strains of this facultatively marine fungus are metabolically different from nonpathogenic, terrestrial strains has also been suggested (Alker et al. 2001).
2. **Environmental changes that affect the immunity of the host may make it susceptible to the disease.** *Aspergillus sydowii* has been detected in healthy corals (Toledo-Hernandez et al. 2008), where the symbiotic fungus may occur as a commensal, a mutualist, or a weak parasite. Sea fans produce antimicrobial compounds. Despite this, infection or pathogenicity may take place under unfavorable environmental conditions (Alker et al. 2001). Temperature plays a crucial role in the virulence of the fungus *A. sydowii* to the sea fan coral *Gorgonia ventalina*. The fungus grows at an optimum temperature of 30 °C suggesting that increased water temperature during the summer is likely to promote its pathogenicity (Alker et al. 2001). This points to a potential danger of climate changes and accompanying warming. Increased temperature (31.5 °C) caused 76% increase in activity of host-derived antifungal compounds, but also increased the pathogen's growth rate, providing an opportunity to the

pathogen to establish itself before the host could defend itself (Ward et al. 2007). With increasing global warming, it is predicted that aspergillosis will continue to play havoc in gorgonian sea fan communities.

3. **Sea fan disease may perhaps also be caused by other factors.** Thus, *Aspergillus sydowii* was not isolated in diseased sea fans collected from 13 reefs in Puerto Rico (Toledo-Hernandez et al. 2008).

As the fungus can be cultured and the host is easy to maintain in laboratory, aspergillosis disease of the sea fan *Gorgonia ventalina* is taken as a model system to understand host–pathogen interactions in context with environmental stressors.

## 9.3 Sponge-Associated Fungi

**Fungi inhabiting marine sponges have attracted a lot of attention because of the potential of sponge-associated microorganisms in producing bioactive substances.**

Sponges are sedentary organisms with highly efficient filter-feeding mechanisms. A sponge weighing about 1 kg in weight can filter an enormous volume of up to 24,000 l of seawater per day. Fungal spores in the water may pass into the animals. Some of these may colonize and grow within sponges, while others may remain as spores. Thus, sponge-inhabiting fungi may be "residents" or "transients" (Li and Wang 2009). "Resident fungi" are sponge-specific and long-term inhabitants of sponges. "Transient fungi" are merely spores that have been ingested by the sponge, do not actively grow therein, and can be recovered in culture.

### *9.3.1 Diversity of Sponge-Associated Fungi*

Marine-derived fungi belonging to a wide diversity of species have been cultured or detected using DNA sequences from sponges at a number of locations in the Atlantic and Pacific (Höller et al. 2000; Namikoshi et al. 2002; Morrison-Gardiner 2002; Li and Wang 2009; Yarden 2014).

**Most resident, facultative marine mycetaen fungi in sponges are terrestrial species belonging to anamorphic forms of Ascomycota. Obligate marine fungi and novel fungi not described so far may also reside in sponges.**

- **Culture-dependent and culture-independent studies on the sponges *Gelliodes fibrosa*, *Haliclona caerulea*, and *Mycale armata* from Pacific waters of Hawaii have indicated the presence of facultative as well as obligate mycetaen marine fungi.**

- Culturing following surface sterilization of tissues yielded 44 isolates of terrestrial ascomycete species belonging primarily to Pleosporales and Hypocreales. *Aspergillus* and *Trichoderma* were particularly prevalent. Mycosphaerellales, Eurotialies, Dothideales, Hypocreales, Diapothales, Xylariales, Pleosporales and Saccharomycetales of Ascomycota, and Aphyllophorales of Basidiomycota are facultatively marine in the sponges. The most common ones belong to *Acremonium*, *Alternaria*, *Aspergillus*, *Cladosporium*, *Fusarium*, *Myrothecium*, and *Penicillium* (Wang et al. 2008).
- Another study, also based on surface sterilization and plating, yielded 86 isolates which were identified based on ITS sequences belonging to 1 basidiomycete and 25 ascomycete genera. These corresponded to Aphyllophorales, Diaporthales, Eurotiales, Hypocreales, Mycosphaerellales, Pleosporales, Saccharomycetales, and Xylariales. Terrestrial species belong to *Aspergillus*, *Eupenicillium*, *Penicillium*. *Ampelomyces*, *Bipolaris*, *Cochliobolus*, *Curvularia*, *Didymella*, *Leptosphaerulina*, *Phaeosphaeria*, *Paraphaeosphaeria*, *Bionectria*, *Fusarium*, *Hypocrea*, *Myrothecium*, *Solheimia*, *Trichoderma*, *and Tubercularia*. Seventy one of the 86 isolates belonged to terrestrial species (Li and Wang 2009).
- A culture-independent technique based on denaturing gradient gel electrophoresis (DGGE) technique, using primer sets designed to detect rRNA gene and ITS sequences of fungi, was used to detect fungi in these three sponges (Gao et al. 2008). These fungi mostly belonged to Ascomycota and Basidiomycota.
- *Aspergillus sydowii*, the causal agent of the sea fan disease, has also been isolated from the sponge *Spongia obscura* in coral reefs of Bahamas (Ein-Gil et al. 2009).

• **Obligate marine fungi may be present in sponges**. The culture-dependent study of sponges from Hawaiian waters (see above, Li and Wang 2009) also yielded obligate marine fungi, which comprised 17% of the 86 fungi cultured from three sponges. Sequences of obligate marine fungi belonging to *Asteromyces cruciatus* and *Phialophorophoma litoralis* have been detected in sponges from the Atlantic and Pacific (Höller et al. 2000). Many of the DNA sequences retrieved from Hawaiian sponges had low identity with those available in GenBank. This shows that **novel fungi inhabit sponges.** Likewise, similarity of sequences with terrestrial species of fungi obtained from other marine samples suggests that these are "marine phylotypes" that belong to facultative marine fungi (Gao et al. 2008).
• **Yeasts are common inhabitants of sponges.** *Malassezia* species are ubiquitous lipophilic yeasts and are found in soils, sediments, and deep-sea habitats and in terrestrial metazoa. This ascomycetous genus, which consists of 13 known species, is usually associated with the skin of humans and various animals such as cats, dogs, goats, and cows. They also infect marine mammals, such as seals or sea lions (Guillot et al. 1998; Nakagaki et al. 2000; Pollock et al. 2000; Nagahama et al. 2011). A great diversity of this yeast has been detected in

sponges using molecular sequences (Gao et al. 2008). It is currently unclear whether the diversity of sponge-associated yeasts represents symbionts and/or parasites (Fell 2012).

**Numbers and diversity of sponge-associated fungi vary depending on individual sponge species and geographical locations.** Some fungi may have a broad host range and occur in a wide variety of sponges, while others may be restricted to one or a few species (Fig. 9.6; Gao et al. 2008; Li and Wang 2009).

- Culturable fungi generally range from zero to 21 genera per sponge species. However, a much higher diversity of marine-derived fungi has been detected using DGGE methods in the marine sponges *Suberites zeteki* and *Gelliodes fibrosa* from Hawaii compared to other sponges (Gao et al. 2008). Fungal communities differed between sponge species *Suberites zeteki* and *Gelliodes fibrosa* and between sponges and the surrounding seawater (Gao et al. 2008).
- *Arthrinium* sp. and *Niesslia* sp. have been isolated exclusively from sponges of the North Sea at Helgoland, Germany, while *Aspergillus* spp. were seen to predominate in sponges from Malta in the Atlantic. *Phoma* spp. were most frequent in Dominica in the Caribbean. *Acremonium*, *Cladosporium*, *Fusarium*, and *Penicillium* were common in many sites in the Atlantic and Pacific (Höller et al. 2000).

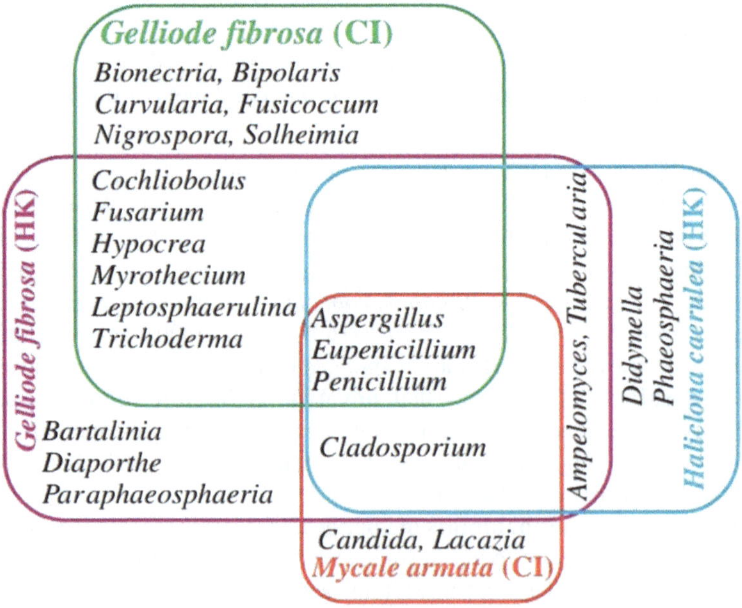


**Fig. 9.6** Association of fungi with three different sponges collected from Coconut Island (CI) and in Hawaii Kai (HK), showing "sponge generalists," "sponge associates," and "sponge specialists" fungi *(Source: Li, Q. and Wang, G. 2009. Diversity of fungal isolates from three Hawaiian marine sponges. Microbiological Research 164: 233–241. With Permission from Elsevier)*

- Based on diversity studies in three Hawaiin sponges, the fungi were classified as "sponge generalists," "sponge associates," and "sponge specialists" (Fig. 9.6; Li and Wang 2009). "Sponge generalists" such as *Aspergillus*, *Penicillium*, and *Eupenicillium* were found in all sponges. Fungal genera such as *Ampelomyces*, *Tubercularia*, and *Cladosporium* were found in more than one sponges and were called "sponge associates." The "sponge specialists" *Didymella*, *Fusicoccum*, and *Lacazia* occurred only in one sponge species. Another important criterion could be the location of the sponge. Thus, the same sponge species from two different locations could harbor different sponge specialists. Some species may be geographically widespread. This interesting concept needs to be studied further.

## 9.3.2 The Symbiotic Association of Fungi with Sponges

**The actual relationship of fungi and sponges, whether commensalistic, mutualistic, or parasitic, has not been clearly elucidated.** Information on these is limited.

**Some fungi have been suspected to cause disease in sponges** even since 1942 (Galstoff 1942). As with scleractinian and gorgonian corals, it is likely that fungi that are resident in sponges may become opportunistic parasites. Alternatively, fungi may invade diseased tissues as secondary colonizers.

**Some yeasts might have a mutualistic association with sponges.** Yeasts are associated with both adult sponge tissue and reproductive structures of three species of the sponge *Chondrilla*. Their presence within sponges has been confirmed beyond doubt by electron microscopy and immunocytochemical labeling of the fungal signature β-1,4-N-acetyl- D-glucosamine residues of chitin walls. These fungi remain in direct contact with the cell's cytoplasm. These endosymbiotic yeasts are maternally transmitted from one generation to another. A large number of yeast cells (ca. 4.4 cells per 10 $\mu m^2$) were transmitted from the soma through the oocytes to the fertilized eggs (Maldonado et al. 2005).

Sponges may possess a molecular mechanism for recognizing fungi. The demosponge *Suberites domuncula* possesses a cell surface receptor that recognizes (1-3)-β-D-glucans (Perovic-Ottstadt et al. 2004). The gene that codes for the 45 kDa protein has also been determined.

Thraustochytrids have been shown by histochemical studies to become abundant in the dying marine sponge *Halichondria panicea* (Richter 1985). **Hence, thraustochytrids may have a saprotrophic role in decomposing dead sponges.** The subsistence of Labyrinthulomycetes in coral mucus and the seawater surrounding the coral reef area also suggests a definite role for saprobic Labyrinthulomycetes in association with these invertebrates.

## 9.4 Fungi in Diverse Coral Reef Invertebrates and Algae

A rich diversity of invertebrates belonging to cnidarians, molluscs, and crustaceans inhabit coral reefs. Fungi are likely to be associated with them in various ways. **For example, primary cell cultures of coral reef invertebrates are invariably contaminated with thraustochytrids, suggesting an intimate relationship of these straminipilan fungi with them.** A group of closely related thraustochytrids have been isolated from primary cell cultures of the colonial tunicate *Botryllus schlosseri* (Rabinowitz et al. 2006).

However, our knowledge of the association of fungi with them is still rudimentary.

Coral reefs are also home to a high diversity of algae. Several fungal pathogens of green, brown, and red algae are known from the Lakshadweep Islands in the Arabian Sea (see references in Raghukumar 2006). Most of these pathogens belong to Chytridiomycota and Oomycetes comprising aquatic fungi that are pathogens of algae and phytoplankton. Littler and Littler (1998) described infection of crustose coralline algae by an unidentified fungus in American Samoa. The pathogen occurred in shallow reef habitat but was not restricted to calm waters. Spreading dense black fungal bands caused death of coralline algae (Raghukumar and Ravindran 2012).

- Fungi live symbiotically with organisms in the coral reef ecosystem.
- Fungi constitute an important component of the coral holobiont and grow epi- and endobiontically in scleractinian corals.
- Many endolithic fungi inhabit calcium carbonate skeletons of living as well as dead coral.
- Most endolithic coral fungi are facultative marine fungi, belonging to terrestrial species.
- A high diversity of Chytridiomycota, Ascomycota, and Basidiomycota with many novel lineages live in hard corals.
- Species of the ascomycete family Koralionastaceae are some of the obligate, marine, corallicolous fungi, inhabiting dead coral skeletons and coral slabs
- Thraustochytrids densely colonize coral surfaces and grow in coral mucus.
- Hyphae of endolithic fungi densely pervade coral skeleton and cause bioerosion.
- Fungi often cause black bands within corals.
- Fungi and corals appear to live in a balanced equilibrium of commensalism or mutualism under normal conditions.
- Fungi may be involved in the cycling of recalcitrant nitrogen metabolism for uptake by zooxanthellae in corals.

(continued)

- Adverse environmental conditions may trigger endolithic fungi to become parasitic in corals.
- Fungi are associated with healthy as well as diseased gorgonian corals.
- *Aspergillus sydowii* is the causal agent of a widespread disease called aspergillosis in the sea fan *Gorgonia*. The disease can cause epizootics.
- Terrestrial species of facultative marine fungi, obligate marine fungi, as well as novel fungal phylotypes have been detected in sponges.
- Most resident, facultative marine mycetaen fungi in sponges are terrestrial species belonging to anamorphic forms of Ascomycota. Besides, obligate marine fungi and novel fungi not described so far may reside in sponges.
- Numbers and diversity of fungi may vary with sponge species and geographical locations.
- Yeasts belonging to *Malassezia* and others are frequent inhabitants of sponges.
- Some yeasts may be associated with sponges through their entire life cycle, probably in a mutualistic manner.

**Future Directions**

1. Diversity of fungi in coral reef organisms is poorly known.
2. The source of endolithic species of fungi, their mechanism of entry into corals, and their method of spread have not been addressed.
3. The nutrition of endolithic coral fungi has not been studied.
4. The relationship between fungi and corals, whether mutualists, commensals, or parasites needs to be elucidated. What triggers fungi to become parasitic in corals?
5. Thraustochytrids appear to be abundant in coral reef organisms, but their diversity, population, and role have not been studied.
6. The prevalence of fungal diseases in gorgonian corals worldwide has to be studied, and triggers of the disease need to be clarified.
7. The relationship between fungi and sponges is not known. Are they commensals or mutualists? How active are fungi in sponges?
8. The role of yeasts in association with sponges requires further study.

# Chapter 10
# Animals in Coastal Benthic Ecosystem and Aquaculture Systems

*The best way to observe a fish is to become a fish*
Jacque Yves Cousteau

Animals are secondary and higher level producers in any ecosystem. Health of animals at each trophic level is important for efficient energy transfer to trophic levels above. Many marine animals are also cultivated in aquaculture for human use. Hence, their health is also of economic importance to human society. Following their death by natural mortality, disease, and incomplete predation by other animals, they are decomposed by bacteria and fungi which recycle nutrients in the ecosystem. The littoral and sublittoral zones of coastal seas are home to a large diversity and population of benthic animals. These include meiobenthos such as nematodes, amphipods, and polychaetes, as well as macrobenthos such as crustaceans, molluscs, and fish. Many of these are cultivated in aquaculture. The balance of health of marine animals when they are alive and their decomposition and recycling of nutrients upon their death are crucial to the marine ecosystem.

**A high diversity of fungi belonging to Mycetae as well as Straminipila are widely associated with animals as symbionts and saprotrophs.** Mutualistic fungi play a role in maintaining the health of animals. Saprophytic fungi decompose dead animals and their fecal matter. Most of what we know about fungi in the marine ecosystem pertains to their role as parasites and pathogens. Many members of marine Oomycetes (Straminipila) as well as terrestrial species of fungi are widely prevalent as animal parasites.

## 10.1 Nonpathogenic Symbiotic Fungi in Animals

**Diverse fungi occur in a variety of invertebrates without causing any external disease symptoms.**

**Obligate as well as facultative marine fungi are frequently found in invertebrates.**



S. Raghukumar, *Fungi in Coastal and Oceanic Marine Ecosystems*,
DOI 10.1007/978-3-319-54304-8_10

- Sea urchins (Phylum Echinodermata) harbor fungi in their guts.
  - The irregular sea urchin *Echinocardium cordatum* (Pennant) collected from a muddy sediment at a depth of 28 m from the Baltic Sea harbors a variety of microorganisms in its guts (Thorsen 1999). An obligately anaerobic fungus belonging to the Neocallimastigomycota appears to be a resident in its guts (Fig. 10.1). Motile spherical cells of 4–6 µm with an extremely long, posteriorly directed flagellum resembling zoospores were found in the gut and coelomic fluid. Thalli, zoosporangia, and zoospores of the chytrid fungus have been found in various parts of its gut. Cells of the chytrid as well as zoospores are abundant in the anoxic anterior cecum of the gut, along with bacteria and protozoa. This fungus may be mutualistic with the symbiont, with a similar function as in terrestrial ruminant animals.
  - Thraustochytrids have been cultured from the guts of another sea urchin, *Lytechinus variegatus* collected from the Gulf of Mexico (Wagner-Merner et al. 1980). Straminipilans also appear to be associated intimately with other invertebrates. The thraustochytrid *Ulkenia visurgensis* (Ulken) Gaertner has been detected using an immunofluorescence detection method in the coelenterons and hydranth of a hydroid from the west coast of India (Raghukumar 1988). The aplanochytrid, *Aplanochytrium (Labyrinthuloides) yorkensis*, was first isolated from the mantle cavity of the oyster *Crassostrea virginica* (Perkins 1973).
- Trichomycetes, a group of mycetaen fungi belonging to Glomeromycota, colonize the digestive tracts of many marine arthropods. These fungi are believed to

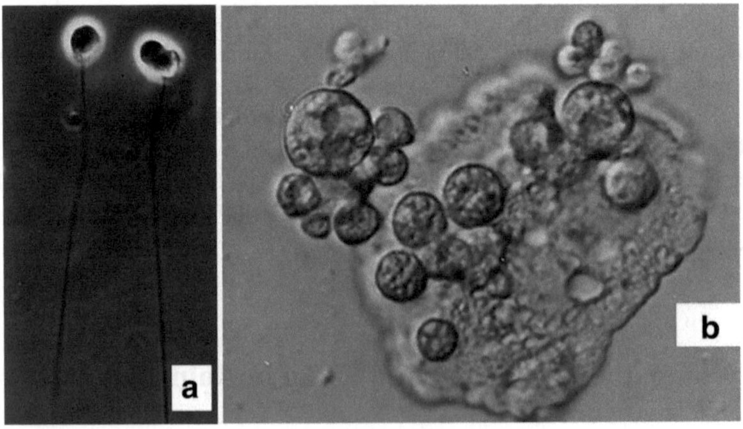


**Fig. 10.1** Some commensalistic or mutualistic fungi in animals. (**a, b**) A fungus resembling anaerobic Neocallimastigomycota in gut of the sea urchin *Echinocardium cordatum*. (**a**) Zoospores with a posterior flagellum from the coelomic fluid. (**b**) Particulate material from the anterior cecum, inhabited by nonmotile spherical cells resembling the fungus (**a** and **b**: Source: Thorsen, M.S. 1999. Abundance and biomass of the gut-living microorganisms (bacteria, protozoa and fungi) in the irregular sea urchin *Echinocardium cordatum* (Spatangoida: Echinodermata). Marine Biology 133:353–360)

live in a mutualistic association with their hosts. Their thallus either consists of branched, septate filaments attached to the host cuticle by a holdfast or may be unbranched, nonseptate, and coenocytic (http://www.nhm.ku.edu/~fungi/). Trichomycetes cannot be cultured and have to be detected by direct microscopy methods. These fungi have been found in a number of marine arthropods, isopods, decapods, and amphipods (Misra and Lichtwardt 2000). One particular study found several members of Eccrinales belonging to Trichomycetes in crustaceans of the littoral region of the San Juan archipelago, Washington (Hibbits 1978).

- A number of marine-derived fungi have been obtained from the sponges *Amphimedon viridis, Axinella corrugata, Dragmacidon reticulata, Geodia corticostylifera, Mycale laxissima*, and *Mycale angulosa* and the ascidians *Didemnum ligulum* and *Didemnum* sp. collected at 5–10 m from the coast of Sao Paulo, Brazil. Culturing following surface sterilization yielded facultative marine fungi belonging to Ascomycota, Basidiomycota, and Mucoromycota (Menezes et al. 2010). Marine-derived fungi have also been detected from sponges along the coast of Hawaii, using DGGE analysis of ITS or SSU rRNA sequences. These fungi included the PCG environmental sequence groups belonging to Ascomycota as well as the Hy-An group belonging to Basidiomycota (Gao et al. 2008). Most fungi belonged to hitherto unidentified taxa. Among the identified ones, most belonged to *Trichoderma, Penicillium, Aspergillus* and *Fusarium*. The ascidian *Didemnum* sp. harbored the highest diversity of filamentous fungi.
- Yeasts have been isolated from a variety of marine animals, such as the Mexico shrimp *Penaeus setiferus*, shrimp eggs, sponges, and other invertebrate material collected from the North Atlantic Ocean, guts of the fiddler crab, *Uca pugilator,* shellfish, and intestine of farmed rainbow trout (*Salmo gairdneri*) (Kutty and Philip 2008). Yeasts appear to be abundant in internal fluids of some animals, such as those of the guts of fiddler crab and the liquid portion of the shellfish. A variety of yeasts have been identified. Red-pigmented yeasts dominated and composed about 90% of the isolates from the rainbow trout. Other yeasts are *Trichosporon cutaneum, Rhodotorula glutinis, Candida parapsilosis, Pichia guilliermondii, Pullularia pullulans, Debaryomyces hansenii, Torulopsis candida, Trichosporon cutaneum, Debaryomyces hansenii, Saccharomyces cerevisiae, Rhodotorula rubra* and *R. glutinis.*

## 10.2 Animal Diseases Caused by Fungi

Parasitism is as much a natural, ecological strategy as commensalism and mutualism are. Natural or man-made environmental changes may induce physiological stress in organisms and make them susceptible to diseases. Under such conditions, organisms which are mildly parasitic or even mutualistic may become highly virulent and cause infections of epidemic proportions that affect the population of the host organisms (Harvell et al. 1999). New diseases have often emerged through host or range shifts of known pathogens.

**Fungal infections are common in both wild and cultivated marine animals** (Porter 1986; Polglase et al. 1986; Shields and Overstreet 2007; Ramaiah 2006; Hyde et al. 1998; Hatai 2012; Marano et al. 2012). Fungi parasitize almost all groups of marine animals in the oceans. In addition to their importance in the marine ecosystem, infectious diseases, including fungal infections, are a major problem for aquaculture industries. Marine fishes, prawns, and crabs are very popular seafood. Finfish and shellfish seed production and culture are important economic activities. Most of our knowledge on fungal parasites of marine animals comes from diseases of commercially useful ones.

## *10.2.1 Fungi that Cause Animal Diseases*

**Fungi belonging to the Oomycetes, Straminipila have an enormous host range and are probably most important as marine invertebrate parasites, especially in the seed production of marine crustaceans such as shrimps and crabs** (Marano et al. 2012; Hatai 2012). This includes species belonging to *Atkinsiella* (Atkinsiellales, Saprolegniomycetes), *Haliphthoros, Halioticida*, and *Halocrusticida* (Haliphthorales), *Lagenidium* (Peronosporales), and *Sirolpidium* (Olpidiopsidales) (Beakes et al. 2014) (Fig. 10.2; Table 10.1).

- **Members of "Haliphthorales" are exclusively marine and occur as parasites of molluscs, crustaceans, or their eggs** (Beakes et al. 2014). Members of this group produce irregularly branched mycelial thalli. Part of the thallus is converted into the reproductive structures, the zoosporangia, while others remain vegetative (eucarpic thallus). Species of this order can be cultured on artificial media. Large segments of the thalli convert into zoosporangia. All genera form very long narrow hypha-like exit tubes from which the biflagellate zoospores escape in uniseriate fashion.
- **Species of *Lagenidium,* particularly *L. callinectes* Couch, infect and kill many invertebrate eggs and embryos**. This oomycete was first discovered in eggs of the blue crab *Callinectes sapidus* from Chesapeake Bay and described by the renowned mycologist John N. Couch in 1942. The fungus was later reported to infect ova of the barnacle *Chelonibia patula. Lagenidium chthamalophilum* Johnson infects ova of the barnacles *Chthamalus fragilis. Lagenidium callinectes* represents the greatest fungal threat in cultivation of marine decapods. Presence of the fungus can be observed as brown or gray patches in the clutch. Dead eggs are opaque and smaller than healthy ones. Infection is generally more on eggs in the periphery of the clutch. The fungus rarely penetrates more than 3 mm of the clutch. Eggs die prematurely. The fungus may infect 100% of eggs and kill all of them under severe conditions. Older clutches are attacked more heavily than recently laid ones. Larvae are also

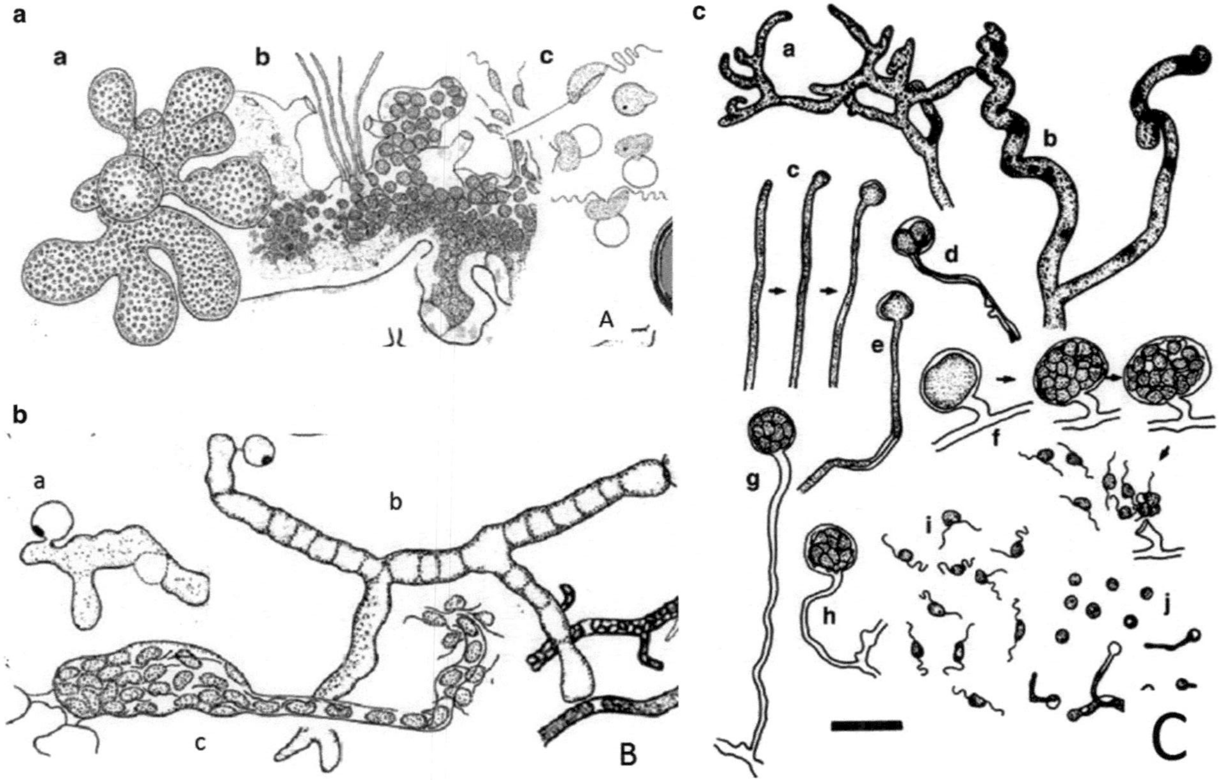


**Fig. 10.2** Oomycete parasites of marine animals. (**a**) *Atkinsiella dubia*. (a) lobate thalli; (b) zoosporangia; (c) biflagellate zoospores. (**b**) *Haliphthoros milfordensis*. (a) Encysted zoospores; (b) lobate thalli; (c) zoosporangia. (**c**) *Lagenidium callinectes*. (a, b) Vegetative thallus. (c–h) Zoosporangium formation. (i, j) Liberation of biflagellate zoospores (Source: **a** and **b**: Beakes, G.W. et al. Systematics of the Straminipila: Labyrinthulomycota, Hyphochytriomycota, and Oomycota. In: Systematics and Evolution, 2nd Edition The Mycota VII Part A. D.J. McLaughlin and J.W. Spatafora (Eds.) Springer-Verlag. **c**: Hatai, K. 2012. Diseases of Fish and Shellfish Caused by Marine Fungi. In: Raghukumar, C. (ed.). Biology of Marine Fungi. Springer Verlag)

**Table 10.1** Examples of diseases caused by Oomycetes in marine animals

| Fungus | Hosts |
|---|---|
| *Lagenidium* | • *L. callinectes*—ova of the blue crab, *Callinectes sapidus, Artemia salina;* ova of the barnacle, *Chelonibia patula*; eggs and zoeae of the marine crabs, *Portunus pelagicus, Cancer magister, Neopanope texana, Pinnotheres ostreum, Portunus trituberculatus, Scylla serrata;* American lobster, *Homarus americanus*<br>• *L. chthamalophilum*—ova of the barnacle, *Chthamalus fragilis*. Cultivated crustaceans, e.g., white shrimp, *Penaeus setiferus*, the Dungeness crab, *Cancer magister*, and the American lobster, *Homarus americanus*<br>• *L. scyllae*—ova and larvae of the mangrove crab, *Scylla serrata*, in Philippines<br>• *L. thermophilum*—eggs and larvae of mangrove crab, *Scylla serrata*, affecting the seed production in Bali, Indonesia, *Penaeus monodon, Crangon vulgaris, Palaemon serratus* (as *Leander serratus*)<br>• *L. marinum*—bivalves *Barnea candida, Acanthocardia echinata* (as *Cardium echinatum*), *Mytilus edulis* |
| *Haliphthoros* | • *H. milfordensis*: endoparasite of eggs of the oyster drill, *Urosalpinx cinerea*; juveniles of the American lobster, *Homarus americanus, H. gammarus*; brine shrimp *Artemia salina;* adults of the white shrimp, *Penaeus setiferus, Farfantepenaeus duorarum* (as *Penaeus duorarum*), *Penaeus japonicus, P. monodon, Litopenaeus setiferus* (as *P. setiferus*); crabs, *Callinectes sapidus, Portunus pelagicus, P. pisum, P. trituberculatus, Pinnotheres* spp., *Scylla serrata;* in abalone, *Haliotis gigantea (sieboldii), Japan; Penaeus monodon* in Nha Trang, Vietnam; gill lesions of juvenile kuruma prawns, *Penaeus japonicus*, with black gill disease.<br>• *H. philippinensis:* larvae of the jumbo tiger prawn, *Penaeus monodon* in Philippines; crab *Scylla serrata;* oyster *Crassostrea virginica;* hard clam *Mercenaria mercenaria* (as *Venus mercenaria*) |
| *Halodaphnea (Halocrusticida)* | • *H. hamanaensis*—Eggs and larvae of *Scylla serrata*<br>• *H. parasitica*—Rotifer *Brachionus plicatilis*, crab *Portunus trituberculatus*<br>• *H. awabi*—Abalone (*Haliotis gigantea (sieboldii)*),<br>• *H. okinawaensis*—zoea of the crab *Portunus pelagicus, Eroicheir japonicus*; crab *Scylla serrata*<br>• *H. panulirata*—Philozoma of spiny lobster (*Panulirus japonicus*)<br>• *H. baliensis*—crab *Scylla serrata* |
| *Atkinsiella dubia* | • Abalone, *Haliotis gigantea (sieboldii)*<br>Crabs—eggs of pea crab, *Pinnotheres pisum* in England, eggs of *Gonoplax rhomboids,* gills of swimming crab, *Portunus trituberculatus,* Japan, *Eriocheir japonicus, Gonoplax angulata, Hyas* sp., *Macropodia* sp., *Oregonia, Portunus depuratus, Typton spongicola* |
| *Halioticida noduliformans* | Abalone—*Haliotis medae*, Japan, gills of wild mantis shrimp, *Oratosquilla oratoria* in Tokyo Bay, Japan |
| *Sirolpidium* | Eggs and larvae of the prawn *Penaeus monodon* |

attacked. The fungus penetrates the exoskeleton, grows rapidly within the body of the larva, and replaces the internal tissues, particularly the striated muscles (Polglase et al. 1986). An isolate from the blue crab grew better on simple sugars (e.g., fructose, glucose) than on complex carbohydrates and polysaccharides and required vitamin B1 (Bahnweg and Bland 1980). Most strains are obligately marine, but isolates from the American lobster and the Dungeness crab *Cancer magister* do not require NaCl (Bahnweg and Gotelli 1980). Infection starts through germination of encysted zoospores that form a germ tube which penetrates the host and grows rapidly. The vegetative thallus consists of coenocytic, intramatrical hyphae. These grow rapidly, ramify in the host, and replace the host tissues. They reproduce by means of zoospores produced in zoosporangia. The zoosporangium is first delimited by a septum from the vegetative hyphae. Sporangia develop long discharge tubes. Following this, cytoplasm is discharged (5–30 min) into a gelatinous vesicle. Flagellar formation precedes cleavage, and the flagella can be observed actively beating inside the sporangium (Gotelli 1974). Cleavage is rapid and spore release occurs within 10 min of sporogenesis. From 20 to 200 zoospores are produced by a single sporangium. The pyriform zoospores, 10 by 13 μm, have two flagella arising from a groove that spans the length of the spore (Shields and Overstreet 2007).

**Many terrestrial species of fungi, such as species of *Fusarium*, are opportunistic pathogens and have been reported to be associated with shell disease of marine crustaceans and lobsters.**

## *10.2.2 Infections of Shrimps and Prawns*

**Fungal diseases frequently pose a huge problem in prawn hatcheries.**

- ***Penaeus monodon***, the highly rated giant tiger prawn or the Asian tiger shrimp, is the second largest cultivated aquaculture prawn in the world. Eggs and larvae of the prawn in the hatchery of an aquaculture facility in Thailand, were reported to have been infected by ***Lagenidium thermophilum*** K. Nakam., Miho Nakam., Hatai & Zafran in August 2000 (Muraosa et al. 2006). A new species, ***Haliphthoros philippinensis*** Hatai, Bian, Batic. & Egusa, was isolated from larvae of the jumbo tiger prawn, *Penaeus monodon* in Philippines (Hatai et al. 1980). This species grows in a wide range of temperature, even up to 36 °C. The anamorphic, opportunistic, terrestrial fungus *Fusarium* may infect all stages of the prawn (Ramaiah 2006). In Vietnam, a new *Fusarium* infection, caused by ***F. incarnatum*** (Roberge) Sacc., occurred in black tiger shrimp, *Penaeus monodon* (Khoa and Hatai 2005). Infected shrimps showed typical signs of black gill disease and mortalities about a month prior to harvest. Optimal temperature for the fungus ranged from 20 to 30 °C. The fungus grew very well at 35 °C, but not at 5 and 40 °C.

***Haliphthoros milfordensis*** Vishniac was isolated from larvae of *Penaeus monodon* in Nha Trang, Vietnam (Chukanhom et al. 2003).

- **The kuruma prawn, *Penaeus japonicus*,** is another highly prized cultured prawn. Many fungi cause the "black gill disease" of juveniles of this prawn (Fig. 10.3a, b). The oomycete *Halipthoros milfordensis* was reported from black gill lesions at a private farm in August 1989 in Japan. The fungus *Fusarium* seems to be an important agent of the disease. *Fusarium* infection was first reported from black gill disease of pond-cultured kuruma prawn in Japan in 1972. Subsequently, the fungus has been implicated in a number of such infections. The fungi *Fusarium solani* (Mart.) Sacc., *F. moniliforme* J. Shield, and *F. oxysporum* Schltdl have also been isolated from kuruma prawn with black gill disease in Japan. Among these, *F. solani* appears to be an important pathogen (Fig. 10.3c). The pathogenicity of anamorphic fungi *Plectosporium oratosquillae* and *Acremonium* sp., isolated from mantis shrimp in kuruma

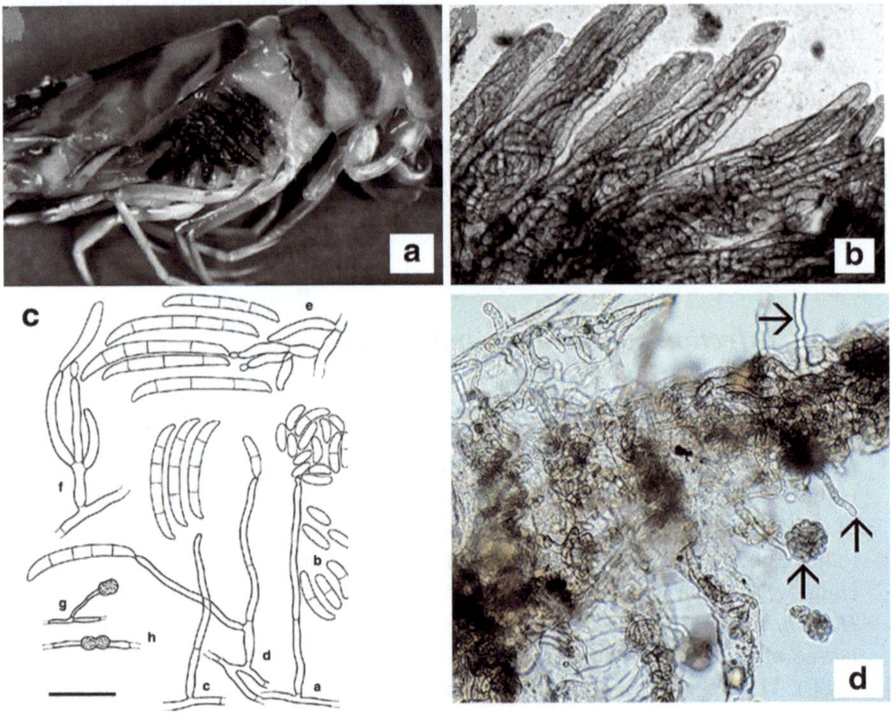


**Fig. 10.3** Fungal parasites of marine crustaceans. (**a**, **b**) The black gill disease of the kuruma prawn, *Penaeus japonicus*. (**a**) The typically black gills. (**b**) Hyphae of the parasite *Haliphthoros milfordensis*. (**c**) Conidia of *Fusarium solani*, causal agent of black lesions in *Penaeus japonicus*. (**a–c**: Source: Hatai, K. 2012. Diseases of Fish and Shellfish Caused by Marine Fungi. In: Biology of Marine Fungi (Ed.: C. Raghukumar, Springer). (**d**) Mycelium of *Lagenidium callinectes* pervading larvae of *Penaeus setiferus*. Zoosporangia with zoospores have emerged (*arrows*) (Courtesy: Dr. D.H. Lightner, Aquaculture Pathology Laboratory, University of Arizona)

prawn *Penaeus japonicus*, has been established by intramuscular injection of conidial suspensions. Cumulative mortality in kuruma prawn reached 100% when injected with a conidial suspension of *Acremonium* sp. The prawns showed typical black gills, and the clinical sign was similar to that of prawn naturally infected with fungus. The prawn is an important cultured crustacean in Japan and lives in the same environmental conditions.

- **The mantis shrimp, *Oratosquilla oratoria***, is an economically important and delicious culinary crustacean species. The famous Japanese "sushi" is made from the meat of mantis shrimp. This shrimp is a dominant benthic species in coastal Japan. Gill lesions from the wild mantis shrimp from Tokyo, Japan, harbored the oomycete *Halioticida noduliformans*. The fungus grows in an aerobic environment. In culture, it grows well at 15–25 °C, with optimal temperature of 20 °C.

  **Two anamorphic fungi, *Plectosporium oratosquillae* and *Acremonium* sp., cause mortality of the mantis shrimp** by infecting the gills which turn brown due to discoloration. Animals injected with conidia of the two species result in brown spots in the gill filaments similar to the clinical sign of mantis shrimp naturally infected with the fungi and subsequently in death. Hyphae and conidia are found in the gill filaments and heart, and the hyphae become encapsulated by hemocytes in the gill filaments and the base of gills. The result confirmed that these two anamorphic fungi were pathogenic to mantis shrimp. The antifungal agent voriconazole is an efficient antifungal agent against *Acremonium* sp. (Duc et al. 2010).
- **Adult northern shrimp, *Pandalus borealis*, cultured at the Japan Seafarming Association (JASFA) was infected by *Salilagenidium myophilum*** (Hatai & Lawhav.) M.W. Dick (*Pythium myophilum, Lagenidium myophilum*) which grew in the abdominal muscles and swimmeret. Lesions became filled with hyphae. The fungus could grow at temperatures of 5–25 °C. It also infected larvae of the coonstripe shrimps, *Pandalus hypsinotus*. The disease caused 100% mortality at Hokkaido in Japan (Nakamura et al. 1994).
- **The cultivated white shrimp *Penaeus setiferus* is often infected by *Lagenidium callinectes* and *Haliphthoros milfordensis*.**
- A species of the yeast ***Kluyveromyces*** was isolated from the heart tissue of subadult penaeid shrimp ***Penaeus chinensis*** during tissue culture (Tong and Miao 1999).
- The yeast ***Metschnikowia bicuspidata*** (Metschn.) T. Kamienski, a pathogenic yeast of aquatic invertebrates, infected aquaculture-reared, disease-free ***Artemia*** (Moore and Strom 2003).

### *10.2.3 Infection of Crabs and Lobsters*

- **The gazami crab, Japanese blue crab** or horse crab, *Portunus tuberculatus*, is the most fished pelagic crab species in the world. It is being cultivated in China. An explosive epidemic disease, now called milky or "emulsification" disease, has appeared in cultured *Portunus trituberculatus* since 2001 in Zhoushan, Zhejiang Province, China, leading to high mortality of this crab and great economic loss in this area. Infected crabs are emaciated and a milky liquid flows out from cut legs. The causal organism is the yeast *Metschnikowia bicuspidata.* The yeast can produce similar symptoms in the muscle, heart, and hepatopancreas when inoculated artificially. The use of "killer yeasts" to control the disease has been suggested. "Killer yeasts" can be applied to control growth of pathogenic yeasts in humans, animals, and plants. Such yeasts produce low molecular mass proteins or glycolipid toxins that kill the target pathogen. A strain of *Pichia anomala* (E.C. Hansen) Kurtzman was subsequently identified for the purpose (Wang et al. 2007b). Gills of this swimming crab was seen to be infected also by *Atkinsiella dubia* (D. Atkins) Vishniac in Japan. Heavy mortalities reaching 100% among gills of swimming crab have been observed. Infected zoeal larvae were filled with numerous aseptate hyphae. Zoea of this crab is also infected by *Haliphthora milfordensis.*

- Another pelagic crab, the blue **swimmer crab or flower crab, *Portunus pelagicus***, is vulnerable to a fungal disease. Eggs and zoeae of these crabs may be infected by ***Lagenidium thermophilum and Halocrusticida okinawaensis*** (K. Nakam. & Hatai) K. Nakam. & Hatai.
- **The mud crab or the mangrove crab, *Scylla serrata***, is a dominant crab of Indo-Pacific mangroves and is now widely cultivated. **Several fungi cause diseases in this crab.** In 1997, fungal diseases occurred in the eggs and zoeae of the mangrove crab, *S. serrata,* at the Gondol Research Station for Coastal Fisheries, Bali, Indonesia. The mortality rate reached almost 100% in the larvae. The infected larvae were whitish in color and filled with numerous aseptate hyphae. *Lagenidium callinectes, L. scyllae* Bian et al., *Haliphthoros milfordensis, Halocrusticida baliensis* Hatai, Roza & T. Nakay sp. nov., and *H. (Atkinsiella) hamanaensis* Bian & Egusa have all been reported as parasites in eggs and larvae of this crab (Hatai et al. 2000; Hatai 2012). A new fungus named *Lagenidium thermophilum* was isolated from the diseased mangrove crabs in 1993 at Gondol Research Station for Coastal Fisheries, Bali, Indonesia. This thermotolerant fungus grows rapidly between 15 and 45 °C and has an optimum at 30–40 °C (Nakamura et al. 1995). *Haliphthoros philippinensis* Hatai, Bian, Batic. & Egusa and *H. milfordensis* are known to abort eggs of captive *Scylla serrata* before hatching in several hatchery runs at the Aquaculture Department of Southeast Asian Fisheries Development Center in lloilo, Philippines (Leaño 2002).
- A 100% mortality of larvae of the **Japanese mitten crab, *Eriocheir japonicus***, is known to have been caused by infections, the main parasite being ***Atkinsiella dubia***.
- Live eggs of the subtidal yellow **rock crab, *Cancer anthonyi,* become infected by the chytrid fungus *Rhizophydium littoreum*** Amon, which otherwise is generally known to be a saprobe on its dead eggs. About 14–52% infections have been recorded throughout the year at sites along the west coast of North America. The chytrid probably is a facultative parasite with a low degree of virulence.

- **Eggs of the blue crab, *Callinectes sapidus*, pea crab, *Pinnotheres pisum*, and the Dungeness crab, *Cancer magister*, are prone to attack by *Lagenidium callinectes*.** Up to 50% of the eggs in the former succumb to the infection. Eggs of pea crab *Pinnotheres pisum* and the angular crab *Gonoplax rhomboides* are also parasitized by *Atkinsiella dubia* (Atkins) Vishniac. This fungus as well as *Halocrusticida hamanaensis* infect crustacean eggs freshly detached from ovigerous females (Porter 1986). The embryos of the crabs *Dyspanopeus texana, Panopeus herbstii*, and *Pinnotheres ostreum* are also susceptible to infection by *Lagenidium callinectes* (Polglase et al. 1986; Shields and Overstreet 2007).
- **The Black Mat Syndrome in the Tanner Crab *Chionoicetes bairdi* is caused by the biotrophic, ascomycetous fungus *Trichomaris invadens*** Hibbets, Hughes et Sparks (Polglase et al. 1986). The carapace turns black because of the growth of the fungus and presence of ascocarps. The hyaline hyphae penetrate the exoskeleton and grow in the epidermis. In advanced cases, the hyphae invade the connective tissue of the animal, presumably preferring to grow along lines of low physical weakness. The symptoms of the disease indicate that this disease could severely affect the health of the crab. As is typical of biotrophic parasites, the fungus cannot be grown in the absence of the host and has not been cultured so far.
- **The American lobster, *Homarus americanus***, is cultured in California for food. The animals are **host to the pathogens *Lagenidium callinectes* and *Haliphthoros milfordensis***. The fungi penetrate and fill larvae with mycelia giving a white, opaque appearance. Appendages or body filled with white mycelia, vegetative fruiting structures are visible under dissecting microscope. The anamorphic ascomycete fungus *Fusarium* also causes a disease in American lobsters. The fungus is an opportunistic parasite and invades the animal through damaged or dead tissues.
- **Philozoma of the diseased spiny lobster, *Panulirus japonicus*, in Japan has been reported to harbor the parasite *Halocrusticida (Atkinsiella) panulirata*.** The spiny lobster *Palinurus elephas* is known to be infected by a fungus which extensively invades and destgroys the shell (see Kohlmeyer and Kohlmeyer 1979).
- **Hermit crabs have been reported to be infected by *Fusarium*** species.
- Exoskeleton of the crab *Carcinus maenas* is infected by *Periconia prolifica* that causes 'burnspots' and black nodules in the hepatopancreas (see Kohlmeyer and Kohlmeyer 1979).

### *10.2.4 Bivalves and Gastropods*

- The bivalve, the abalone, is a greatly valued traditional food. **Mariculture of the northern abalone, *Haliotis kamtschatkana*, and the red abalone, *H. rufescens*, has been attempted** in view of their declining natural populations. This has been beset with serious problems, including a fungal disease.

**One of these is a disease caused by the aplanochytrid *Aplanochytrium (Labyrinthuloides) haliotidis*** (S.M. Bower) C.A. *Leander* & D. Porter (Fig. 10.4). The high mortality of up to 100% of juvenile abalones caused by this disease in 1980s is considered to be one of the reasons why the mariculture facility on Vancouver Island, British Columbia, in the West coast of Canada was closed down (Bower 1987, 2000; Bower and Meyer 2005).

The fungal infection causes destruction of head and foot tissues of the juvenile abalones which are less than 4 mm in shell length and younger than

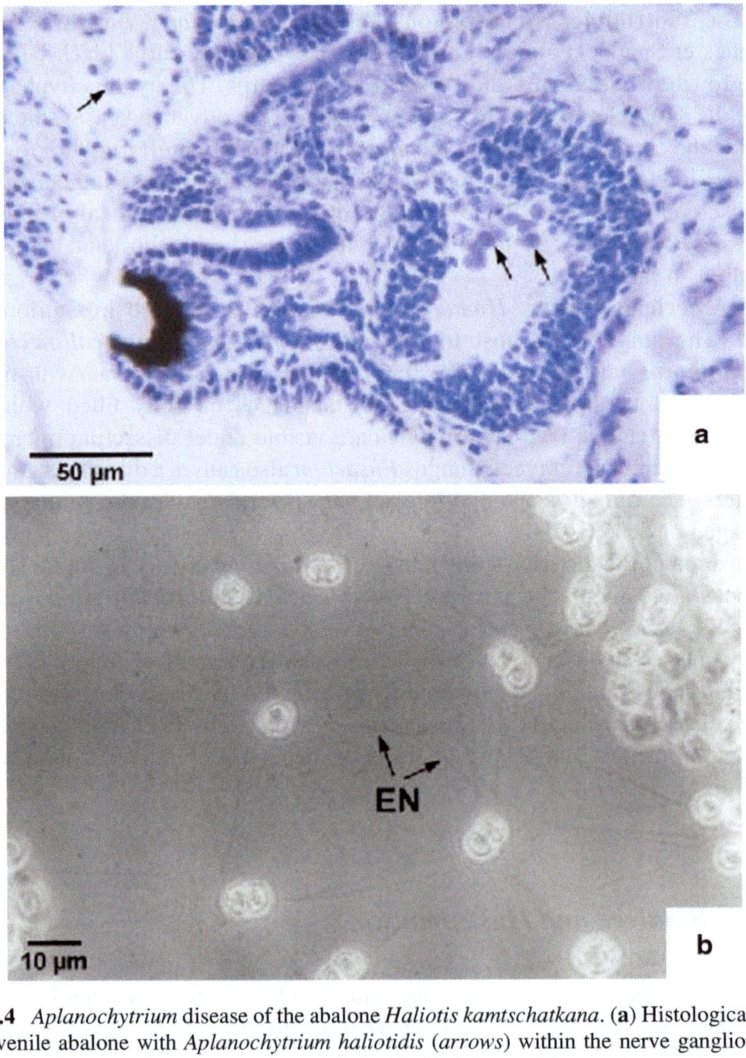


**Fig. 10.4** *Aplanochytrium* disease of the abalone *Haliotis kamtschatkana*. (**a**) Histological section of a juvenile abalone with *Aplanochytrium haliotidis* (*arrows*) within the nerve ganglion and in surrounding tissues. (**b**) Cells of *Aplanochytrium haliotidis* in culture showing ectoplasmic net elements (EN, *arrows*) (Courtesy: Dr. Susan M. Bower)

6 month. Over 90% of the 100,000 small abalones in the British Columbia raceway succumbed to infection within 2 weeks of detection in a raceway. Abalones greater than 5 mm in shell length were more resistant to infection and mortality compared to those which were about 25 mm in shell length. Cells of the aplanochytrid could be detected in internal tissues. The fungus might be capable of infecting other bivalves if it gains access to internal tissues. Thus, small Pacific oysters (*Crassostrea gigas*) with damaged shells became infected, although those with intact ones were not affected. The scallop (*Patinopecten yessoensis)* was also resistant to infection.

The aplanochytrid grows vegetatively by binary divisions of cells. Mature cells are released into the water upon death. They produce biflagellate zoospores which may again infect live abalone. *Labyrinthuloides haliotidis* grew well on a wide variety of nutrient media and could also grow on pine pollen, which is used as a standard bait to culture labyrinthulomycetes.

Various control measures were considered. Exposure to chlorine, the antifungal compound cyclohexamide, and ozone treatment of incoming water have been recommended. Many of these are not practical for various reasons. A major precaution is to physically separate broodstock from the progeny, disinfecting fertilized eggs and rearing of the progeny in treated (disinfected) water until they reach a shell size of more than 5 mm. Quarantine practices are also recommended.

A few other species of abalone are also susceptible to fungal diseases. *Haliotis sieboldii* held in aquaria with circulating sea water adjusted to 15 °C by a cooling system in Japan has been infected by the oomycete *Haliphthoros milfordensis*. Tubercle-like swellings are produced on the mantle and melanized lesions on the peduncle. The fungus grew at a temperature range of 4.9–26.5 °C, with optimum of 11.9–24.2 °C. It grew best in shrimp extract medium at 25 °C.

*Haliotis midae* imported from the Republic of South Africa, *Haliotis rufescens* imported from the Republic of Chile and the United Mexican State, and *Haliotis sieboldii* collected in Japan were seen to have been infected by the oomycete *Halioticida noduliformans* Muraosa & Hatai. Here too white nodules were produced on the mantle.

- Many oysters such as *Ostrea edulis*, *Crassostrea gigas*, *Saccostrea cucullata*, *Crassostrea cucullata*, and *C. angulata* are prone to a wart disease caused by the mycetaen fungus *Ostracoblabe implexa* (Golubic et al. 2005; Raghukumar and Lande 1988). The fungus pervades the shells of the oysters. It is characterized by spindle-shaped swellings along the hypha. The fungus is probably an opportunistic pathogen, which normally resides in the shells without causing a disease. However, in the case of a disease, the hyphae grow through the shell, eventually penetrating the inner surface. The disease is limited to the shell, first appearing as small round white spots, which are slightly raised and have a clear center. These spots coalesce. "Conchiolin warts" are formed on the inner surface of the shell and can cause severe thickening of the shell margin. If the area beneath the adductor muscle is infected, attachment is weakened and closing of the shell is compromised due to excessive shell production in that area. Distortion can render the oyster unmarketable. Proliferation of the fungus is restricted to waters where temperatures exceed 22 °C for more than 2 weeks. Hence, shallow beds are more severely affected than deep growing sites. Infections have been reported from Europe, India, and Canada (both Pacific and Atlantic coasts).

- **Hatchery-related mortalities of the oyster *Crassostrea virginica* by the oomycete *Sirolpidium zoophthorum* Vishniac have been reported** along eastern U.S. Larvae and post-metamorphic juveniles measuring up to 400 μm in diameter are affected. The fungus spreads throughout the soft tissues, resulting in their disintegration. The long discharge tubes of the zoosporangia protrude outside the shell and release motile biflagellate zoospores. The disease is transmitted through zoospores. Over 90% of larvae can be killed within 2 days. Zoospores of *S. zoophthorum* can germinate and grow on nutrient agar in the absence of bivalve larvae. The disease is believed to be because of poor husbandry. The disease also occurs in natural populations of the oyster.
- **Eggs of the oyster drill, *Urosalpinx cinerea*, have been reported to become infected with *Haliphthoros milfordensis*.**
- **The edible "hard clam" or quahog, *Mercenaria mercenaria,* caused by a thraustochytrid, now termed the Quahog parasite or QPX (Quahog Parasite Unknown),** is widespread in North and Central America. A serious disease of the clam emerged in Provincetown, Massachusetts, USA, in 1993. About 90% of the clams were affected, threatening to shut down the clam industry (Powell 2005). Hatchery-reared and commercially harvested clams throughout the northeastern coast of North America have been affected. Infected clams are weak. Their shells become partially opened, allowing the lodging of sand inside. Tan nodules develop within the clams. Such clams are more likely to be consumed by predators, leading to mortalities. As the siphon sucks in seawater, particles that are not taken into the gut are discarded. These "pseudofaces" may harbor QPX, initiating the infection at the clam's siphon. The QPX thraustochytrid is believed to be a saprobe in the surrounding water and may be an opportunistic pathogen. In culture, the thraustochytrid grows best between 20 °C and 23 °C. Production of abundant mucus by the parasite is believed to promote inflammatory response of the clam and also protect the thraustochytrid from phagocytic hemocytes and humoral antimicrobial agents. Mortalities peak in summer and early fall (Garcia-Vedrenne et al. 2013).
- The fungi ***Curvularia* sp., *Exserohilum rostratum*** (Drechsler) K.J. Leonard & Suggs, and an unidentified species were regularly isolated from fungal lesions in juveniles of the **boring clam *Tridacna crocea***, which is cultured in Australia (Norton et al. 1994).
- Natural populations of the **nudibranch *Tritonia diomedea*** from the North American Atlantic coast, as well as laboratory, often contract a disease called the yellow spot disease or ringworm. Symptoms consist of large yellow spots in the subepidermal tissues, up to 1.5 mm in diameter. This may lead to erosion of surface layers. Cells of a **thraustochytrid** are found in the amoebocytes of the nudibranch and is probably the cause of the disease (McLean and Porter 1987).

### 10.2.5 Cultured Fish

**A number of commercially cultured fish are infected by opportunistic parasites belonging to terrestrial species of fungi** (Hatai 2012).

- **The devil stinger, *Inimicus japonicus* is parasitized by the anamorphic fungus *Ochroconis humicola*.** Symptoms of the disease include necrotic surface lesions that become sloughed off, leaving the trunk muscles exposed in the form of a crater. Hyphae of the fungus pervade the lesions. This fungus also infects the red sea bream, *Pagrus major*, and marbled rockfish, *Sebasticus marmoratus*. Body lesions are caused on the surface in both cases. Yet another anamorphic fungus *Fusarium oxysporum* also infects the red sea bream. Kidneys of the fish have been noticed to become swollen and discolored, although the disease did not cause external symptoms.
- **Juveniles of the cultured striped jack, *Pseudocaranx dentex*, at a fish farm in Japan also became infected by *Ochroconis humicola*** *during* April 2004. Moribund fish had a distended abdomen. About 25% of the fish died. Fungal hyphae were found in the musculature, spleen, and kidney. Resistant fish showed inflammatory reaction involving mycotic granulomas and granulation tissues. The isolate grew at 10–30 °C, but not at 35 °C. The isolate could grow up to 9% NaCl indicating that *O. humicola* could grow in an environment with a wide range of salinity. *Exophiala* sp. is another parasite of the striped jack.
- **In Japan, infection by *Exophiala xenobiotica* occurred in cultured striped jack, *Pseudocaranx dentex***, in 2005. One hundred out of 35,000 fish died per day and mortalities continued for 1 month. Diseased fish showed swelling of the abdomen and kidney distension. Fungal hyphae and conidia were found in gill, heart, and kidney.
- **Caged rainbow trout *Salmo gairdneri* becomes infected by a thraustochytrid.**

### 10.2.6 Other Animals

- Cephalopods are often parasitized by thraustochytrids. The thraustochytrid *Ulkenia amoeboidea* (Bahnweg and Sparrow) Gaertner as well as a labyrinthulid are associated with a fatal ulcerative dermal necrosis in the lesser octopus, *Eledone cirrhosa* (Polglase et al. 1986). Amoebocyte in the octopus only partially encapsulates the pathogens, resulting in a more serious disease compared to that in the nudibranch *Tritonia diomedea* (see above), where the encapsulation appears to be complete. Thraustochytrids are also found in gill lesions of the squid *Illex Illecebrosus*. **Many fungi have caused epizootics in mariculture, resulting in high mortalities and economic losses. Several crustaceans, such as shrimps, crabs, lobsters, as well as bivalves and fish have been affected seriously. Most of these diseases are caused by**

**oomycetan fungi and facultative marine fungi. A number of trials have been carried out on antifungal substances in aquaculture.** Malachite green has been tried in shrimp culture to reduce zoospore motility and infection. Trifluralin and captan reduced mortality rates when exposed for 96 h and caused minimal larval mortalities. Furanace was effective against *H. milfordensis* in shrimp aquaculture.

- **The littoral marine nematode *Rhabditis marina* Bastian becomes infected by three species of oomycetes**: *Gonimochaete latitubus, Haptoglossa heterospora*, and *Myzocytiopsis vermicola. Haptoglossa* is an obligate parasite of rhabditid nematodes (Marano et al. 2012). *Myzocytiopsis vermicola* is parasitic in adults and larvae, but does not infect eggs of the nematode.
- **The rotifer, *Brachionus plicatilis,*** bred in a concrete tank as food supply for seed production of crustaceans and fishes has been reported to be infected by ***Halocrusticida (Atkinsiella) parasitica.***
- The whole body of the **amphipod *Podocerus brasiliensis*** was found to be invaded by ***Rhodotorula minuta***. Seki and Fulton showed that the tissues of living **marine copepods (*Calanus plumchrus*) were attacked by *Metschnikowia*** sp. Fize et al. (1970) reported a *Metschnikowia* sp. parasitizing living copepods (*Eurytemora velox*) in southern France (Kutty and Philip 2008).
- Many captive marine mammals develop a variety of mycotic infections (see Chap. 11).

Fungal pathogens of animals are likely to continue as a major problem, because many of them are difficult to eradicate and cause death of their hosts (Hyde et al. 1998).

**However, no effective strategies are available even now to counter fungal diseases in aquaculture practices.**

## 10.3 Saprobic Fungi in Marine Animals

Soft tissues of dead animals are decomposed much more rapidly compared to plant material, since they are relatively less recalcitrant to decomposition. Under these circumstances, bacteria have a greater advantage compared to filamentous, mycetaen fungi, which require time to colonize and establish themselves in decomposing tissues. Hence, the role of such fungi in decomposition of soft animal tissues may be negligible compared to bacteria. However, the role of straminipilan fungi in such decomposing tissues is not known.

Compared to soft tissues, parts of animal tissues that are less labile to degradation may be vulnerable to fungal degradation. Some of these are the calcareous shells and chitinous exoskeleton of invertebrates and bones of vertebrates. Fungi play a role in decomposition of all these three animal remains. Calcium carbonate shells are secreted by molluscs (gastropods and bivalves). The calcium carbonate is embedded in a matrix of the horny protein conchyolin made up of quinone-tanned proteins. Chitinous exoskeletons are secreted by crustaceans. Chitin is a polymer of N-acetylglucosamine. Bones of marine vertebrates, including fish and mammals, are a matrix made up of calcium phosphate and collagen.

**Calcareous shells secreted by marine animals such as molluscs, barnacles, and corals become colonized by endolithic fungi upon the death of the animals** (Fig. 10.5; Kohlmeyer and Kohlmeyer 1979; Golubic et al. 2005; Raghukumar 2008). Endolithic organisms in calcareous shells comprise cyanobacteria, green algae, red algae, and fungi. Many such shells are often cast on intertidal beaches where they can be collected and studied (Fig. 10.5a). Endolithic fungi were discovered more than 130 years ago by the French phycologists Edouard Bornet and Charles Flahault, who described the fungus *Ostracoblabe implexa* in oyster shells (Sect. 10.2.4). Willy Hőhnk in Germany reported the presence of endolithic fungi from shells in the 1930s. Filaments of endoliths can be directly detected under a stereoscopic microscope.

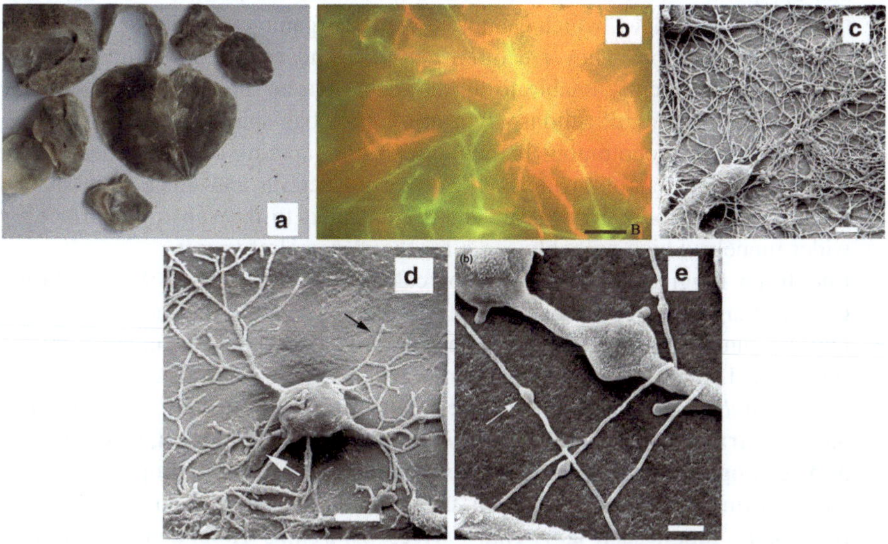


**Fig. 10.5** Endolithic fungi in calcareous shells of molluscs. (**a**) Shells of the windowpane oyster *Placuna placenta* and others darkened possibly of colonization by endolithic fungi and algae (S. Raghukumar). (**b**) Epifluorescence microscopy of autofluorescing cyanobactria (*red*) and Calcofluor stained *green* fungal mycelia in windowpane oyster shells (Source: C. Raghukumar 2008. Fungal Diversity 31:19–35. With permission of Dr. Kevin Hyde.) (**c–e**) Resin casts of calcareous animal shells. (**c**, **d**) Dense networks of tapering fungal borings in a bivalve shell fragment from shallow coastal waters. Bar represents 10 μm. (**e**) The fungus *Ostracoblabe implexa* in calcareous shells spread next to a much larger tunnel of the endolithic *green* alga *Phaeophila dendroides*. (**c–e**) (Reprinted from Trends in Microbiology 13:229–235, Golubic, S. et al. Dense networks of tapering fungal borings permeate a bivalve shell fragment in shallow coastal waters. 2005. With permission from Elsevier)

- **Filamentous fungi and algae pervade and grow within calcareous animal shells.** It is often difficult to distinguish between filaments of green algae, cyanobacteria, and fungi. Chelation and removal of the calcium carbonate shells using EDTA is a way of detecting fungal hyphae. Fungi and algae can be simultaneously distinguished by epifluorescence microscopy. Algae can be distinguished by red autofluorescence and the fungi by a blue fluorescence following staining with the optical brightener Calcofluor (Raghukumar 2008; Fig. 10.5b).
- **The resin-cast method** provides a clear picture of the branching pattern of endolithic organisms in shells (Fig. 10.5c, d). The method yields **three-dimensional plastic casts of the endoliths or their empty borings that were previously within the shell. SEM observations of the surface features of these resin casts reveal details not visible with conventional light microscopy** (Golubic et al. 1975).
- Photosynthetic algae and fungi display a difference in their colonization of shells (Golubic et al. 2005). Fungi prefer to grow within organic lamellae of the shells and apparently make use of the organic matrix for the nutrition. On the other hand, endolithic algae, which cannot utilize the organic material, seem to prefer the mineral portion. They are often confined and crowded within particular crystallites, apparently unable to digest and penetrate the surrounding organic lamellae. Fungal ramifications often arise from bag-shaped swelling and frequently display dichotomous branching and fine tapering of the hyphae. These features are not common in endolithic cyanobacteria and algae. Bag-shaped swellings of endolithic fungi are often connected to the substrate surface by wider tunnels, which probably serve for the dispersal of spores.
- Endolithic fungi cause bioerosion of calcareous shells (Sect. 9.1.2). During growth, fungi produce tunnels of uniform diameter within the shells.
- Diverse fungi, including obligately marine mycetaen fungi, facultative marine fungi, and straminipilan fungi colonize calcareous shells. The fungus *Ostracoblabe implexa* is an example (see Sect. 10.2.4; Fig. 10.5e). **The obligately marine ascomycete, *Pharcidia balani* (Winter) Bausch, is a common and cosmopolitan endolith in various barnacles, snails, and limpets.** Shells of wood boring shipworms found in the borer tunnels are often colonized by lignicolous fungi, such as *Arenariomyces trifurcatus* Höhnk, *Arenariomyces triseptalas* Kohlm., *Corollospora maritima* and *Corollospora pulchella* Kohlm., *Antennospora quadricornuta*, *A. salina*, *Hydea pygmaea*, and *Humicola alopallonella* Meyers & R.T. Moore. These fungi reside in the wood, from where they invade the calcareous shells (Kohlmeyer and Kohlmeyer 1979). Shells of barnacles and bivalves that are brought to the laboratory and incubated for several months may yield ascocarps and basidiocarps of ascomycetes and basidiomycetes, respectively, developing on them (Ananda et al. 1998). These presumably arise from the endolithic hyphae of the fungi that inhabit the shells. Most of these ascomycetes belong to the so-called "arenicolous fungi" such as *Corollospora maritima* that produce fruiting bodies adhering to sand grains,

while drawing nutrition from another source. The most common ones found along the west coast of India were the ascomycetes *Corollospora maritima* Werdermann, *C. angusta* Nakagiri el Tokura, *C. cinnamomea* Koch, and *Arenariomyces parvulus* Koch. Shells of balanids and the gastropod *Turritella* had the highest density of ascocarps, while cuttlebone samples showed the highest diversity.

- **A thraustochytrid with the characteristics of the genus *Schizochytrium* was found in carbonate shell fragments** that were baited in coastal US waters (Porter and Lingle 1992). Thraustochytrids were particularly abundant in fragments of mussel shells, which have more organic matrix than clam or oyster shells. The thraustochytrid has an unusual morphology of elongated, tapered, and sometimes branched thalli that were divided into many vegetative cells. Observations suggested that the thraustochytrid was responsible for the formation of tunnels and cavities, thus causing bioerosion.

The exoskeleton of tunicates is made of tunicin, which is an animal cellulose. **The ascomycete *Antennospora quadricornuta,* isolated from tunicates has been shown to grow well on pure tunicin and degrade it** (Kohlmeyer and Kohlmeyer 1979).

Chitinous exoskeleton of dead crabs harbor thraustochytrids (Bongiorni et al. 2005b). Two thraustochytrids isolated from this substrate produced a wide variety of enzymes, including esterase, esterase lipase, lipase, leucine, cystine and valine arylamidase, acid and alkaline phosphatase, naphthol-AS-BI-phosphohydrolase, β-galactosidase and β-glucosidase, N-acetyl-β-glucosaminidase, α-mannosidase, and L-aminopeptidase. Interestingly, these enzymes were produced as ectoenzymes that were adsorbed to the cell surface, rather than exoenzymes that were secreted into the environment. The presence of chitinase (N-acetyl-β-glucosaminidase) indicate that thraustochytrids may play an important role in the process of degradation of this biopolymer present in the carapace of crustaceans. Crab exoskeleton collected from intertidal beaches of the west coast of India did not initially reveal any fungi (Ananda et al. 1998). However, after prolonged moisture chamber incubation for several months, a high density of ascocarps of two obligate marine fungi, *Corollospora angusta* Nakagiri & Tokura and *C. maritima*, were noticed. This indicates that fungal mycelia may be present in chitinous shells of dead animals. These grow subsequently and reproduce on these shells.

- A number of fungi are found as symbionts and saprotrophs in marine animals.
- Fungal infections are common in both wild and cultivated marine animals.
- Many fungi such as species belonging to thraustochytrids, yeasts, and Trichomycetes live as commensals in invertebrates.
- A number of marine-derived fungi have been detected in sponges.

(continued)

- Fungi belonging to the Oomycetes are probably the most important parasites of marine animals and often affect seed production of marine crustaceans such as shrimp and crabs.
- Fungal diseases affect eggs and larvae of a number of economically important crustaceans, such as the giant tiger prawn, the kuruma prawn ("black gill disease"), the cultivated white shrimp, the mantis shrimp, the gazami crab, Japanese blue crab or horse crab, the mangrove crab, the American lobster, and the Japanese mitten crab.
- Many other species of shrimps, prawns, crabs, and molluscs are also affected.
- The most important oomycete parasites of marine invertebrates are *Lagenidium callinectes and other Lagenidium species; Haliphthoros milfordensis;* as well as other species of the genus, *Salilagenidium myophilum, Sirolpidium zoophthorum, Atkinsiella dubia, and Halocrusticida* spp. Yeasts belonging to *Kluyveromyces, Metschnikowia bicuspidata*, and *Pichia anomala* have been reported as parasites. Many terrestrial species of fungi, such as species of *Fusarium*, are opportunistic pathogens and have been reported to be associated with shell disease of marine crustaceans and lobsters.
- Many molluscs are also affected by fungal parasites. Cultured abalone, *Haliotis rufescens*, is affected by the aplanochytrid *Aplanochytrium haliotidis*. The edible "hard clam" or quahog, *Mercenaria mercenaria*, is parasitized by the QPX thraustochytrid.
- A number of commercially cultured fish are infected by opportunistic parasites belonging to terrestrial species such as *Ochroconis humicola* and *Exophiala xenobiotica*
- Many other straminipilan parasites occur in invertebrates such as cephalopods and nematodes
- No effective strategies are available even now to counter fungal diseases in aquaculture practices.
- Saprobic fungi may be much more common in recalcitrant body parts such as exoskeletons of invertebrates, rather than on easily degradable soft tissues of dead animals.
- Calcareous shells secreted by marine animals such as molluscs, barnacles, and corals become colonized by endolithic fungi.
- The resin-cast method provides a clear picture of the branching pattern of endolithic organisms in shells. Endolithic fungi cause bioerosion of calcareous shells. Fungi produce tunnels of uniform diameter within shells.
- Diverse fungi, including obligately marine mycetaen fungi, facultative marine fungi, and straminipilan fungi, colonize calcareous shells.
- Only a few fungi are known in chitinous exoskeleton.

## Future Directions

1. Animal guts may harbor mutualistic fungi, particularly yeasts, Trichomycetes, and thraustochytrids. Their diversity and role are not known.
2. Most studies on fungal diseases are from aquaculture animals. The importance of fungal diseases in wild animals is not known.
3. Do fungi with an r-strategy, such as yeasts and thraustochytrids, have a role in decomposition of soft tissues of animals?
4. Molluscan shells are almost invariably colonized by fungi. What is their diversity, nutrient substrates, and role in recycling organic matter?

# Chapter 11
# The Pelagic Ecosystem

*In one drop of water are found all the secrets of the Ocean*
Kahlil Gibran

In terms of surface, oceans cover 70% of the earth. More importantly, the pelagic ecosystem comprises 80% of the biosphere on earth in terms of volume. Food web dynamics of the pelagic ecosystem includes a variety of organisms (Fig. 11.1). Their interactions vary temporally and spatially (Lalli and Parsons 1997; http://courses.washington.edu/ocean101/Lex/Lecture26.pdf).

**Life in the pelagic ecosystem is sustained almost entirely by primary production from single-celled phytoplankton** (Fig. 11.2). Primary production is restricted to the upper, well-lit euphotic zone that may extend up to 80 m in well-lit tropical waters. Primary productivity ranges from 130 g C $m^2$ $yr^{-1}$ in the oceans to 640 gC $m^2$ $yr^{-1}$ in coastal upwelling regions, where nutrients from deeper layers are brought up to the surface. Upwelling may be seasonal or perennial depending on the location. Total global primary production from all ocean areas is approximately $53 \times 10^9$ metric tons per year, most of which (nearly 89%) is from the pelagic.

**Large, fast-growing diatoms**, followed by dinoflagellates, are the most important of these in coastal waters and in nutrient-rich parts of the ocean (Fig. 11.1a). On the other hand, **picoplankton** such as *Synechocystis* and *Prochlorococcus* **constitute the major primary producers in nutrient-poor oceanic waters** (Fig. 11.1b).

**In nutrient-rich waters, the mesozooplankton, comprising copepods and other crustaceans, cnidarians, chordates, chaetognaths, and larvae of larger animals, are the most important grazers of primary consumers** (Fig. 11.1c). About 40% of the primary production is grazed in coastal waters. Fish that feed on mesozooplankton constitute the secondary consumers. These may in turn be eaten by larger, carnivorous fish, the tertiary consumers.

**Unicellular microbial protists such as flagellates, dinoflagellates, ciliates, acantharids, radiolarians, foraminiferans, and the microzooplankton are the primary consumers of picoplankton in oceanic waters.** About 56% of picophytoplankton is consumed by these "microzooplankton" (Fig. 11.1d, e).



S. Raghukumar, *Fungi in Coastal and Oceanic Marine Ecosystems*,
DOI 10.1007/978-3-319-54304-8_11

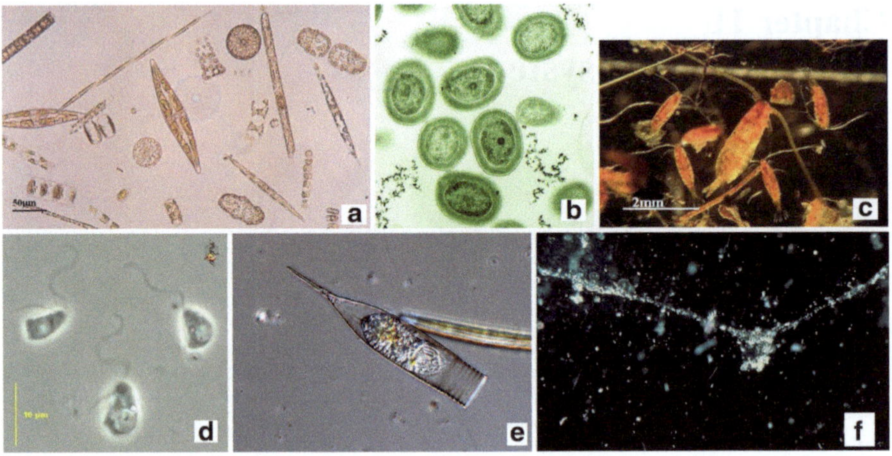


**Fig. 11.1** Important organisms of the pelagic ecosystem. (**a**) Diatoms, the major constituents of phytoplankton (Courtesy: Marine Scotland/Scottish Government). (**b**) *Prochlorococcus*, a major component of oceanic picoplankton (Courtesy: Dr. Sallie Chisholm, Massachusetts Institute of Technology). (**c**) Copepods that constitute mesozooplankton (Courtesy: Marine Scotland/Scottish Government). (**d**) *Cafeteria roenbergensis*, the most common flagellate in pelagic waters (Photomicrograph by David Patterson, Linda Amaral Zettler and Virginia Edgcomb; http://eol.org/pages/912371/overview; Courtesy Creative Commons). (**e**) A tintinnid ciliate that constitutes the microzooplankton (Source: Phyto'pedia—The Phytoplankton Encyclopaedia project of the University of British Columbia, UBC Department of Earth, Ocean and Atmospheric Sciences. With permission). (**f**) Marine aggregates or marine snow (Courtesy: Dr. Susumu Honjo, Woods Hole Oceanography Institute, USA)

**Non-living organic material, such as dead phytoplankton and their exudates, dead zooplankton, empty larvacean houses, pteropod feeding webs, and their fecal pellets, coalesce to form marine aggregates, or "marine snow"** (Fig. 11.1f).

The large size of the aggregates, ranging from half to several millimeters, and their high density cause them to sink beyond the euphotic zone and down into the deep water column. **The process by which organic matter produced in the euphotic zone is transported to deeper layers is called the biological pump.** Marine aggregates are ideal substrates for saprotrophic microbial growth and decomposition. A plume of organic matter trails behind a sinking aggregate as a result of diffusion of DOM from the aggregate into the surrounding waters resulting from microbial action (Alldredge and Youngbluth 1995; Simon et al. 1990; Smith et al. 1992; Kiørboe 2001).

**Sinking marine aggregates are sites of intense microbial activities** (Kiørboe 2001). Microorganisms colonizing these particles may solubilize and remineralize them to various extents. The DOM released in the form of a trail in this process in turn becomes sites of microbial growth. Thus, both particulate aggregates and DOM in pelagic waters support microbial growth. DOC concentrations may vary widely, supporting a range of copiotrophic to oligotrophic forms.

**Dissolved Organic Carbon (DOC) is constantly liberated into the pelagic owing to several processes.** Living phytoplankton exude copious amounts of

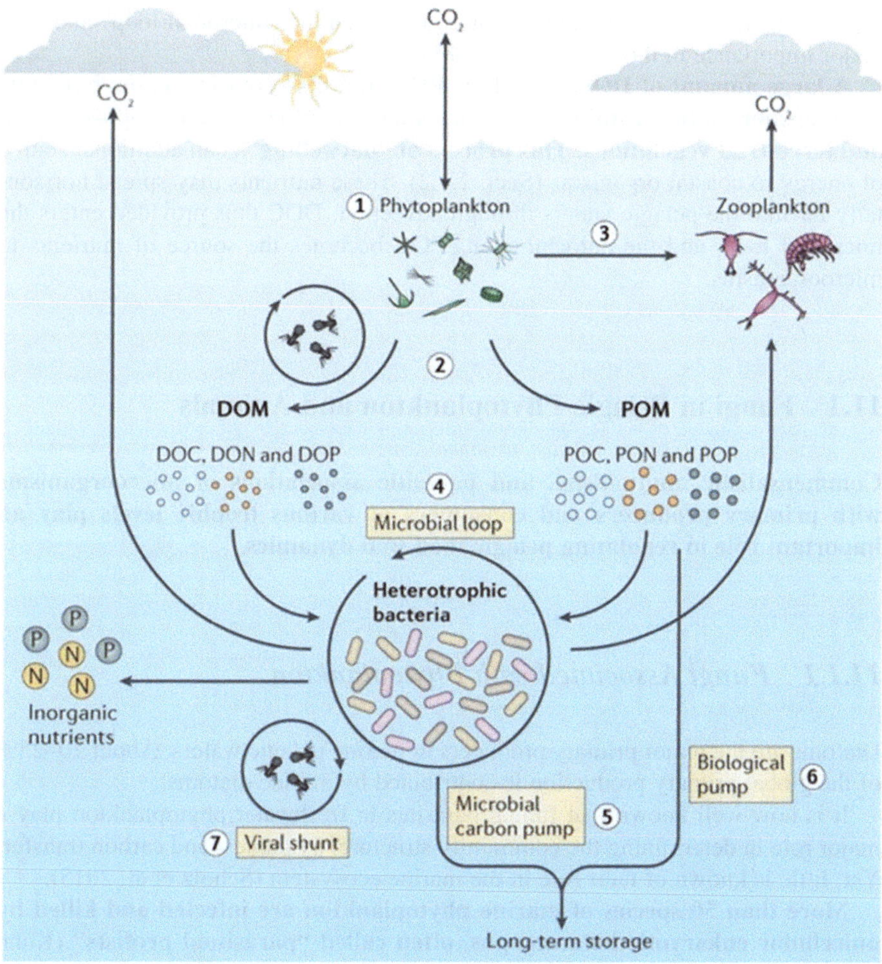


**Fig. 11.2** The marine carbon cycle (Reprinted by permission from Macmillan Publishers Ltd.: Nature Reviews Microbiology, 12, 686–698. Buchan, A. et al. Master recyclers: features and functions of bacteria associated with phytoplankton bloom, copyright 2014)

primary production into the environment as DOC. "Sloppy feeding" of the zooplankton on phytoplankton and viral lysis of phytoplankton are other sources of DOC. Microbial degradative action on dead particulate organic matter (POM), including marine aggregates, also releases substantial amounts of DOC. The DOC is recycled in the pelagic by free-living, planktonic microorganisms, particularly the bacteria. These utilize the DOC, resulting in buildup of their biomass. These are then fed upon by unicellular, microzooplankton such as ciliates and flagellates. The microzooplankton are then consumed by the larger, mesozooplankton. The process by which DOC is made available to larger organisms through microbial intervention is termed the "microbial loop" and is of major importance in the pelagic ecosystem.

**A large amount of DOC as well as POC and dissolved inorganic nutrients are transported to coastal pelagic waters from land run off, as well as coastal and terrestrial vegetations. This process of "outwelling"** is an additional source of energy to coastal organisms (Sect. 2.4.2). These nutrients may spread horizontally far into the pelagic waters through advection. DOC thus provided enters the microbial loop, and the autochthonous POC becomes the source of nutrients to microorganisms.

## 11.1 Fungi in Pelagic Phytoplankton and Animals

**Commensalistic, mutualistic, and parasitic associations of microorganisms with primary producers and consumers of various trophic levels play an important role in regulating pelagic food web dynamics.**

### *11.1.1 Fungi Associated with Phytoplankton*

Diatoms are the major primary producers in marine pelagic waters. About 20–25% of the global primary production is contributed by marine diatoms.

It is now well known that fungal parasites in freshwater phytoplankton play a major role in determining the community structure, dynamics, and carbon transfer. Yet, little is known of their role in the marine ecosystem (Scholz et al. 2015).

**More than 50 species of marine phytoplankton are infected and killed by unicellular eukaryotic heterotrophs, often called "parasitoid protists"** (Kühn et al. 2004; Scholz et al. 2015; http://sbroscoff.fr/Phyto/PICODIV/Meeting_Roscoff_2002/Stefanie_Parasitoid_Protists.pdf). These parasites include euglenozoa, dinoflagellates, cercomonads, and plasmodiophorids, as well as fungi belonging to **oomycetes (Straminipila) and chytrids (Mycetae).** Some of these parasitic outbreaks can decimate more than 90% of the diatom population, yet their role in pelagic ecosystem dynamics is poorly understood.

Fungi belonging to the Chytridiomycota, as well as several members of Straminipila, which produce zoospores that are typically adapted to aquatic conditions are often found to infect diatoms (Hanic et al. 2009; Wang and Johnson 2009; Gleason et al. 2012a; Table 11.1). Staining with calcofluor and direct examination of phytoplankton samples have revealed endobiontic fungi within (Li et al. 2010).

**Parasitic eukaryotes enter the diatom cells by making use of the gaps in their ornamented siliceous frustules of diatoms, called areolae, or through the**

**Table 11.1** List of fungal parasites in marine diatoms and dinoflagellates (Wang and Johnson 2009)

| Host Diatom | Stramenipilan parasite | Chytrid parasite |
|---|---|---|
| *Actinocyclus normanii* | *Ectrogella perforans* | |
| *Bacteriastrum hyalinum* | *Ectrogella* sp., | *Rhizophydium* sp. |
| *Bellerochea malleus* | *Ectrogella perforans,* | Unidentified |
| *Cerataulina pelagica* | *Ectrogella* sp., *Pirsonia eucampia, P. formosa, P. mucosa* | |
| *Chaetoceros danicus* | *Ectrogella perforans* | |
| *C. densus* | | *Rhizophydium* spp. |
| *Coscinodiscus centralis* | *Lagenidium* sp. | |
| *C. concinnus* | *Lagenisma coscinodisci, Pirsonia mucosa* | |
| *C. granii* | *Lagenisma coscinodisci, Pirsonia diadema* | |
| *C. wailesii* | *Lagenisma coscinodisci* | |
| *Chaetoceros* sp. | | *Olpidium rostriferum; O. phytophagum* |
| *Coscinodiscus* sp. | *Ectrogella perforans* | |
| *Cylindrotheca closterium* | | *Rhizophydium* sp. |
| *Detonula pumila* | *Ectrogella perforans; Pirsonia diadema* | |
| *Eucampia zodiacus* | *Ectrogella perforans, Pirsonia eucampia, P. Formosa, P. mucosa, P. strain 99–2* | |
| *Guinardia delicatula* | *Ectrogella perforans, Pirsonia formosa, P. guinardiae, P. mucosa, P. verrucosa, Pirsonia strain 98, 99–1, 99–2, 99–5* | |
| *G. flaccida* | *Pirsonia guinardiae, P. formosa, P. mucosa, P. strains 99–1, 99–2, 99–5* | |
| *G. striata* | *Pirsonia formosa* | |
| *Gymnodium* sp. | | *Verticillium sp.* |
| *Helicotheca tamesis* | | Unidentified |
| *Lauderia annulata* | *Ectrogella perforans, Pirsonia eucampiae* | |
| *Leptocylindrus danicus* | *Pirsonia eucampiae, Pirsonia formosa, P. mucosa, Pirsonia strains 99–1, 99–2* | |
| *Licmophora* spp. | *Ectrogella perforans* | |
| *Lithodesmium undulatum* | *Ectrogella perforans* | Unidentified |
| *Navicula* sp. | *Ectrogella or Olpidiopsis* | Unidentified |
| *Nitzschia* | | *Olpidium andreei* |
| *Odontella sinensis* | | Unidentified |
| *Palmeria hardmaniana* | *Lagenisma coscinodisci* | |

(continued)

**Table 11.1** (continued)

| Host Diatom | Stramenipilan parasite | Chytrid parasite |
|---|---|---|
| *Porosira glacialis* | *Ectrogella* sp. | |
| *Pseudonitzschia fraudulenta, P. pungens* | | *Olpidium marinum* |
| *Pseudonitzschia* sp. | | *Olpidium rostriferum; O. phytophagum; Rhizophydium* sp. |
| *Rhizosolenia imbricata* | *Pirsonia formosa, P. mucosa, Pirsonia strains 99–1, 99–2* | |
| *Rhizosolenia pungens* | *Pirsonia formosa* | |
| *R. setigera* | *Pirsonia formosa, P. lenis* | |
| *R. similoides* | *Pirsonia formosa, P. strains 99–2, 99-S* | |
| *Skeletonema costatum* | | Unidentified |
| *Skeletonema* sp. | *Schizochytrium* sp. | |
| *Schroederella schroederi* | *Ectrogella perforans* | |
| *Stephanopyxis turris* | *Ectrogella perforans, Pirsonia formosa* | |
| *Striatella unipunctata* | *Ectrogella perforans* | |
| *Synedra* sp. | *Ectrogella perforans* | |
| *Thalassionema nitzschioides* | *Ectrogella perforans, Pirsonia formosa* | |
| *Thalassiosira hendeyi* | *Pirsonia punctigerae* | |
| *T. minima* | *Ectrogella* sp. | |
| *T. cf. pacifica* | | Unidentified |
| *T. punctigera* | *Ectrogella perforans, Pirsonia punctigerae* | |
| *T. rotula* | *Ectrogella perforans, Pirsonia punctigerae, cf. Thraustochytrid* | Unidentified |
| *Triceratium alternans* | *Ectrogella* sp. | |

**gap between the girdle bands and overlapping valve.** Chemical cues may attract some parasites (Kagami et al. 2007). Species of the marine chytrids belonging to *Rhizophydium* are chemotactic towards amino acids and carbohydrates (Muehlstein et al. 1988).

- **The biotrophic, obligate parasite, the oomycete *Lagenisma coscinodisci* Drebes in the diatoms *Coscinodiscus* spp. and *Palmeria hardmania*, is one of the best studied diatom parasites** (Fig. 11.3a). The fungus was first reported in 1962 by Parsons at Hecate Strait, British Columbia, infecting the important marine, centric diatom *Coscinodiscus centralis*. T.W. Johnson described the

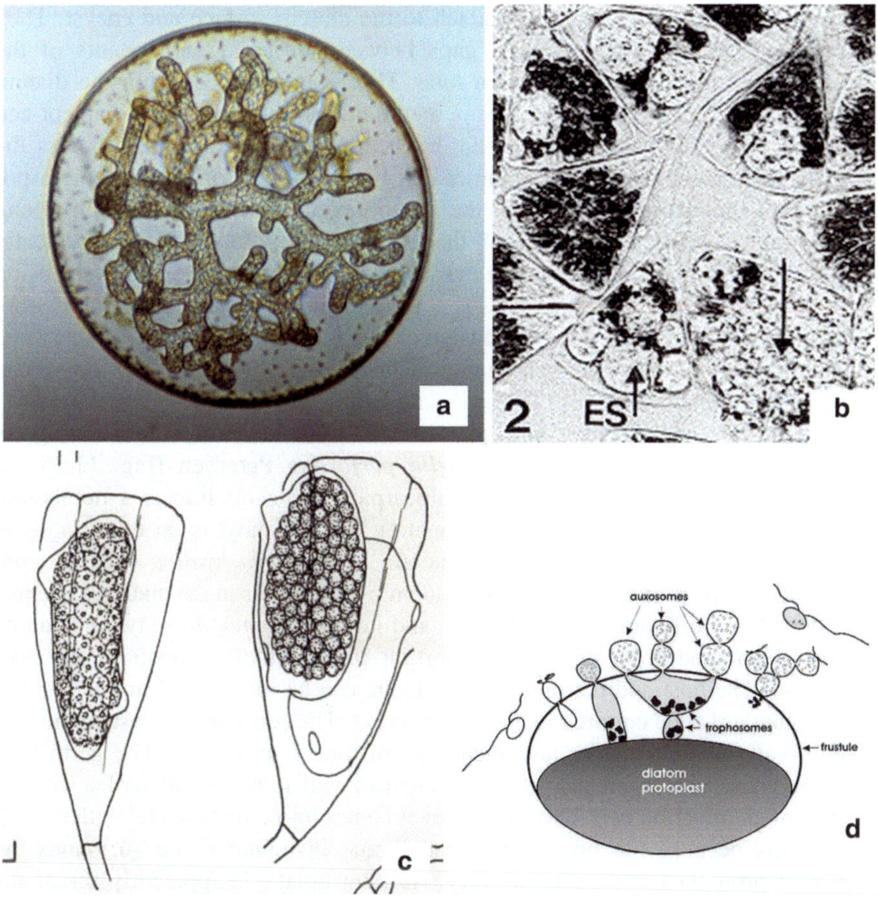


**Fig. 11.3** Some fungal parasites of marine diatoms. (**a**) The fungus *Lagenidium coscinodiscii* in *Coscinodiscus* (Courtesy: Dr. Stefanie Kuehn). (**b**, **c**) The fungus *Ectrogella* in *Licmophora*. (**b**) *Ectrogella perforans* in *Licmophora hyalina*. *Arrows* indicate spherical zoosporangia and zoosporangia with zoospores (Chandralata Raghukumar). (**c**) *Licmophora* cells with a fully formed zoosporangium (*left*) and a zoosporangium about to release zoospores (*right*) (Source: Johnson, T.W. 1966. Journal of Elisha Mitchell Science Society 82: 25–29. Courtesy: North Carolina University). (**d**) Schematic diagram of developmental stages of *Pirsonia* and *Pseudopirsonia*. Reprinted from Protist, Vol. 155, Stefanie Kühn et al., Phylogenetic Position of the Parasitoid Nanoflagellate *Pirsonia* inferred from Nuclear-Encoded Small Subunit Ribosomal DNA and a Description of *Pseudopirsonia* n. gen. and *Pseudopirsonia mucosa* (Drebes) comb. Nov., pp. 143–156., 2004, with permission from Elsevier)

fungus as infecting the same diatom in Puget Sound. Gerhard Drebes recorded the fungus again in 1966 from Helgoland waters in Germany and described it in detail in 1968 (Gotelli 1974). Gerhard Drebes and E. Schnepf from Helgoland, Germany, carried out a number of remarkable studies on ultrastructure and infection by this and other plankton parasites. Zoospores of this stramenipilan fungus *Lagenisma coscinodisci* attach to the diatom surface and encyst. They overcome the frustules through gaps between different components of the frustule by means of an infection tube. The fungus grows within the diatom cell in the space between the plasma membrane and the cell wall (Schnepf and Drebes 1977; Schnepf et al. 1978a, b). It draws its nutrition from across the plasmalemma of the host. The entire cell becomes transformed into a zoosporangium (holocarpic mode), and the zoospores are discharged through a long discharge tube emerging out of the dead diatom. Zoospores are typically straminipilan and biflagellate. Sexual reproduction takes place through fusion of swarmers. The zoospores pass two cyst stages before infecting the diatom again. Sexual reproduction is by fusion of two haploid zoospore cysts to form a diploid oospore following karyogamy (Schnepf et al. 1978a, b). The fungus can be cultured axenically.

- **A number of diatoms are parasitized by the oomycetes *Ectrogella licmophorae* Scherffel and *Ectrogella perforans* Petersen** (Fig. 11.3b, c). The unicellular, endobiontic, and holocarpic *Ectrogella* infects a number of diatoms, but the most detailed information that we have is on this fungus in the pennate, stalked, and attached diatom, *Licmophora hyalina* (Raghukumar 1980a, b). The disease may assume epidemic proportions in the diatom (Sparrow 1969). Natural populations of healthy and infected *Licmophora hyalina* can be easily obtained from surfaces supporting the epiphytic growth of the host diatom. The host and the biotrophic parasite can be maintained in dual culture in a standard algal culture medium, such as F/2. The zoospores encyst on the host cell wall. A branched or unbranched germ tube is produced. The germ tube forms an appressorium and penetrates the host wall at the site of areolae with the help of an infection peg. The fungus never comes into direct contact with the cell contents because the host plasma membrane invaginates and surrounds the fungal protoplast at the site of entry. Nutrient uptake is apparently across the host plasma membrane. Several membranes may be laid later. When the fungal thallus is mature, all the membranes except the fungal plasmalemma break down, and it secretes an amorphous wall around itself. Subsequently, the host organelles break down and the cell dies. Cytoplasmic contents of the holocarpic fungus divide into zoospores which are liberated.
- **One of the most interesting reports of fungal parasites in oceanic phytoplankton is that of chytrids that infect marine diatoms in the upwelling ecosystem of the Humboldt current system off central Chile** (Gutiérrez et al. 2016). The diatoms *Skeletonema*, *Thalassiosira*, and *Chaetoceros* have been found infected by chytrids. A large number of zoosporangia were found attached to *Thalassiosira* and *Skeletonema* during the active upwelling season of austral spring when these diatoms were abundant. The estimated biomass of attached chytrids ranged from 0.02 to 4.2 mg C $L^{-1}$ and that of their zoospores from 3.7 to 10.1 mg C $L^{-1}$. Fungal hyphae were also detected associated with the infections, suggesting the presence of saprophytic fungi along with affected diatoms.
- **Members of the genus *Pirsonia* infect many diatoms** (Fig. 11.3d). ***Pirsonia* is an interesting organism that blurs the borderline of the straminipilan fungi,**

**the Hyphochytriomycetes**. The genus was described by E. Schnepf et al. in 1990 for nanoflagellate parasitoids of diatoms in the North Sea. The genus comprises six species (Kühn et al. 2004; Schweikert 2015). *Pirsonia* has a phagotrophic mode of nutrition, whereby it ingests the cellular contents of the diatom. Although it differs from osmoheterotrophic fungi in this manner, the genus is worth a discussion because it is most closely related to the straminipilan fungi Hyphochytriomycetes (see Sect. 1.2.2) such as *Hyphochytrium catenoides* and *Rhizidiomyces apophysatus*. The biflagellate zoospore or flagellate attaches to the diatom frustule and squeezes a pseudopod through the frustule. The pseudopod then creates the trophosome, which performs phagocytosis and takes up parts of the diatom protoplast and digests it in a central vacuole. Nutrients are transported into the part of the cell remaining outside the frustule named the auxosome. The auxosome grows, divides, and produces numerous offspring as long as the nutrient supply continues. The first cells that detach are named flagellate mother cells. These give rise to motile flagellates after two divisions. Up to 60 flagellated cells may be generated from a single cell infection in less than a day depending on the species. Multiple infections can occur in the environment and are common under culture conditions, resulting in joint trophosomes of different individuals. Flagellates can infect new diatom cells immediately after detachment from the auxosome. Dormant stages have been reported for two *Pirsonia* species only, namely, *P. guinardiae* and *P. formosa*. These cysts may enable the flagellate to overcome times of absence of host diatoms in the environment.

- **Although a number of fungal parasites are known in marine diatoms, their role in regulating phytoplankton populations is not clear.** It has been reported that some parasitic fungi can infect up to 75% of the total natural marine phytoplankton populations (Tillmann et al. 1999). The fast-growing fungal parasites can consume the major part of the host phytoplankton within the water column, imposing a large impact on trophic web structure by directly contributing to phytoplankton lysis or death (Walsh 1983). The relative role of fungal parasites in controlling energy flow and food web dynamics compared to zooplankton remains to be studied (Tillmann et al. 1999). A conceptual model where fungi contribute to controlling the dynamics of phytoplankton populations, as well as the release of organic matter and the transfer of organic carbon through the pelagic trophic web in coastal upwelling ecosystems, has been proposed (Gutiérrez et al. 2016). It is believed that such infections can impact population dynamics of phytoplankton, and therefore evolution of spring blooms in the coastal upwelling ecosystem off central Chile. Parasitic chytrids may also cause substantial release of organic matter from primary producers into the waters.
- **Infection by planktonic fungal parasites can be important in regulating the dynamics of harmful algal blooms in coastal environments. The case of the diatom *Pseudonitzschia* species, some of which produce the toxin domoic acid, is a case in point.** Domoic acid is a neurotoxin, the causative agent of amnesic shellfish poisoning. The presence of *Pseudonitzschia* spp. cells,

parasitized possibly by an oomycete, has been reported in waters of Gulf of Mexico and Hood Canal in Washington State, Fanny Bay in Vancouver Island, British Columbia, Canada, and at the Brudenell River site in Prince Edward Island (PEI), Canada (Hanic et al. 2009). Hasle et al. (1996) reported an unexplained decrease in *P. multiseries* abundance, possibly by parasitic fungi in the Skagerrak between 1991 and 1993. Infection of *Pseudo-nitzschia* cells may be widespread. Two eukaryotic parasites were found infecting the bloom-forming marine diatom *Pseudonitzschia pungens* (Grunow ex Cleve) Hasle in Prince Edward Island, Canada. The most common was an oomycete; the other was a chytrid. Cell infection frequencies ranged from 0.6% to 15.9% during 1992–1995, at four sampling sites (Hanic et al. 2009). It is possible that these fungal infections might have caused the decline of the diatom bloom since 1987.

## *11.1.2 Fungi in Mesozooplankton, Fish, and Mammals*

**Fungi may occur as commensals, mutualists, or parasites of mesozooplankton, fish, and mammals, which constitute secondary, tertiary, and higher level producers of the pelagic ecosystem.**

**Only a few fungal parasites have been reported in marine animals.** However, this is more likely the result of inadequate studies, rather than a paucity of fungal diseases, considering that fungi are common parasites in many benthic animals, aquaculture animals, and fish. According to the famous marine microbiologist, Claude Ephraim ZoBell, "marine animals do have infectious disease, perhaps much more extensively than indicated by the fragmentary literature on the subject" (Seki and Fulton 1969). This may be particularly true of fungal parasites.

- **Yeasts, particularly belonging *Metschnikowia* species, are common as parasites of various animals in the marine pelagic.** *Metschnikowia krissii* has been reported as a pathogen of the copepod *Calanus plunchrus* in the Strait of Georgia, Canada, in water depths up to 200 m (Seki and Fulton 1969). Infected copepods were motile even when their tissues were infected. About 5% of the copepods from below 200 m were infected. Movement of the animals decreased with progression of the disease, followed by secondary infections by bacteria and protozoans. The yeasts were present on the copepod surface, in the gastrointestinal tract, or within the animal tissues. Asci of the yeast contained needle-shaped ascospores. Copepods were killed between 5 and 10 days of infection. Bottom samples revealed exoskeletons of the copepod, which presumably died of the disease. Pathogenicity by the species may be due to the production of extracellular enzymes that hydrolyze chondroitin sulfate and proteins. Another species of the yeast, *Metschnikowia kamienski*, is known to cause infections of the copepod *Eurytemora velox* off the French coast (Fize et al. 1970). *Metschnikowia bicuspidata* is a pathogen of marine invertebrates and fishes.

- **The straminipilan fungus *Leptolegnia baltica* has been reported to cause mass mortality of the copepod *Eurytemora hirundoides*** in the Gulf of Bothnia during the early 1950s by Willy Höhnk and Vallin (Gleason et al. 2012a).
- **A wide range of superficial and systemic mycoses have been reported from captive as well as stranded marine mammals in Central California coast**. These included California sea lions, southern sea otters, Pacific harbor seals, Dall's porpoise, and northern elephant seal, indicating that fungal diseases may be common in them (Huckabone et al. 2015). A number of animals revealed cytological, cultured, or histologically confirmed locally invasive or systemic mycoses. These included coccidioidomycosis, zygomycosis, cryptococcosis, and a systemic infection by an ascomycete *Scedosporium apiospermum*. *Coccidioides* infection was the most common. Even a pulmonary aspergillosis has been reported in pinnipeds and dolphins. Many yeasts such as *Candida* spp., *Torulopsis* spp., *Trichosporon* spp., and *Malassezia* spp. appear to be opportunistic pathogens, causing superficial dermatitis. For example, *Malassezia pachydermatis* has been reported to cause dermatitis in a captive South American Sea Lion (Nakagaki et al. 2000). The fact that many of these were found in stranded animals indicates that fungal infections in the wild are common in the pelagic ecosystem.

**Yeasts and labyrinthulomycetes live symbiotically in pelagic animals.**

- **Yeasts are commonly found on skin, gills, mouth, gut, and feces of fish** (Kutty and Philip 2008). The yeast *Metschnikowia zobellii occurs* in high numbers in guts of many fish. The yeast was found in guts of 8 of 11 fish from the coast of California. They were particularly abundant in the fishes *Atherinops affinis littoralis* and *Trachurus symmetricus*, where up to 5730 viable cells per g gut contents were estimated (Van Uden and Branco 1963). These numbers were much higher than those found for the same yeast in the surrounding seawater. It is possible that the fish gut is the niche for this yeast, rather than seawater. It is also possible that the high numbers in the guts came from the feed of the fish on which the yeast was abundant. *Debaryomyces hansenii* is another yeast that is very frequently found in fish. The yeasts *Leucosporidium antarcticum* and *Metschnikowia australis* have been isolated from guts of krill in the Southern Ocean (Donachie and Zdanowski 1998).
- ***Aplanochytrium kerguelense* is a dominant species in mesozooplankton at depths of surface to 1000 m in the equatorial Indian Ocean.** About 8% of mesozooplankton from this region yielded this straminipilan fungus in culture (Damare and Raghukumar 2010). In situ hybridization techniques based on internal transcribed spacer (ITS) sequences suggested that aplanochytrids might be present within the body of zooplankton belonging to chaetognaths. It is possible that aplanochytrids live symbiotically within zooplankton. This species is also common in zooplankton of coastal waters (Damare et al. 2013).

## 11.2 Fungi in the Water Column

The significant role played by bacteria in heterotrophic processes of the open ocean has come to light just in the last 50 years. Despite the equally important role that fungi play in the terrestrial ecosystem, we know little about the role they play in the pelagic ecosystem. However, efforts to understand this aspect have commenced of late.

The role of fungi in the pelagic can be understood only by considering their fundamental biological characteristics that circumscribe their ecological niches. **Three major groups of fungi are involved in heterotrophic activities in pelagic waters. These are (1) filamentous fungi, (2) unicellular yeasts, and (3) unicellular straminipilan fungi belonging to aplanochytrids and thraustochytrids.**

Fungi are larger in size than bacteria and have a lower affinity to dissolved organic carbon (Newell 1984, 1996a; Raghukumar 1990). Thus, they may not be as successful as bacteria when nutrient levels approach oligotrophic conditions. However, a number of picoplanktonic eukaryotes have been discovered of late and there may yet be many a surprise (Raghukumar 2007). **Filamentous fungi are ideally suited to grow on or within solid particulate matter such as marine aggregates,** but not as plankton, freely suspended in the water column. The term "mycoplankton" often used by many authors for filamentous fungi cultured from seawater, therefore, is a misnomer. **Yeasts are adapted to the aqueous environment.** They can grow on the surfaces of particulate organic matter or suspended in the water column where dissolved organic matter concentrations are high. **The straminipilan labyrinthulomycetes, particularly thraustochytrids and aplanochytrids, appear to be the most dominant group of fungi in the pelagic water column, occupying a wide variety of niches.** These are suited to penetrate particulate organic matter by virtue of their ectoplasmic nets, while the vegetative cells remain outside, or they may live endobiontically within. They may also live planktonically as with yeasts where nutrient levels are high.

**A consideration of the role of fungi in any single pelagic heterotrophic process has to take into account the combined role of filamentous mycetaen fungi, yeasts, as well as the straminipilan fungi, particularly the labyrinthulomycetes.** However, such holistic information is not yet available.

**Presence of fungi in the water column has been conclusively demonstrated by culturing, metagenomics, and direct detection.**

- Culture methods and metagenomics have frequently been used for fungi in the water column (Wang et al. 2012). Culture methods for mycetaen fungi yield colonies from non-active spores transported from terrestrial sources or active mycelia of such fungi and, therefore, has a serious limitation. On the other hand, culturing and metagenomics are reliable when used for obligate marine fungi, such as the straminipilan fungi, and demonstrate their autochthonous presence in the samples.
- Direct detection methods have conclusively demonstrated the presence of fungi in water samples. Detection of filamentous, mycetaen fungi is possible by

various methods (see Sect. 15.3). Detection of fungal mycelia using these techniques unequivocally shows active growth of mycetaen fungi. Thraustochytrids in the water column can be detected by the Acriflavine direct detection (AfDD) epifluorescence method (Raghukumar and Schaumann 1993). This method also includes aplanochytrids. Direct detection methods help in confirming the presence and in estimating biomass. They do not help in unravelling the diversity of fungi.

**Mycetaen fungi, comprising both mycelial fungi and yeasts, as well as straminipilan fungi, including thraustochytrids and aplanochytrids, appear to be active and abundant in the water column and display seasonal dynamics. Thraustochytrids are the most well-studied group of fungi in the coastal and oceanic pelagic.**

### *11.2.1 Diversity of Fungi in the Water Column*

**A number of thraustochytrids and aplanochytrids have been cultured from the pelagic in different geographical regions using pine pollen and brine shrimp baiting techniques** (Table 11.2). These are obligate marine organisms, and culture methods are adequate to demonstrate that they are active in the pelagic. Pioneering studies on this aspect were carried out by Alwin Gaertner in Germany during the 1960s. He cultured, isolated, and enumerated thraustochytrids from the Weser Estuary, Fladen Ground areas of the North Sea, North Sea waters of Shetland and Faeroer islands, the Barents Sea, and the Atlantic. These studies also revealed that a large diversity of thraustochytrids was present in coastal and oceanic waters (Raghukumar 2002).

- **Thraustochytrids occur in coastal and oceanic waters in tropical, temperate, and Antarctic regions.** Coastal waters of the North Sea, eastern Atlantic, Seto Inland Sea, and the Arabian Sea have been studied for thraustochytrids. Among the oceanic waters, the North Sea, Atlantic, the Arabian Sea, equatorial Indian Ocean, the Antarctic, and the North and South Pacific regions have been explored.

  Many species such as *Oblongichytrium multirudimentale*, *Thraustochytrium aureum*, and *Ulkenia visurgensis* are cosmopolitan in coastal waters. *Ulkenia amoeboidea* was originally described from Antarctic waters, but subsequently found in the North Sea.

  *Aurantiochytrium limacinum* and *A. mangrovei* appear to be restricted to tropics. *Thraustochytrium antarcticum, T. kerguelense*, and *T. rossii* are examples of cold water, Antarctic species.

  The aplanochytrid, *Aplanochytrium kerguelense* (called *Labyrinthuloides yorkensis* or *L. kerguelensis* in earlier literature), is a common inhabitant of oceanic waters. This species is the dominant stramenipile in the Arabian Sea and the equatorial Indian Ocean (Raghukumar 1985; Damare and Raghukumar

**Table 11.2** Thraustochytrids found in pelagic waters

| Species | Geographical Location |
|---|---|
| *Aplanochytrium kerguelense* Bahnweg et Sparrow | Antarctic Ocean. Equatorial Indian Ocean, Arabian Sea, Northern Pacific Gyre near Hawaii |
| *Aurantiochytrium limacinum* (D. Honda & Yokochi) R. Yokoyama et D. Honda | Seto Inland Sea, Japan |
| *Aurantiochytrium mangrovei* (Raghuk.) R. Yokoyama et D. Honda | Seto Inland Sea, Japan |
| *Botryochytrium radiatum* (A. Gaertn.) R. Yokoyama, B. Salleh et D. Honda | Helgoland waters (as *Ulkenia radiata*), Langkawi Is., Malaysia |
| *Oblongichytrium minutum* (A. Gaertn.)R. Yokoyama et D. Honda | North Sea (as *Schizochytrium minutum*) |
| *Oblongichytrium multirudimentale* (S. Goldst.) R. Yokoy. et D. Honda | North Sea; Ligurian Sea coast, NW Mediterranean, (as *Thraustochytrium multirudimentale*), Arabian Sea |
| *Oblongichytrium octosporum* (Raghuk.) R. Yokoy. et D. Honda | North Sea (as *Schizochytrium octosporum*); Ligurian Sea coast, NW Mediterranean |
| *Schizochytrium aggregatum* S. Goldst. et Belsky | Ligurian Sea coast, NW Mediterranean |
| *Schizochytrium minutum* Gaertn. | North Sea, Fladen Ground |
| *Thraustochytrium aggregatum* Ulken | North Atlantic |
| *Thraustochytrium antarcticum* Bahnweg et Sparrow | Antarctic Ocean |
| *Thraustochytrium aureum* Goldst. | North Sea, New Zealand waters |
| *Thraustochytrium benthicola* Raghuk. | North Sea, Fladen Ground |
| *Thraustochytrium kerguelense* Bahnweg et Sparrow | Antarctic Ocean |
| *Thraustochytrium kinnei* Gaertn. | New Zealand waters, North Atlantic, Seto inland sea, Japan |
| *Thraustochytrium motivum* | Ligurian Sea coast, NW Mediterranean |
| *Thraustochytrium pachydermum* Scholz | New Zealand waters |
| *Thraustochytrium roseum* | North Sea |
| *Thraustochytrium rossii* Bahnweg et Sparrow | Antarctic Ocean |
| *Thraustochytrium striatum* J.Schneider | North Sea |
| *Parietichytrium sarkarianum* (A. Gaertn.) R. Yokoyama, B. Salleh et D. Honda | Weser estuary (as *Ulkenia sarkariana*), Penang Is., Malaysia, Iriomote Is., Japan |
| *Sicyoidochytrium minutum* (Raghuk.) R. Yokoyama, B. Salleh et D. Honda | North Sea (as *Ulkenia minuta*), Langkawi Is., Malaysia |
| *Ulkenia amoeboidea* (Bahnweg and Sparrow) Gaertner | Antarctic Ocean, North Sea, Hiroshima (Japan) |
| *Ulkenia profunda* Gaertn. | North Atlantic; Seto Inland Sea, Japan |
| *Ulkenia visurgensis (Ulken)* Gaertn | North Sea |

2010). *Aplanochytrium kerguelense* seemed to be most prevalent also in Pacific waters off Hawaii (Li et al. 2013). A species phylogenetically closest to *Aplanochytrium (Labyrinthuloides) minutum* has been detected in surface seawater of Blanes Bay, Catalan coast, of the Northwest Mediterranean (Massana et al. 2004a, b).

- **In addition to culturing, metagenomic studies have also revealed the presence of labyrinthulomycetes in the pelagic.** A molecular survey based on SSU rRNA gene in the upwelling region of Arraial do Cabo, at the Cabo Frio, Rio de Janeiro state, Brazil, revealed that stramenipiles comprised about 5% of all eukaryotes in coastal, anthropogenically affected waters and 10–20% of oceanic waters. Of the stramenopiles, 5–10% of the diversity consisted of oomycetes. Labyrinthulomycetes constituted a high percentage of 5–40% of the diversity. Most of these were found in the oceanic, upwelling waters at a depth of about 50 m. Hyphochytriomycetes in these waters comprised about 10% (Cury et al. 2011).

**Yeasts are common in both coastal and oceanic waters** (Table 11.3; Gadanho et al. 2003; Kutty and Philip 2008; Fell 2012). Pioneering work on marine yeasts carried out by A.E. Kriss in Russia, J.W. Fell in the USA, and others led to numerous other studies on their diversity and distribution in coasts and oceans.

- **Marine yeasts belong to both Ascomycota and Basidiomycota**. Yeasts have been isolated in culture from a number of coastal waters, as well as from all major oceans, including tropical, temperate, and polar waters. Yeast diversity is

**Table 11.3** Examples of yeasts in pelagic water column

| Species | Geographic location |
|---|---|
| *Blastobotrys parvus* (Fell & Statzell) Kurtzman & Robnett | Southern Ocean |
| *Metschnikowia australis* (Fell & I.L. Hunter) Mend.-Hagler, Hagler, Phaff & Tredick | South Shetland Islands |
| *Candida atlantica* (Siepmann) S.A. Mey. & Simione | Atlantic Ocean |
| *Metschnikowia zobellii* (Uden & Cast.-Branco) van Uden | |
| *Sakaguchia dacryoidea* (Fell, I.L. Hunter & Tallman) Y. Yamada, K. Maeda & Mikata, *Pseudozyma aphidis* (Henninger & Windisch) Boekhout, *Rhodosporidium babjevae* Golubev, *R. diobovatum* S.Y. Newell & I.L. Hunte, *and Debaryomyces hansenii* (Zopf) Lodder & Kreger | Atlantic, South of Portugal; Portugal, above the Alvares Cabral trench head |
| *Candida marinus* (Uden & Zobell) Golubev | Pacific, Great Barrier Reef in Australia |
| *Metschnikowia bicuspidaa* (Metschn.) T. Kamienski | |
| *Metschnikowia krissii* (Uden & Cast.-Branco) Uden | |
| *Candida torresi*, *Kluyveromyces aestuarii* (Fell) Van der Walt | Torres Strait, Australia |
| *Leucosporidium scottii* Fell, Statzell, I.L. Hunter & Phaff, *and L. antarcticum* Fell, Statzell, I.L. Hunter & Phaff | Antarctic Ocean |

high in coastal waters and decreases in the oceans. Most marine yeasts among Ascomycota belong to terrestrial species such as *Candida, Cryptococcus, Debaryomyces, Pichia, Hansenula, Rhodotorula, Saccharomyces, Trichosporon*, and *Torulopsis*. Occurrence of many of these species may be the result of run-off from the surrounding terrestrial vegetation. Many yeasts, which are predominantly marine or even endemic to marine habitats, occur in the water column (Fell 2012; Table 11.3). **The most important genera of true marine yeasts belong to *Metchnikowia, Kluyveromyces, Rhodosporidium, Candida, Cryptococcus, Rhodotorula*, and *Torulopsis*.** The ascomycetous genus *Metschnikowia* has significant endemic marine representatives. Species of *Leucosporidium* and *Rhodosporidium* are the common basidiomycetous yeasts. Two members of the genus *Leucosporidium, L. scottii*, and *L. antarcticum* were repeatedly isolated in the Antarctic Ocean (Fell et al. 1969).

- **Many yeasts show distinct geographical distributions**. *Blastobotrys (Sympodiomyces) parvus* seems restricted to warm Antarctic waters in the vicinity of the polar front and northward into the Subantarctic and Antarctic Intermediate waters. *Candida natalensis* has been found in a narrow longitudinal zone in the Indo-Pacific from the polar front southward. *Candida norvegica* is restricted to a narrow geographical zone southward from the polar front and has been found in water masses with temperatures below 5 °C. The yeast *Leucosporidium antarcticum* is a psychrophilic species found in ice packs (Fell 2012). Four species of *Metschnikowia* species have been isolated from cold-water habitats in the Pacific and Southern Oceans. *Kluyveromyces aestuarii* appears to be an inhabitant of subtropical or tropical waters. A metagenomic study of fungi in the West Pacific Warm Pool revealed that most of the yeast *Rhodotorula mucilaginosa* was the most abundant among OTUs of the subphylum Pucciniomycotina, while those belonging to *Malassezia* were the most abundant among those of Ustilaginomycotina (Wang et al. 2014).

**Filamentous, mycetaen fungi are commonly detected in the pelagic by culture or metagenomic methods.** Roth et al. (1964) were the first to carry out detailed studies on fungi in seawater. Most recent studies on fungal diversity in the pelagic habitat have used metagenomic methods. **Metagenomic studies have revealed novel fungal lineages, in addition to terrestrial species.**

- Forty-two basidiomycetes and 4 ascomycetes were detected in coastal Hawaiian waters at a depth up to 100 m, using DGGE and SSU rRNA-based clone library analysis. Of the basidiomycetes, 27 were novel phylotypes, with less than 98% similarity with available 18S rDNA sequences (Gao et al. 2010). Fungal communities displayed distinct lateral and vertical variations in diversity and composition. Coastal, surface waters at 0–100 m depth had a greater diversity and species richness of fungi compared with the open ocean. Both diversity and species richness in coastal waters were further positively correlated to phytoplankton when vertical profiles of the pelagic waters were compared.

- A 6-year study of coastal waters from Plymouth, UK, using fungi-specific high-throughput sequencing and quantitative PCR analysis revealed a high diversity of Ascomycota, Basidiomycota, and Chytridiomycota, with several orders within these phyla (Taylor and Cunliffe 2016).
- Molecular survey based on SSU rRNA gene in the Brazilian coastal region of the Cabo Frio revealed that 15–35% of eukaryotic OTUs from coastal, anthropogenically affected waters belonged to fungi. A greater percentage of 20–50% of fungal OTUs were found in oceanic samples. Fungal diversity was greater at depths of 20–50 than at surface. Of the fungal OTUs, 80–90% could not be phylogenetically classified, while the rest belonged to Chytridiomycota, Ascomycota, and Basidiomycota (Cury et al. 2011).
- ITS (Internal transcribed region)-based studies of several locations at the West Pacific Warm Pool near Hawaii yielded over 400 fungal phylotypes, predominantly belonging to Ascomycota and Basidiomycota (Wang et al. 2014). Although many were marine-derived fungi belonging to terrestrial species, the majority of OTUs could not be attributed to any known fungi. Overall, the fungal community displayed the highest diversity near Hawaii coast but similar among open ocean stations along the transect.
- **It is possible that many fungi exist as unicellular picoeukaryotes** in the size range of less than 3 μm in size (Gleason et al. 2008). A high diversity of marine picoeukaryotes has been discovered in recent years. These are believed to play an important role in marine microbial dynamics. An analysis of 18S rDNA sequence-based metagenomic studies of picoeukaryote diversity in the euphotic, well-oxygenated surface layer of the ocean from Pacific, Atlantic, Indian, and Southern Oceans and the Mediterranean and North Seas revealed 16 fungal OTUs, representing 0.8% of total clones. A high degree of novelty was also revealed in the fungal clones (Massana et al. 2004b; Massana and Pedrós-Alió 2008).
- In general, **diversity of filamentous, mycetaen fungi in the pelagic, as detected by metagenomic as well as culture methods, constitutes less than 5% of the operational taxonomic units recovered from marine environments** (Richards et al. 2012). However, considering the vast diversity of organisms in the pelagic, fungal diversity could still be very high. **Most of these marine-derived fungi belong to terrestrial species**. Further studies are required to find out if these cultures and sequences resulted from active mycelia.

### *11.2.2 Abundance and Biomass of Fungi in the Water Column*

**The standing crop or biomass of fungi, estimated from their abundance, gives a clue to their relative importance in the ecosystem compared to other organisms.** Culture methods underestimate microbial populations. For example, direct detection methods for bacterial enumeration in seawater yield population densities that are several orders of magnitude greater than cultural estimations, a phenomenon termed the "great plate count anomaly" (Munn 2011).

**The straminipilan fungi, the thraustochytrids, have been the most well-studied group of fungi in the water column.** Alwin Gaertner devised a modified Most Probable Number technique used by bacteriologists, whereby thraustochytrids were cultured and enumerated by using the pine pollen baiting technique (Gaertner 1968). His studies showed that culturable numbers of thraustochytrids varied from <1 to 640 cells $L^{-1}$ seawater. These studies also revealed that the population densities varied in space and time (Raghukumar 2002). However, culture studies are limited by the "great plate count anomaly" as mentioned above.

**Thraustochytrids and aplanochytrids in the water column can be enumerated fairly accurately using the Acriflavine direct detection (AfDD) epifluorescence method** (Raghukumar and Schaumann 1993; Fig. 11.4a). Thraustochytrid populations can be converted into biomass, based on a conversion value of $1.65 \times 10^{-4}$ μg C and $0.158 \times 10^{-4}$ μg N $cell^{-1}$ of 10 μm diameter (Kimura et al. 1999) or more conservatively, $20.6 \times 10^{-12}$ g C and $1.98 \times 10^{-12}$ g N for a smaller cell size of 5 μm diameter. The AfDD method has provided us much insight into the dynamics of labyrinthulomycetes in the water column (Naganuma et al. 1998; Kimura et al. 1999; Raghukumar et al. 2001; Leaño and Damare 2012).

**Filamentous fungi have been estimated in the pelagic using culture methods.** However, as mentioned earlier, colonies in cultures can arise from single spores, cluster of spores, single cells or mycelial bits of varying sizes. Hence, their populations cannot be estimated by culturable numbers. Biomass of mycetaen fungi can be estimated by biovolume measurements based on direct detection of fungal hyphae or ergosterol estimations (see Sect. 15.3). A popular method is an epifluorescence microscopy method based on staining with Calcofluor White by0020which fungal hyphae can be detected, their biovolumes estimated and extrapolated to biomass (Fig. 11.4b). Biomass fungal carbon is estimated using the conversion factor of 1 pg C $(\mu m^3)^{-1}$ volume (Gutiérrez et al. 2011). Another method is to estimate ergosterol contents in a sample and convert these to fungal biomass.

Fungi and bacteria coexist in the marine environment. In order to understand the role of fungi in the ecosystem, it is important to consider the relative roles of both bacteria and fungi in heterotrophic processes. Bacterial numbers are estimated using the epifluorescence microscopy based on Acridine Orange Direct Counts (AODC method). Bacterial biomass is estimated by using conversion values of $20.0 \times 10^{-15}$ g C (Ducklow 2000) and $4.0 \times 10^{-15}$ g N $cell^{-1}$ (being 20% of C).

There are three important parameters in studying fungal carbon contribution to pelagic waters.

1. The individual biomass carbon and nitrogen contribution of mycetaen and straminipilan fungi.
2. Their relative contribution with respect to bacteria.
3. The relative contribution of each of these to the total organic carbon in the water column.

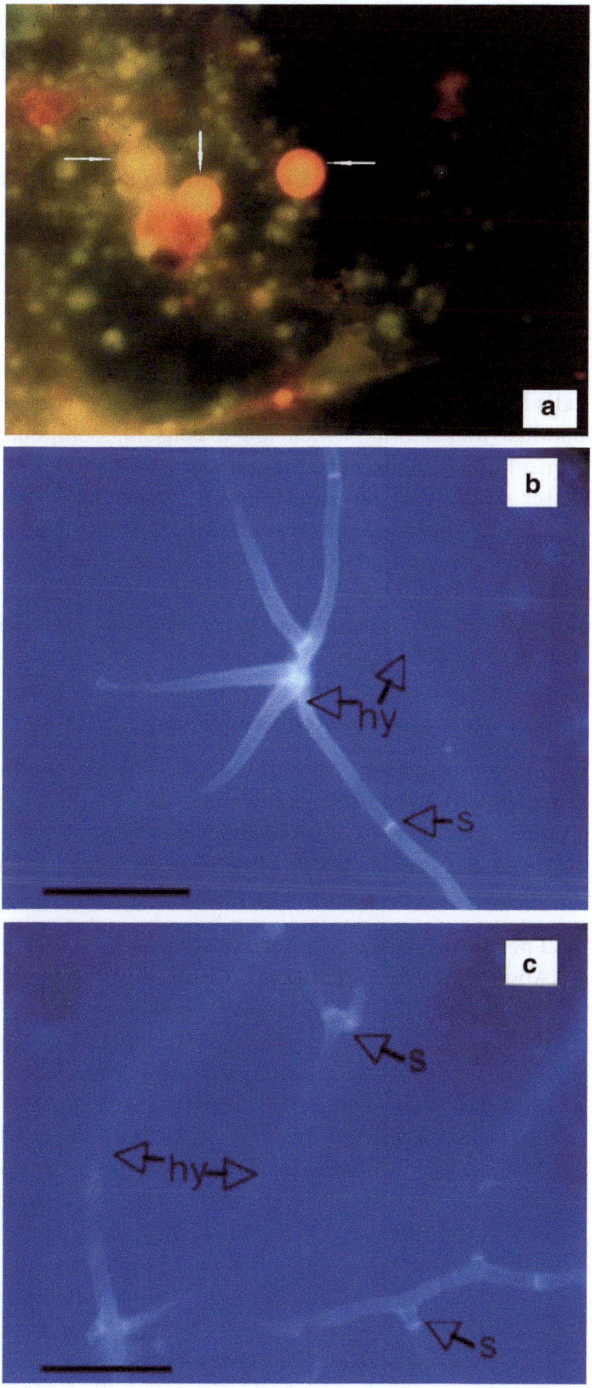


**Fig. 11.4** Fungi in water column. (**a**) Epifluorescence microscopy photographs of thraustochytrid cells stained using the AfDD technique. Besides the larger cells (*arrows*), numerous smaller cells are present outside (S. Raghukumar). (**b**, **c**) Epifluorescence microscopy photographs of fungal hyphae stained with Calcofluor (Source: M.H. Gutie'rrez et al. 2011. The role of fungi in processing marine organic matter in the upwelling ecosystem off Chile. Marine Biology 158: ©Springer-Verlag 2010)

All of the above have not been addressed together in any study. However, **information is available on many of the above individual parameters and an insight into fungal dynamics in the pelagic has begun to emerge.**

**Microbial heterotrophic activities in the water column depend upon the availability of DOM and POM generated by biological production, which is strongly influenced by the seasons.** Seasonal variations are characterized by alterations in levels of solar radiation, water temperatures, and input of nutrients through upwelling and eddies, as well as by outwelling from land run-off and coastal vegetations. Filamentous fungi, yeasts, and thraustochytrids have been independently studied by specialists in these groups from different geographical regions.

**Culturable numbers of thraustochytrids vary widely from $<1$ to $600 \times 10^3$ cells $L^{-1}$seawater** (Table 11.4). AfDD direct counts yield much higher numbers. In contrast to bacteria, which are a constant component of the pelagic microbial consortium, thraustochytrids are not always present. **They are present in the water column in extremely low numbers below detection levels, or they can attain substantial population densities.** Thus, they may vary from $<1$ to 21,000 to 674,000 cells per liter seawater. The maximal numbers attainable vary by a factor of about 30. Based on these values, it can be estimated that **thraustochytrid populations can attain maximum biomass ranging from 3.465 to 15.8 μg C and 0.33 to 1.51 μg N $L^{-1}$water.** Biomass values using the AfDD technique are conservative estimates because AfDD technique will not be able to detect small cells or zoospores with a thin cell wall.

**Thraustochytrid populations show distinct seasonal changes. Thraustochytrid numbers are generally low during periods of phytoplankton blooms and during periods of low nutrients. They achieve peak densities towards the end of productive seasons** (Naganuma et al. 1998; Kimura et al. 1999, 2001; Raghukumar et al. 2001).

- **The Arabian Sea** is characterized by a summer premonsoon period during April/May. The mixed layer of the water column at this time is shallow and lies at about 25 m. Nitrate levels, phytoplankton biomass, and POC are low, while bacterial abundance is high. Thraustochytrids ranged from $<1 - 230 \times 10^3$ cells $L^{-1}$. Thraustochytrids during this period are infrequent and were noticed only in 29% of the total 41 samples taken during this time (Raghukumar et al. 2001). Subsequently, high levels of primary productivity are achieved in the Arabian Sea during the southwest or summer monsoon during June to September and the northeast or winter monsoon between November and February (Burkill et al. 1993; Shetye et al. 1994; Madhupratap et al. 1996; Morrison et al. 1998).

  Thraustochytrids are abundant at the end of the southwest monsoon in September (Fig. 11.5). The range of their population was the same as that in summer ($<1 - 230 \times 10^3$ cells $L^{-1}$), but patches of high densities were found more frequently than in summer, up to 93% of the samples revealing these fungi. Populations reached up to $1313 \times 10^3$ cells $L^{-1}$ at one station. Thraustochytrid abundance ranged from 3.7 to $183 \times 10^3$ cells $L^{-1}$ water at the end of the winter

**Table 11.4** Abundance of thraustochytrids and bacteria in the water column at different geographical locations

| Geographical location | Thraustochytrids $L^{-1}$ water | µg Thraustochytrid C $L^{-1}$ water | Bacteria $L^{-1}$ water | µg Bacterial C $L^{-1}$ water | Reference |
|---|---|---|---|---|---|
| Seto Inland Sea | $2.1 \times 10^3$–$56 \times 10^3$ (10 µm cell size) | 0.3465–9.24 | $0.05 \times 10^9$ $-1.0 \times 10^9$ | 1.01–20.2 | Naganuma et al. (1998) |
| Seto Inland Sea, Japan | $2.5 \times 10^3$–$45 \times 10^3$ (10 µm cell size) | 0.4125–7.425 | $0.15 \times 10^9$ $-1.1 \times 10^9$ | 3.03–22.22 | Kimura et al. (1999) |
| Seto Inland Sea and the Hyuga Nada area, Japan | $0.09 \times 10^3$–$21 \times 10^3$ (10 µm cell size) | 0.015–3.465 | $1.1 \times 10^9$–$66 \times 10^9$ | 22.22–1333 | Kimura et al. (2001) |
| Arabian Sea | $<1 - 230$ (1313) $\times 10^3$ (5 µm cell size) | <0.00003–4.74 (27) | 0.3–$1.5 \times 10^9$ | 6.06–30.3 | Raghukumar et al. (2001) |
| Equatorial Indian Ocean | $<1 \times 10^3$–$750 \times 10^3$ (5 µm cell size) | <0.00003–15.8 | $0.183 \times 10^9$–$0.8 \times 10^9$ | 0.004–16.1 | Damare and Raghukumar (2008) |
| North Pacific Subtropical gyre, near Hawaii | $1.89 \times 10^3$–$630 \times 10^3$ (5 µm cell size) | 0.039–13 | $0.145 \times 10^9$–$0.7 \times 10^9$ | 2.93–14.14 | Li et al. (2013) |

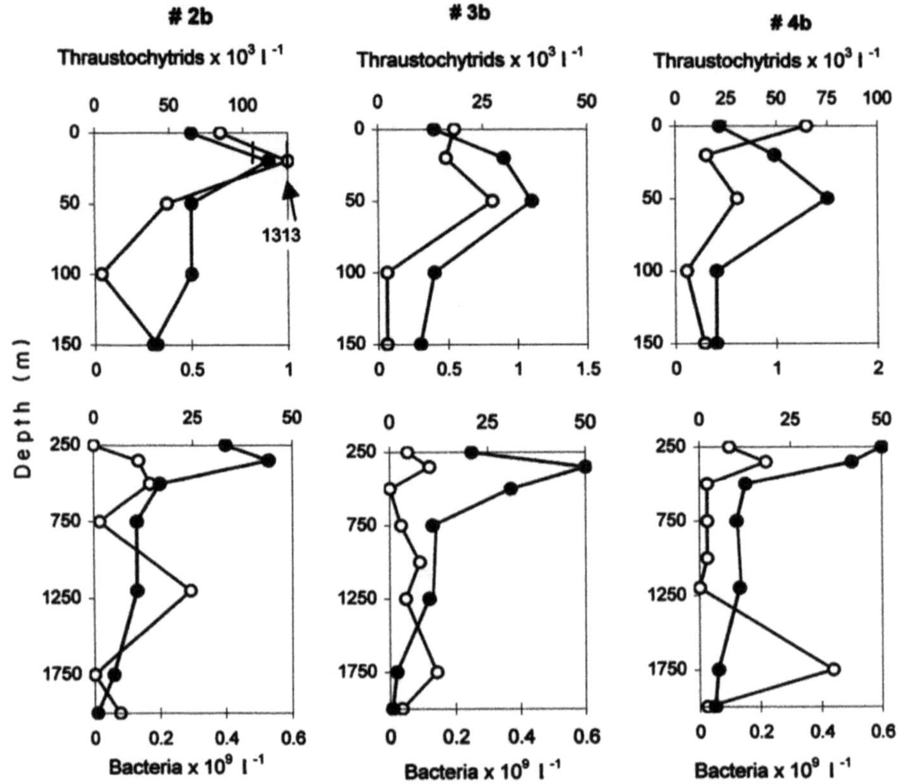


**Fig. 11.5** Vertical distribution of thraustochytrids and bacteria in the water column of the Arabian Sea during September. Note the patchy distribution of thraustochytrids (Source: Raghukumar S et al. 2001. Dynamics of thraustochytrid protists in the water column of the Arabian Sea. Aquatic Microbial Ecology 24: 175–186. © Springer-Verlag 2010)

monsoon in February/March, more or less similar as in summer and the late summer monsoon. However, all samples yielded thraustochytrids (100% frequency). Thus, it appears that **pockets of nutrients supporting thraustochytrids, probably POM, are present in the water column at all seasons. However, abundance of these nutrient pockets may be seasonally variable**, certain seasons being characterized by a high frequency of these. Thraustochytrids in the Arabian Sea are not related to chlorophyll and are only occasionally related to bacteria. They appear to be more related to POC.

- **Thraustochytrid populations, as estimated using pine pollen baiting culture method, showed distinct seasonal changes in the Fladen Ground area of the North Sea. This region shows a distinct, temperate water pattern of seasonality.** Numbers of thraustochytrids were extremely low or below detection levels during the low nutrient and low temperature conditions of winter (November to February). Abundant phytoplankton blooms occur during spring when light availability is high enough for phytoplankton to use the nutrients brought up during the winter. Although thraustochytrids were more during this time than in winter, their numbers were still few at this time (Gaertner and Raghukumar 1980; Raghukumar and Gaertner 1980). Thraustochytrids have been seldom microscopically observed in association with living phytoplankton (Gaertner 1979). Peak thraustochytrid numbers in the Fladen Ground area of the North Sea were found only at the end of the phytoplankton blooms in the early summer of June. **Contrary to thraustochytrids, yeast populations may be related to phytoplankton blooms.** Meyers et al. (1967) reported cell numbers of up to 3000 cells/liter in the North Sea during a period of *Noctiluca* blooms (Fell 2012).

- **The North Pacific Subtropical Gyre** (NPSG) contains two layers within the euphotic zone, an upper mixed layer down to 50 m and a lower one below that (Karl 1999). The least abundance of thraustochytrids in the NPSG adjacent to Hawaii occurred during spring when chlorophyll values are high in the lower layers and in winter when they are high in the upper layer (Li et al. 2013). Thraustochytrids were also low during the low productive summer from March to August and were observed in only 39% of the samples. Maximal numbers reached only $2.95 \times 10^3$ and $1.75 \times 10^3$. Thraustochytrids reached maximum densities of $1.89 \times 10^3$ cells $L^{-1}$ to $630 \times 10^3$ cells $L^{-1}$ during the fall season of November at the end of the productive season. Thraustochytrids were found at 100% frequency of samples. Their numbers were never related to chlorophyll in any of the seasons.
- **Seto Inland Sea, Japan,** is a semi-enclosed, coastal sea. It is severely affected by human activities and is highly eutrophicated. Thraustochytrid numbers were positively correlated to chlorophyll in these waters, suggesting that these protists depend to some extent on dissolved organic matter or particulate detritus from phytoplankton (Naganuma et al. 1998).
- **Population densities of individual species of thraustochytrids using quantitative polymerase chain reaction** (qPCR) methods will provide better insight into dynamics of thraustochytrids in the water column. Genus-specific primer sets targeting seven genera have successfully detected the abundant presence of *Aurantiochytrium* in the range of $1.12 \times 10^4$ to $1.31 \times 10^4$ cells $L^{-1}$ and *Oblongichytrium* in the range of $1.02 \times 10^4$ to $3.14 \times 10^4$ cells $L^{-1}$ in coastal waters of Japan (Nakai et al. 2013).
- Culturable numbers of thraustochytrids in seawater and sediments of Ligurian Sea coast, NW Mediterranean, Italy, were not related to chlorophyll (Bongiorni and Dini 2002).
- We may draw certain conclusions about thraustochytrid dynamics in the open water column. **Thraustochytrid abundance appears to be related to particulate organic carbon (POC) both in coastal and oceanic waters.**
  - **The high abundance of thraustochytrids at the end of the summer monsoon period in the Arabian Sea during September appears to be associated with senescent phytoplankton and detritus formed therefrom and appear to be related to particles** following phytoplankton degradation

at the end of the biologically productive seasons. They were also significantly related to POC in the upper 150 m of the water column. Thraustochytrids densely colonized mucoid particles in the water column, reaching a density of $1313 \times 10^3$ cells $L^{-1}$. Dense swarms of salps make their appearance in the northern Arabian Sea during the winter period. These animals are highly efficient filter feeders and rapidly churn out fecal pellets, which appear to be excellent substrates for these fungi (Raghukumar and Raghukumar 1999). Thraustochytrid biovolumes were twice those of bacteria during certain stages of decomposition of salp fecal pellets.

- **Thraustochytrids in Hawaiian waters are often attached to marine snow particles** and have a patchy distribution in the water column (Li et al. 2013). It appears that in these waters, POC formed from phytoplankton-based detrital aggregates during the fall control their population, rather than phytoplankton biomass.
- **Thraustochytrid abundance in Seto Inland Sea was significantly correlated primarily to POC concentration** and secondarily to chlorophyll a concentration. Their abundance is more influenced by non-phytoplankton, non-bacterioplankton POC (Kimura et al. 2001).
- **Cells of *Aplanochytrium kerguelense* have been detected in marine aggregates** obtained from the equatorial Indian Ocean (Damare and Raghukumar 2010). Coastal isolates of this aplanochytrid produced abundant proteases and lipases, which might play a role in decomposition of POC in the pelagic (Damare et al. 2013).
- **Thraustochytrids in Greenland and the Norwegian seas were found to be associated with non-ATP-related or the non-phytoplanktonic particles** such as post-bloom phytodetritus and terrigenous matter non-ATP-related (Naganuma et al. 2006).
- **Thraustochytrids and aplanochytrids grow and survive on laboratory-generated aggregates**. They grew well in the presence of diatom polysaccharides and in the absence of phytoplankton cells (Damare and Raghukumar 2012).
- Thus, they may utilize allochthonous, outwelled detrital material in coastal waters, as well as autochthonous material in oceanic waters.

• Thraustochytrids, in general, seem to grow poorly when phytoplankton are under active, bloom conditions. Phytoplankton, particularly diatoms, defend themselves through production of antimicrobial compounds and may deter thraustochytrid growth.

**Mycetaen fungi contribute biomass to the pelagic microbial consortium.**

**Filamentous fungi have often been estimated in the pelagic using culture methods.** The first extensive investigation of fungal populations was made in the northwestern subtropical Atlantic (Roth et al. 1964). All eulittoral water samples contained relatively high fungal populations with a minimum of ten different fungal species, ranging from 187 to 296 colonies per liter. Water samples obtained more distant from land and in deep waters displayed a more than tenfold decrease in average fungal population density and about a 50% reduction in the number of species. A typical milliliter of seawater contains over $10^3$ CFUs (Gao et al. 2008).

**Actively growing fungal hyphae can be detected in the water column using Calcofluor white direct detection methods** ((Fig. 11.4b, c; Gutiérrez et al. 2011). Single or grouped septate hyphae have been found on particulate material in upwelling waters of the south Pacific using Calcfluor staining and epifluorescence microscopy. The size of fungal structures ranged between 1 and 4 μm in diameter and from 10 to 400 μm length. Biomass measurements based on mycelial biovolume have shown that **mycelial fungi contribute 0.01–40 μg C $L^{-1}$C in these waters.**

**Cryptic endobiotic, filamentous, mycetaen fungi in particulate matter in the pelagic might have been overlooked in our preoccupation of looking at planktonic microorganisms in the water column.** Thus, filamentous, mycetaen fungi have generally been considered to be of no significance in the pelagic. However, a few recent studies have shown that such fungi do indeed occupy an important ecological niche within particles present in the water column. Thus, actively growing fungal mycelia have been detected in productive waters of the Pacific (Gutiérrez et al. 2010, 2011; Wang et al. 2012).

- **The Humboldt Current System in the South Pacific along the west coast of South America** is one of the most productive ecosystems in the world. It supports a thriving microbial community in surface waters of the upwelling ecosystem off central-southern Chile during spring and summer (September to March), while productivity is low during winter (April to August) (Gutiérrez et al. 2011). As with thraustochytrids, fungal biomass, determined by means of detecting fungal hyphae (Fig. 11.4b, c), is much more variable seasonally compared to bacteria, but could obtain substantial values at certain times (Fig. 11.6). **Highest fungal biomass occurred in these waters during the**

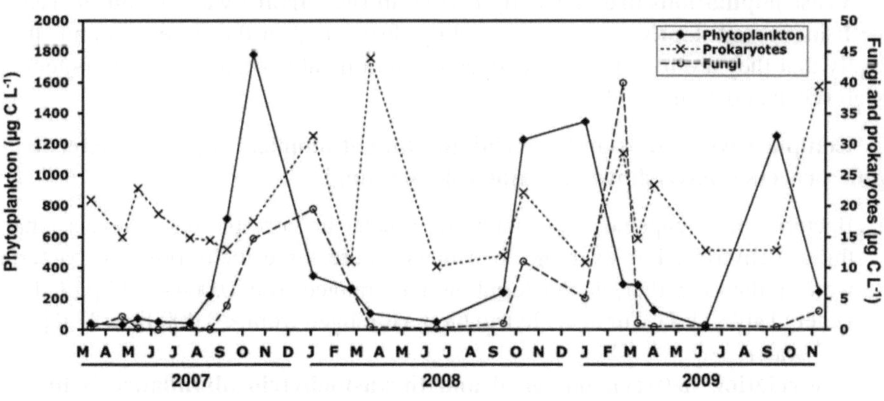


**Fig. 11.6** Seasonal variations of biomass of mycetaen fungi, prokaryotes, and phytoplankton in the coastal upwelling waters off Chile (Source: Gutierrez, M.H. et al. 2011. The role of fungi in processing marine organic matter in the upwelling ecosystem off Chile. Marine Biology 158: 205–219. ©Springer-Verlag 2010)

**austral spring and summer at the depth of maximum chlorophyll fluorescence. The biomass amounted to 40 μg C $L^{-1}$** and decreased with depth in the water column. Fungal biomass drastically declined during the low productive austral winter (April to August) to values of 0.01–2 μg C $L^{-1}$. This was also reflected in the fungal biomarker fatty acid 18:2ω6. **The vertical distribution of fungal biomass in the water column was closely related to that of phytoplankton biomass** (Gutiérrez et al. 2010, 2011). Fungi were not planktonic, but were associated with particles. They appear to be related to increments in organic matter available during active upwelling and phytoplankton growth in the photic zone. Fungi were most abundant in particle size fraction of 25–90 μm during the productive spring and summer. Bacteria were associated with particle sizes of 0.2–3.0 μm, being apparently planktonic. Large particles with which fungi were associated also showed a high level of extracellular enzymatic hydrolysis activity compared to other particle sizes, thus indicating an important ecological role for mycelia fungi. On the contrary, enzymatic activity during the low productive winter season was mainly distributed in the prokaryote size class.

**Yeasts are regularly found in high numbers in coastal waters, but only occasionally in oceanic waters.** Yeasts are enumerated by culture techniques. Yeast populations vary from tens to thousands of cells/liter of water. Sewage and land runoff may result in high numbers of yeasts in estuarine and coastal waters. In contrast, organically poor oceanic waters may contain much fewer yeasts. However, these too may contain pockets of high organic nutrients resulting from plankton blooms, surface slicks, current boundaries, eddies, and thermoclines. Such localized niches may sustain high populations of a few thousand cells of yeasts per liter (Kutty and Philip 2008; Fell 2012; Gadanho et al. 2003). *Leucosporidium antarcticum* was found in Antarctic waters adjacent to the ice pack in concentrations of 1–22 cells/liter and at depths to 3000 m (see Jones and Fell 2012).

**Yeast populations are generally higher in the surface waters and decrease with increased depth.** Yeasts are patchily distributed in the water column. It is likely that they are found in microzones where utilizable organic matter content is high (Gadanho et al. 2003).

**Comparative studies on bacterial and fungal biomass in pelagic heterotrophic processes have shown a major role for fungi.**

- Bacteria are capable of attaining much higher biomass levels than thraustochytrids. For example, in waters where thraustochytrids and bacteria were studied together, the bacterial biomass ranged from 0.004–1333 μg C $L^{-1}$ water (Table 11.4). Thraustochytrid biomass ranged from <0.00003 to 27.0 μg C $L^{-1}$ water.
- **The relation between bacterial and thraustochytrid abundance is highly variable.** The two groups were positively related in the water column of Seto Inland Sea, fish farm sediments in Japan, and the equatorial Indian Ocean (Kimura et al. 2001; Bongiorni et al. 2005a; Damare and Raghukumar 2008b; Damare and Raghukumar 2010). However, the relationship between the two

groups was highly variable in the Arabian Sea waters (Raghukumar et al. 2001). Hence, the two groups may share the same or different nutrient sources in the water column at different periods of time or habitats. This is an aspect that deserves further study.

- **It has been suggested that thraustochytrids might play the role of "microbial scavengers."** This may be akin to the terrestrial ecosystem where lions and hyenas, both carnivores, coexist. Thus, although thraustochytrids (hyenas) are capable of degrading an organic substrate entirely by themselves, their heterotrophic activity does not start till bacteria (lions) rapidly colonize the substrate, attain a threshold population density, and cease to degrade the substrate further (Raghukumar and Damare 2011).
- **Abundance of thraustochytrids in the water column displays a "boom or bust" phenomenon. These fungi are often absent in the water column, unlike the ubiquitous bacteria. However, they do rival bacteria in terms of biomass quite often.**

Peak thraustochytrid biomass in various waters that have been studied so far ranges from 3.465 to 27.0 μg C and 0.33 to 2.6 μg N $L^{-1}$ water (Table 11.4). Bacterial biomass attained values of 1.01 to 1333 μg C $L^{-1}$ and 0.2 to 266 μg N $L^{-1}$ in these waters. The biomass contributions of thraustochytrids and bacteria in those locations where the former attained maximum values are given in Fig. 11.7. During such instances, their relative contribution to the microbial biomass with respect to

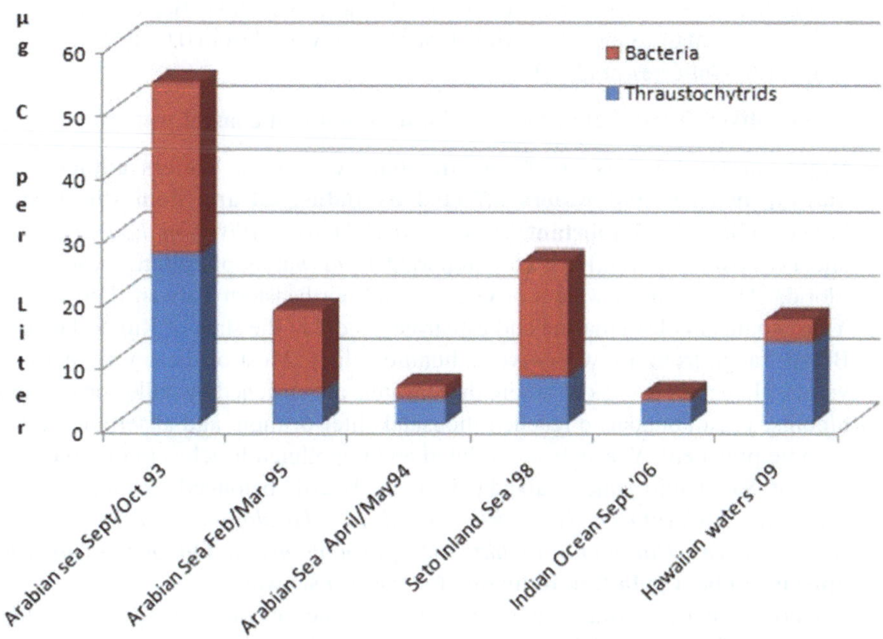


**Fig. 11.7** Thraustochytrid and bacterial biomass in the water column of different locations when thraustochytrids reached their maximum abundance

bacteria amounted at least to 38–50% (Naganuma et al. 1998; Kimura et al. 1999; Raghukumar et al. 2001; Damare and Raghukumar 2008b; Li et al. 2015). Their contribution far surpassed that of bacteria in a sample in the Arabian Sea, the Central Indian Ocean, and Hawaiian waters. Despite possibilities of errors in estimation, **thraustochytrids may apparently contribute substantially to microbial biomass under certain conditions**. Available information also indicates that even when thraustochytrid biomass was low, their carbon contribution to the water column might now and then exceed that of bacteria. Overall, thraustochytrid contribution to the total microbial biomass, comprising both bacteria and thraustochytrids, ranged from 3.4 to 29% in coastal waters of Japan; 0–27%, 0–50%, and 0–68%, respectively, during three different seasons in the Arabian Sea; 0–79.1% and 0–80% in Hawaiian waters. Thraustochytrids also formed a significant component below 200 m and comprised 34.5 to 56% of bacterial carbon biomass in different seasons (Raghukumar et al. 2001). These values point out to the tremendous potential importance of these protists in mineralization processes and in the food web.

- Biomass of filamentous mycetaen fungi have been noticed to contribute significant amounts to microbial biomass carbon in the upwelling Pacific waters off Chile, ranging from 0.01–40 μg C $L^{-1}$ (Gutiérrez et al. 2011). Bacterial biomass in these waters ranged from 13 to 40 μg C $L^{-1}$. Bacterial biomass at the time of maximum fungal biomass of 40 μg C $L^{-1}$ was only 29 μg C $L^{-1}$. On the other hand, bacterial biomass was 2–4 orders of magnitude higher than fungal biomass during the periods when the biomass of the latter was low. Bacterial biomass ranged from 10 to 23 μg C $L^{-1}$ while that of fungi varied from 0.01 to 2 μg C $L^{-1}$ (Fig. 11.6; Gutiérrez et al. 2011).

  **Fungal diversity and abundance is high in polluted coastal waters.**

- **High counts of yeasts are found in polluted waters. Waters adjacent to bathing beaches and waters affected by industrial and domestic wastes harbor clinically important yeasts.** *Candida tropicalis, Candida albicans*, and *Trichosporon* spp. have been reported from waters of bathing beaches of Florida, USA; coastal waters of Greece; and northeastern Taiwan (Fell 2012). Yeast counts in clean marine and estuarine waters in the state of Rio de Janeiro, Brazil, range from a few to several hundreds/liter. Most of these were aerobic yeasts. However, in regions where these waters are enriched by pollution or algal blooms, yeast abundance reached thousands/liter or more and anaerobic yeasts became prevalent. Yeasts from polluted and unpolluted beaches in the southern area of Sao Paulo state, Baixada Santista, Brazil, belonged to nine genera, *Candida, Cryptococcus, Rhodotorula, Torulopsis, Trichosporon, Debaryomyces, Hansenula, Pichia*, and *Sporobolomyces*. **The genus *Candida* appears to be a pollution indicator for coastal seawater.**
- Oil pollution of coastal waters may be chronic or catastrophic. The latter severely affects coastal life. Many yeasts are efficient oil degraders and are found in abundance in waters and sediments polluted with oil. *Debaryomyces*

*hansenii* was a common yeast in North Sea waters before oil production started and continued to be so later. The hydrocarbanoclastic *Candida guilliermondii* became abundant in these waters and sediments following oil production.

- **Thraustochytrid populations increase with organic matter input through pollution.** Excessive enrichment of coastal waters with organic nutrients is detrimental to the benthic ecosystem and affects both meiofaunal and macrofaunal abundance and communities (Bongiorni et al. 2005a). Populations of bacteria, thraustochytrids, and heterotrophic nanoflagellates increased in water under cages of fish farms where uneaten food pellets fed to the fish became deposited. In the case of oligotrophic locations of the Mediterranean Sea, thraustochytrid numbers increased from an average of $0.96 \times 10^3\ L^{-1}$ water in control plots to $3.35 \times 10^3\ L^{-1}$ water in plots under fish cages in soft bottom sediments. Their numbers increased even more drastically from 1.47 to $13.45 \times 10^3\ L^{-1}$ water in areas with seagrass beds. Thraustochytrids apparently could utilize refractory plant detritus very successfully in the seagrass beds. Thraustochytrids were also more dominant compared to nanoflagellates in the seagrass sediments, than in soft bottom ones.

**Thraustochytrids and aplanochytrids in the water column might serve as important sources of nutrition for zooplankton.**

Patches of very high density of thraustochytrids in the water column constitute "hot spots" of nutrition for microbivorous protists (Damare and Raghukumar 2008a, b). The highly unsaturated fatty acids (HUFAs) (also called polyunsaturated fatty acids or PUFAs), eicosapentaenoic acid (EPA 20:5 ω3), docosahexaenoic acid (DHA 22:6 ω3), and arachidonic acid (ARA 20:4 ω6) have been linked to species growth, reproductive success, and neural development in both zooplankton and fish, many of which cannot synthesize them. Pelagic animals obtain these PUFAs through their diet (Perhar et al. 2013). Almost 98% of the lipids in the zooplankton body are obtained through dietary intake. Diatoms are a rich source of PUFAs. However, they are highly variable in their content. Labyrinthulomycetes in the water column, namely, thraustochytrids and aplanochytrids are a potential source of PUFAs for pelagic animals (Raghukumar 2009). It is likely that thraustochytrids and aplanochytrids present in the water column play an important role in the marine food web by serving as sources of these fatty acids for the zooplankton (Kimura et al. 1999; Raghukumar 2002). They may do this through two major routes.

1. Planktonic or particle-attached thraustochytrids and aplanochytrids may be directly fed upon by mesozooplankton (Damare and Raghukumar 2010; Damare et al. 2013). Experimental studies have shown that mesozooplankton from coastal waters can feed on cells of *Aplanochytrium kerguelense*. This aplanochytrid species is also abundant in the oceanic waters of the equatorial Indian Ocean. In situ hybridization technique using a molecular probe for the aplanochytrid has indicated the presence of this species within the guts, as well as surrounding the gut region of mesozooplankton, particularly chaetognaths. It has been suggested that *A. kerguelense* might passively inhabit the guts of these

animals which feed upon them. They may subsequently appear in their feces. Copepods generally have high clearance rates on heterotrophic protists and may even prefer them over algae.
2. Thraustochytrids and aplanochytrids may form an important component of the "microbial loop." In this process, these microorganisms may utilize DOM and multiply. These fungi may then be fed upon by microzooplankton, such as flagellates and ciliates. Ciliates are known to feed avidly on thraustochytrid cells under experimental conditions (Raghukumar and Balasubramanian 1991).
3. Organisms higher up the pelagic trophic level, such as fish, may obtain their DHA from mesozooplankton that had fed upon thraustochytrids and aplanochytrids. When thraustochytrid-fed zooplankton, like rotifers and crustaceans, were served as feed to fish in aquaculture, it was observed that the zooplankton incorporated fatty acids from thraustochytrids in their body. Heterotrophic protists are responsible for trophic upgrading wherein they improve poor algal quality for subsequent use by higher trophic organisms (Damare et al. 2013).
4. Probably another important role of these stramenipilan fungi is to supply essential nutrients such as vitamins, amino acids (e.g., lysine and methionine), sterols, and polyunsaturated fatty acids (PUFAs) to detritivores that cannot synthesize them de novo (Bongiorni 2012). Thraustochytrids might also improve the biochemical characteristics of poor quality detritus for subsequent use by higher trophic levels, a phenomenon dubbed "trophic upgrading."

## 11.3 Fungi in Coastal Sediments

Ecology of benthic sediments in coastal and shallow waters is tightly coupled with dynamics of overlying pelagic waters. Hence, it is not possible to separate the dynamics of coastal waters from those in the sediments. Detritus from the water column and outwelled POM settle to the sediments, considerably enhancing their organic matter contents.

**Fungi are abundant in coastal sediments where they may play an important ecological role.** Fungal abundance, biomass, and activities are likely to be considerably higher in the shallow water benthic than in the water column.

- **Culture-based studies have shown that thraustochytrid numbers are much higher in sediments than in the water column** (Bongiorni 2012). Abundance of thraustochytrids in sediments has been estimated mostly through the pine pollen MPN culturing method. Although this method does not yield accurate results, it does provide us an insight into relative seasonal and spatial differences. Culturable numbers in the water column of Fladen Ground North Sea amounted to a few hundreds per litre water, while their numbers were nearly $70 \times 10^3$ cells per liter sediment (Gaertner and Raghukumar 1980; Raghukumar and Gaertner 1980). Likewise, thraustochytrid culturable numbers in the sandy surf area and the sea bottom at 5 m in the Ligurian Sea coast, NW Mediterranean, Italy, averaged about 150 times more than in the water column. Thus, the benthic population averaged 44 $\times$ $10^3$ $L^{-1}$ and 61 $\times$ $10^3$ $L^{-1}$ respectively, compared to an average of 0.13 $\times$ $10^3$ $L^{-1}$ in the middle of the water column (Bongiorni and Dini 2002). The average abundance of thraustochytrids in sandy shore in the Ligurian Sea (NW Mediterranean) was 42.4 $\pm$ 35.2 $ml^{-1}$ (Santangelo et al. 2000).

- **Yeasts are found in much higher numbers in shallow water sediments than in the water column. Up to 2000 viable cells of yeasts per g have** been reported for marine sediments (Kutty and Philip 2008). Most of the yeast population is confined to the top layers of the sediments. In shallow water sediments off Florida with strong surf wave action and disturbance, yeasts were found in depths up to 9 cm. Yeasts in the sediments of deeper waters at 540 m in the Gulf Stream with less disturbance were confined to the top 2 cm. Yeasts numbered up to several hundred living cells per ml of damp mud from the Kiel Fjord.

- Single-celled phytoplankton, mesozooplankton and higher level consumers, as well as marine aggregates or "marine snow" and DOC generated by these organisms are all sites of intense microbial activities.
- Microorganisms may be associated with primary producers and consumers as symbionts or parasites.
- More than 50 species of phytoplankton are infected and killed by "parasitoid protists" and fungi belonging to oomycetes (straminipila) and chytrids (fungi).
- The oomycetes *Lagenisma coscinodisci* in the diatoms *Coscinodiscus* spp. and species of *Ectrogella licmophorae* are well studied, biotrophic, obligate parasites.
- Planktonic fungal parasites can be important in regulating the dynamics of harmful algal blooms, such as the diatom *Pseudonitzschia* species, some of which produce the toxin domoic acid.
- Fungi may occur as commensals, mutualists, or parasites of mesozooplankton, fish, and mammals, which constitute secondary, tertiary, and higher level producers of the pelagic ecosystem.
- Only a few fungal parasites have been reported in marine animals.
- Many yeasts, such as *Metschnikowia* species, are common as parasites of various animals in the marine pelagic.
- The straminipilan fungus *Leptolegnia baltica* is known to have caused mass mortality of the copepod *Eurytemora hirundoides*.
- Yeasts and labyrinthulomycetes live symbiotically in pelagic animals.
- *Aplanochytrium kerguelense* is a dominant species in mesozooplankton from surface to 1000 m depth in the equatorial Indian Ocean.

(continued)

- Fungi are active and abundant in the water column and display seasonal dynamics.
- Thraustochytrids appear to be the most dominant group of fungi in the pelagic water column, occupying a wide variety of niches.
- A number of thraustochytrids and aplanochytrids have been cultured from coastal and oceanic waters in the tropic, temperate, and Antarctic regions in different geographical regions using pine pollen and brine shrimp baiting techniques.
- Marine yeasts belong to both Ascomycota and Basidiomycota. Yeast diversity is high in coastal waters and decreases in the oceans.
- Many yeasts show distinct geographical distributions.
- Most of the marine-derived fungi belong to terrestrial species.
- Several metagenomic studies have revealed novel fungal lineages.
- Many fungi may exist as unicellular picoeukaryotes.
- Thraustochytrid populations vary widely from <1 to 600 $\times$ $10^3$ cells $L^{-1}$ seawater.
- Yeasts are regularly found in high numbers in coastal waters.
- Actively growing fungal hyphae are found in pelagic POM
- Thraustochytrid dynamics in the pelagic have been studied in detail in the North Sea, the Arabian Sea, coastal waters of Japan, the North Pacific Gyre near Hawaii, and the equatorial Indian Ocean.
- Thraustochytrid populations show distinct seasonal changes. Their numbers are generally low during periods of phytoplankton blooms and during periods of low nutrients. They achieve peak densities towards the end of productive seasons.
- Yeast populations may be related to phytoplankton blooms.
- Thraustochytrid abundance is related to particulate organic carbon (POC) both in coastal and oceanic waters.
- Mycetaen fungi may contribute significantly to the pelagic microbial consortium.
- Highest fungal biomass in Humboldt Current System in the South occurred during the austral spring and summer. The biomass amounted to 40 μg C $L^{-1}$. The vertical distribution of fungal biomass in the water column was closely related to that of phytoplankton biomass
- Yeast populations are generally higher in the surface waters and decrease with increased depth.
- Polluted coastal waters harbor a high diversity and abundance of fungi.
- Thraustochytrid populations increase with organic matter input through pollution.
- Thraustochytrids might play the role of "microbial scavengers," thriving upon nutrients left over by bacteria.

(continued)

- Abundance of thraustochytrids, as well as filamentous fungi, display a "boom or bust" phenomenon. Their biomass is often negligible but occasionally may rival that of bacteria.
- Thraustochytrids and aplanochytrids in the water column might serve as important sources of nutrition for zooplankton.
- Fungal abundance, biomass, and activities are considerably higher in coastal benthic than in the water column.

**Future Directions**

- A number of fungal parasites are known in diatoms. What is their role in regulating phytoplankton populations?
- Thraustochytrids, aplanochytrids, and yeasts are associated with zooplankton in a symbiotic manner. Do they have a mutualistic or commensalistic role?
- A number of thraustochytrids have been cultured from the pelagic. What is the uncultured diversity of straminipilan fungi in open waters?
- What is the role of thraustochytrids in degradation of organic matter in the pelagic and what is their relation to bacteria?
- Terrestrial species of fungi have been detected through cultures and sequences from coastal and oceanic waters. Which of these grow actively in the pelagic?
- What are the spatiotemporal dynamics of mycelial fungi growing in large particulate organic matter? What is their biomass and what is their role?
- Yeasts are patchily abundant. What is their role in the pelagic ecosystem and what governs their distribution and what is their biomass?
- Fungi have been studied to some extent in the water column, but scarcely in marine sediments. What is the diversity, biomass, and role of fungi in coastal sediments?
- What is the comparative role of bacteria and fungi in the water column?
- How important are fungi as food to meso- and microzooplankton?

# Chapter 12
# Extreme Marine Environments

*We need the tonic of wildness...At the same time that we are earnest to explore and learn all things, we require that all things be mysterious and unexplorable, that land and sea be indefinitely wild, unsurveyed and unfathomed by us because unfathomable. We can never have enough of nature.*

Henry David Thoreau, *Walden: Or, Life in the Woods*

Extreme conditions define the limits of life in our planet (Asmaniadou and Lipiatou 2000). Many parts of the ocean experience extreme conditions. Deep-sea organisms face elevated hydrostatic pressure, low temperature, and scant nutrients. Certain parts of the ocean are home to hydrothermal vents where extreme high temperatures beyond 300 °C are found. Many organisms live in environments of constant or seasonal low oxygen conditions. Hypersaline conditions are found in salterns. Despite such adverse conditions, a surprisingly high diversity of organisms thrives in these environments. While some organisms, the extremotolerant ones, merely put up with such conditions and manage to survive and grow, others, the extremophiles, actually prefer such drastic life conditions. Archaea and Bacteria are the most successful in extreme marine environments and have been studied extensively in the last 50 years (Munn 2011). Detailed studies on fungi in extreme marine environments commenced only in recent years.

The ocean is divided vertically into five zones (Fig. 2.1). **The deep sea may be arbitrarily considered to start with the mesopelagic or the twilight zone at 200 m, from which depth downwards light is too feeble for photosynthetic production to take place.** Beyond lie the bathypelagic, abyssal pelagic, and the hadal zone.

## 12.1 The Deep Sea

It is important that we understand the biology in the deep sea because it covers ~65% area of the Earth's total surface (Sverdrup et al. 1942). **The deep sea is a region of high hydrostatic pressure and low temperature.** We experience an



S. Raghukumar, *Fungi in Coastal and Oceanic Marine Ecosystems*,
DOI 10.1007/978-3-319-54304-8_12

atmospheric pressure of 1 bar or 1 kg $(cm^2)^{-1}$ at the sea surface. One bar is equivalent to the ISI unit of 0.1 MPa (mega pascal). Hydrostatic pressure increases by 1 bar or 0.1 MPa for every 10 m depth. We can experience the elevated hydrostatic pressure by feeling it on our ears when we dive in the sea to a depth of even just 10 m. A whale that dives 2000 m experiences a pressure of 200 bars or 20 MPa. **The deep sea is also a region of darkness.** Light is feeble in the mesopelagic and totally absent in zones below. **The temperature at about 1000 m is close to 0 °C.**

**Organisms in the deep sea depend almost entirely on the drizzle of food from above** (Fig. 12.1). "Almost," because hydrothermal vents are an exception, as discussed further on. The food that is transported to the sea bottom in a process termed the "biological pump" (Sect. 2.3.2) comprises only a small portion that is produced in the euphotic zone above. Yet, the deep sea is also a highly diverse environment. Life processes in the deep sea are often subjected to a "feast or famine" situation. The environment ranges from desert-like oligotrophy to sudden food falls such as whales, wood, or kelp (Jørgensen and Boetius 2007). One might also find oases in the form of hydrothermal vents and seeps.

The main source of oxygen in these habitats is the surface water that carries dissolved oxygen generated by phytoplankton in the euphotic zone, as well as exchange with the surrounding waters. Deep-sea currents generated due to the descending movement of surface water are the main force bringing oxygenated water to the lower depths.

Distance from land and accessibility to the deep sea make study of deep-sea organisms difficult. **Availability of oceanographic research vessels, sophisticated deployment and retrieval mechanisms, and appropriate samplers are important requirements that make deep-sea biology a highly specialized research subject**. Samples are collected using conventional oceanographic samplers as well as manned deep-sea submersibles and automated underwater vehicles. The manned submersibles Shinkai 2000 and Shinkai 6500 and the unmanned submersible Kaiko belonging to the "Deep-sea Microorganism Research Group, Japan Marine Science and Technology Center" are some examples.

Numerous studies have focused on deep-sea bacteria (Li et al. 1999; Takai and Horikoshi 1999; Delong and Pace 2001; Sogin et al. 2006). On the contrary, **eukaryotic microbes have begun attracting attention only in recent years.** A large diversity of micro-eukaryotes, novel marine alveolates, stramenipiles, heterokonts, and dinoflagellates have been discovered from the deep-sea regions (López-García et al. 2001, 2003, 2007; Edgcomb et al. 2002).

Numerous studies have been carried out on fungi in deep-sea waters and sediments in many oceans, leading to the recognition that fungi are an important component of the deep sea (Raghukumar et al. 2010; Richards et al. 2012; Nagahama and Nagano 2012).

F.J. Roth and others were the first to culture deep-sea fungi. They isolated terrestrial species of fungi from 4500 m water samples of subtropical Atlantic Ocean. Subsequently, Willy Höhnk of Germany reported the presence of fungi from a depth of 4610 m (Höhnk 1969). These observations were further confirmed

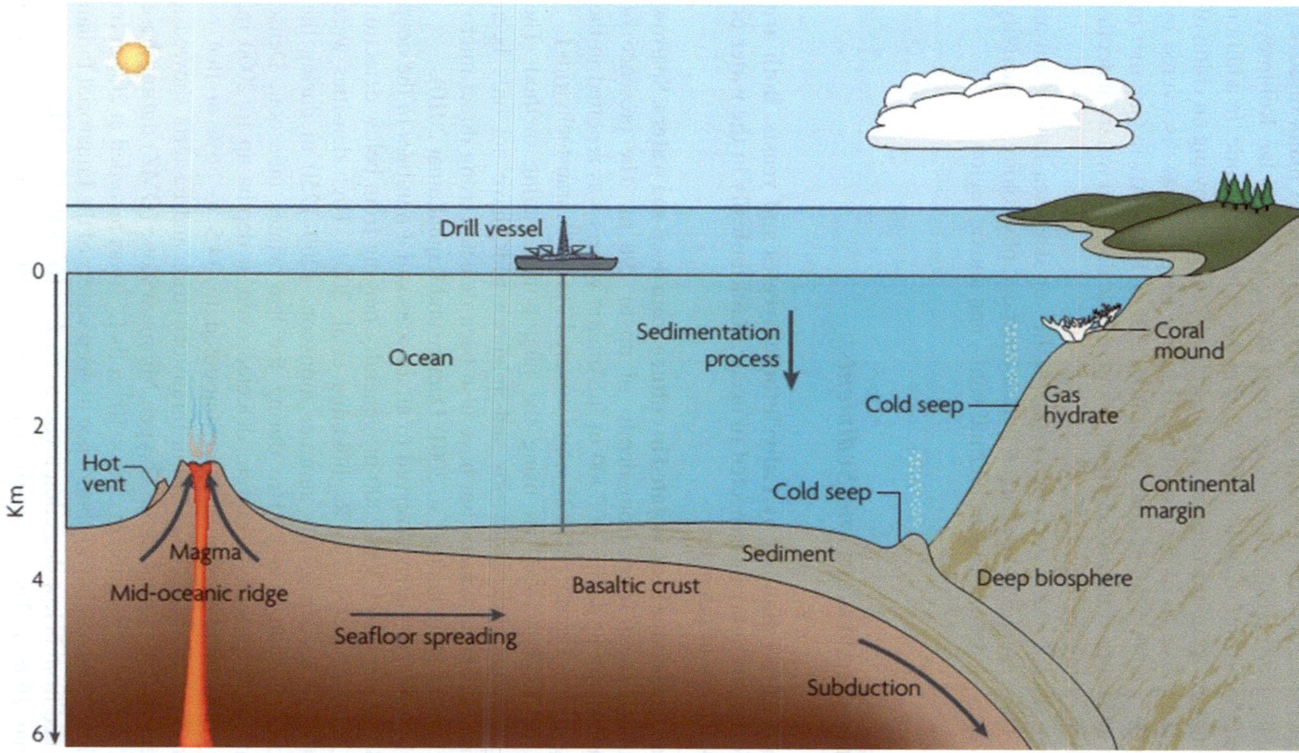


**Fig. 12.1** Vertical section of the seabed and seafloor structures (Reprinted by permission from Macmillan Publishers Ltd.: Nature Reviews Microbiology, 5: 770–781. Barker Jørgensen and Antje Boetius. Feast and famine—microbial life in the deep-sea bed. © 2007 Nature Publishing Group)

by Poulicek et al. in 1986 from shells of molluscs at 4830 m depth in the Atlantic. Raghukumar et al. (1992) were able to culture fungi from surface-sterilized calcareous shells collected from 300 to 850 m in the Bay of Bengal. A major discovery was that of obligately marine, lignicolous fungi in the deep sea. Jan Kohlmeyer (1977) reported them from wood at 1615–5315 m depth in the Atlantic. In addition to four new fungi growing in wood, he also described a fungus growing on chitin of a hydrozoan. The presence of thraustochytrids in the deep sea was discovered by F.K. Sparrow, Jr. who in 1969 isolated them from 298 m. Later, Alwin Gaertner in 1982 cultured a number of thraustochytrids from Atlantic deep-sea waters at depths of 2952–4000 m, using the pine pollen method.

The last two decades have seen numerous studies on deep-sea fungi from all major oceans. These studies have addressed their diversity, physiology, and activity in deep-sea waters, sediments, hydrothermal vents, and anoxic regions.

### *12.1.1 Deep-Sea Waters and Sediments*

**Deep-sea waters are inhabited by Labyrinthulomycetes and yeasts. Both are unicellular forms and may be expected to occur planktonically in the water or attached to surfaces of particles.**

- **Thraustochytrids have been isolated in culture from deep-sea waters.** Sparrow (1969) isolated these fungi from depths of up to 298 m. The presence of thraustochytrids up to depths of 3900 m in Atlantic waters was reported in the pioneering work of Gaertner (1982), who recovered 2 to 1000 thraustochytrids $L^{-1}$ at depths of 2952 to 4000 m using the pine pollen baiting method. The straminipilan fungi, thraustochytrids and aplanochytrids, have not only been cultured from deep-sea waters, but have also been detected using direct microscopy methods (Raghukumar et al. 2001; Damare and Raghukumar 2010).
- Thraustochytrids have been observed to attain substantial numbers in the deep Arabian Sea at depths of 200–2000 m, generally ranging from below detection levels to $38 \times 10^3$ cells $L^{-1}$ (Raghukumar et al. 2001). High densities were noticed at 1750 m during late summer monsoon, 1200–1250 m during the summer pre-monsoon, and 800 m during the late winter monsoon. Dense populations of thraustochytrids were detected even at depths up to 2000 m. Thraustochytrid biomass contribution amounted to 34.5 to 56% of that of bacteria in some of the samples. Both thraustochytrids and bacteria showed a distinct peak at 250–500 m in the Oxygen Minmum Zone (OMZ) during these 2 seasons. A large population of $266 \times 10^3$ cells $L^{-1}$ was recorded at 200 m on one occasion. Thraustochytrid abundance in deep waters of the Equatorial Indian Ocean ranged from below detection levels to $76.7 \times 10^3$ cells $L^{-1}$ (Damare and Raghukumar 2008b). They reached nearly $700 \times 10^3$ $L^{-1}$ at 200 m in one location.

- López-García et al. (2001) have demonstrated the presence of Labyrinthulomycetes at 250–3000 m in the Atlantic deep sea, using molecular signatures of 18S rRNA genes.
- **Deep-sea thraustochytrids apparently depend on POM generated in the mixed layer for their nutrition.**
  - Transparent exopolymer particles (TEPs) are found throughout the water column of the Arabian Sea up to 1000 m towards the end of the southwest monsoon and the Equatorial Indian Ocean. TEPS generally averaged 60 mg alginic acid equivalents (AA eq) $L^{-1}$ in surface waters and 10 mg AA eq. $L^{-1}$ in waters at 1000 m (Raghukumar et al. 2001). It is likely that thraustochytrids are associated with such particles.
  - Enormous quantities of fecal pellets are produced by the salp *Pegea confoederata* during the late winter monsoon period of December and January. These are densely colonized by thraustochytrids and reach depths of 1000 m in 20 h, at a rate of ~50 m $h^{-1}$ (Raghukumar and Raghukumar 1999). A thraustochytrid was isolated from fecal pellets of this salp from surface waters of the Arabian Sea. When grown under experimental conditions of 100 bars and 10 °C corresponding to 1000 m, the thraustochytrid produced proteases that were active under these conditions. This suggests that thraustochytrids are likely to play a significant role in mineralization of particulate organic detritus in deeper parts of the ocean.
- Thraustochytrids and aplanochytrids appear to be abundant not only in the deep-sea water column but also in association with deep-sea zooplankton. **The aplanochytrid, *Aplanochytrium kerguelense*, is the dominant species associated with the equatorial Indian Ocean zooplankton from surface, down to 1000 m deep-sea waters** (Damare and Raghukumar 2010).
- **Yeasts belonging to Ascomycota and Basidiomycota are frequently found in deep-sea waters** (Fell 2012; Singh and Raghukumar 2014)**. Basidiomycetous yeasts appear to be more frequent in the deep sea, compared to ascomycetous yeasts** (Nagahama et al. 2001; Kutty and Philip 2008; Singh et al. 2010). The common basidiomyceteous yeasts belong to species of *Cryptococcus, Malassezia, Rhodosporidium, Rhodotorula, Sporidiobolus*, and *Sporobolomyces*. Common ascomycetous yeasts in the deep sea include species belonging to *Candida, Debaryomyces, Kluyveromyces, Kodamaea, Metschnikowia, Pichia, Saccharomyces*, and *Willopsis*.
- Twenty three of 32 fungal phylotypes recovered from deep-sea water samples of the Atlantic and Antarctic polar front belonged to ascomycetous yeasts (Saccharomycotina) and basidiomycete yeasts (Ustilagomycotina) (Bass et al. 2007).
- ***Cryptococcus* and *Rhodotorula* are dominant deep-sea yeasts.** These two genera were predominant in deep-sea waters from Loma Trough, off California, where the total yeast count varied from 0 to 1920 viable cells $L^{-1}$ (Kutty and Philip 2008). Forty-two per cent of sequences in an RNA-based library from subsurface sediments of the Peru Trench were closely related to *Cryptococcus*

(Edgcomb et al. 2011). *Cryptococcus curvatus* dominated molecular sequences in deep-sea microbial eukaryotic communities (Takishita et al. 2006); a related *Cryptococcus* species (*C. surugensis*) was cultured and described from deep-sea sediments (Nagahama et al. 2003). *Cryptococcus skinneri* sequences have been recovered from deep-sea sediment samples from 1200 to 10,000 m off Japanese islands, including a sample from the deepest ocean depth, the Mariana Trench.
- Out of the 515 clones representing 45 OTUs detected in the East Indian Ocean at about 4000 m, 227 clones and 17 OTUs belonged to yeasts. The most common ones were *Cryptococcus* (63 clones), *Galactomyces candidum*, *Sterigmatomyces halophilus*, *Candida*, *Trichosporon moniliiforme*, and *Rhodotorula slooffiae* (Zhang et al. 2014).

**Fungi belonging to Mycetae may be a very important component of eukaryotic microbes in deep-sea sediments. Large, particulate organic matter in sediments is an ideal substrate for mycelial fungi in the deep sea** (Raghukumar et al. 2010). In spite of the extreme conditions that exist in the deep sea, actively growing fungal hyphae or ascocarps resulting from their growth are found in wood, as well as in particulate organic matter such as molluscan shells and exoskeletons of crustaceans (Kohlmeyer and Kohlmeyer 1979; Kohlmeyer and Volkmann-Kohlemyer 1988; Raghukumar and Raghukumar 1998).

- **Obligate marine fungi have been detected inhabiting deep-sea substrates. One such substrate is wood that has sunk to the deep-sea floor.** Lignocellulosic material from land in the form of decaying wood and leaves may be carried into the oceans via rivers during monsoons in the tropics or spring runoff in high latitudes. For example, a stupendous amount of $8.4 \times 10^9$ kg of woody debris was transported to the oceans off Asia during the Typhoon Morakot in 2009 (Chap. 4; West et al. 2011). Such wood become saturated with seawater and sinks to coastal and oceanic deep ocean floor. "Islands of wood" are often found in the deep sea and constitute oases of organic material on the typically food-limited deep-sea floor (Turner 1973). These are important substrates for wood borers, cellulolytic bacteria, heterotrophic bacteria, chemolithotrophic bacteria, and anaerobic, sulfur bacteria. **Five lignicolous mycetaen species on sunken woody substrata have been described so far** (Table 12.1). Four of these were discovered by Jan Kohlmeyer in 1977 at depths of 1615 to 5315 m in the Atlantic and Pacific sediments on such wood. One more has been described in recent times from the deep-sea floor at depths below 1600 m (Dupont et al. 2009). Such wood may also be inhabited by fungi that were already present under terrestrial conditions and had survived deep-sea conditions. Little is known of such facultative marine fungi. **Another interesting obligate marine fungus inhabiting the deep sea is *Abyssomyces hydrozoicus* Kohlm., which has been found in hydrozoan, hydrorhiza, and hydrocaulon attached to stony corals.**
- **Terrestrial species of Ascomycota and Basidiomycota** are the most common marine-derived fungi detected using **culture or metagenomic methods** (Raghukumar et al. 2010; Nagahama and Nagano 2012). Culturing of deep-sea

**Table 12.1** Obligate marine, lignicolous fungi from wood in deep-sea floors

| Fungus | Group | Depth (m) | Location | Reference |
|---|---|---|---|---|
| *Bathyascus vermisporus* Kohlm. | Ascomycota | 1615 and 1720 | Pacific, Near California | Kohlmeyer (1977) |
| *Oceanitis scuticella* Kohlm. | Ascomycota | 3975 | Atlantic, Gulf of Angola | Kohlmeyer (1977) |
| | | 630–1273 | Pacific Ocean off Vanuatu Island | Dupont et al. (2009) |
| *Allescheriella bathygena* Kohlm. | Anamorphic fungus | 1720 | Atlantic (near Bahams Islands) | Kohlmeyer (1977) |
| *Periconia abyssa* Kohlm. | Anamorphic fungus | 3975 and 5315 | Atlantic, Gulf of Angola and Iberian deep sea | Kohlmeyer (1977) |
| *Alisea longicolla* J. Dupont & E.B.G. Jones | Ascomycota | 630–791 | Pacific Ocean off Vanuatu Island | Dupont et al. (2009) |
| *Abyssomyces hydrozoicus* Kohlm. | Ascomycota | 631–641 | Atlantic, near south Orkney Islands | Kohlmeyer (1977) |

sediments of the Central Indian Basin from 5000 m and ca. 4000 m in the eastern Indian Ocean yielded a number of terrestrial Ascomycota and Basidiomycota, such as *Acremonium, Aspergillus, Aureobasidium, Capronia, Cerrena, Chaetomium, Cladosporium, Curvularia, Eurotium, Exophiala, Fusarium, Hortaea werneckii, Nigospora oryzae, Penicillium, Phoma, Sagenomella, Trametes versicolor,* and *Tritirachium* (Damare et al. 2006a; Singh et al. 2010, 2012a; Zhang et al. 2014). *Penicillium lagena* (Delitsch) Stolk & Samson was retrieved in culture from a depth of 10,500 m sediment samples from the Mariana Trench in the Pacific Ocean (Takami 1999). A number of terrestrial fungi, namely, *Cladosporium, Scopulariopsis, Aspergillus,* and *Penicillium,* were isolated from calcareous sediments at depths up to 860 m in the Bay of Bengal, and their spores were shown to germinate at elevated hydrostatic pressure (Raghukumar and Raghukumar 1998).

- Terrestrial species of fungi have also been detected from a number of deep-sea habitats by recovering their molecular sequences using eukaryotic and fungal-specific primers and amplification (Singh et al. 2012c). According to a rough estimate, sixteen taxa, comprising 14 fungal classes and two uncultured clone groups, have been recognized from the deep-sea resources by culture-independent methods till 2012 (Nagahama and Nagano 2012). Eurotiomycetes and Saccharomycetes were the most frequently detected fungal taxa from deep-sea environments by culture-independent methods, followed by Dothideomycetes and Sordariomycetes. Recovery of fungal sequences using 18S rRNA primers has indicated that many of these fungi are metabolically active and occur in high frequencies in these sediments (Singh et al. 2011, 2012a, c).

Species belonging to *Aspergillus* and *Penicillium* are most frequent using culture-dependent as well as culture-independent methods.

**Zygomycota and Chytridiomycota are seldom encountered.** The absence of recovery in culture or sequences belonging to these two groups suggests that these classes of fungi are not abundant in the deep-sea environment, or additional sampling might be necessary to capture their presence. Several OTUs clustered with the sequences of uncultured fungal clones (Singh et al. 2011, 2012a).

**Detection of hitherto uncultured and novel fungi in the deep sea by culture-independent methods clearly points out that fungi are native inhabitants of this extreme environment** (Nagahama and Nagano 2012; Manohar and Raghukumar 2013).

- The number of novel OTUs obtained from the deep-sea habitats is higher compared with other marine habitats. For example, 27 out of 48 fungal phylotypes recovered from Pacific Ocean deep-sea sediments showed low similarities ($\leq 97\%$) with available fungal sequences in the GenBank (Xu et al. 2014).
- 18S rDNA primer-based study has shown that the majority of active fungi from 1200 to 10,000 m off Japanese islands, including a sample from the deepest ocean depth, the Mariana Trench, were novel sequences within the phylum Ascomycota that were not closely related to previously identified fungal ITS sequences in public databases (Nagano et al. 2010).
- Seven OTUs detected using 18S rDNA and ITS sequences in the 5000 m sediments of the Central Indian Ocean Basin revealed $<97\%$ similarity with sequences in existing NCBI database, suggesting probability of them being novel (Singh et al. 2012a). Two of these belonged to *Aspergillus* sp. and the yeast *Malassezia* sp.
- Several novel phylotypes (OTUs) cluster together in phylogenetic trees and appear to constitute unique "Environmental Sequence Groups," designated in the form of arbitrary names (Table 12.2). The Chytridiomycota are considered the most ancient among the Kingdom Mycota. Many environmental sequences from deep-sea habitats occur as deep branches within the phylum Chytridiomycota. These might represent yet more ancient and unknown fungal groups.

**Presence of fungal hyphae in deep-sea substrates has clearly shown that fungi actively grow and metabolize in this extreme environment.**

- **As is common with calcareous shells in the intertidal and shallow coastal waters, fungi are also present in shells from deeper water (Höhnk, 1969, see above).** Fungal hyphae have been directly detected in calcareous fragments from sediments at depths of up to 860 m in the Bay of Bengal using the fluorescent brightener Calcofluor under an epifluorescence microscope (Raghukumar and Raghukumar 1998). *Aspergillus restrictus,* isolated from such shells, was grown on 1 g of the calcareous substrate for 25 days at 100 bar pressure. The fungus released 512 μg of calcium during this period, indicating that fungi in such substrates may utilize the organic components of shells and cause leaching of calcium during the process (Raghukumar et al. 1992).

**Table 12.2** Novel environmental phylotypes of fungi that have been discovered from various extreme marine environments

| Environmental Sequence Group | Deep-sea water and sediments | Hydrothermal vents | Anoxic waters |
|---|---|---|---|
| Hy-An Group: branches closest to the basidiomycetous yeast *Malassezia* | Various locations (Bass et al. 2007) | Guaymas vent field (Edgcomb et al. 2002); Mid-Atlantic Ridge hydrothermal area (2264 m) (López-García et al. 2003); Lost-city hydrothermal vent (750–900 m) (López-García et al. 2007); Mid-Atlantic Ridge system (860–2630 m) (Le Calvez et al. 2009); Peru margin and trench (252–5086 m) (Edgcomb et al. 2011); | Bolinas tidal flat (coastal anoxic sediment (Dawson and Pace 2002); L'Atlantic basin (3501 m) (Alexander et al. 2009; Arabian Sea (25–200 m) (Jebaraj et al. 2010); Methane cold seeps, South China Sea (350–3011 m) (Lai et al. 2007); Methane cold seep, Sagami Bay (1080 m) (Nagahama et al. 2011) |
| Cryptomycota (*Rozella* and LKM 11 cluster) | Various locations (Bass et al. 2007); Off Japanese Islands (1200–10,000 m) (Nagano et al. 2010); Sagami Bay (1080 m) (Nagahama et al. 2011) | The East Pacific Rise at the Elsa site and the Mid-Atlantic Ridge at the Menez Gwen site (Le Calvez et al. 2009) | Bolinas tidal flat (coastal anoxic sediment (Dawson and Pace 2002); Kagoshima bay (204 m) (Takishita et al. 2005); Kamikoshiki Island (22 m) (Takishita et al. 2007); Sagami Bay (1174–1178 m) (Takishita et al. 2007; Nagahama et al. 2011); Gotland deep, Baltic Sea (200–240 m) (Stock et al. 2009); Arabian Sea (20 to 200 m) (Manohar and Raghukumar 2013) |
| DSF Group-I: closest known relative is *Metschnikowia bicuspidata*. | Western Pacific, Isu-Ogasawara Trench (9700 m) (Nagano et al. 2010) | | Methane seep, Sagami Bay (1174–1178 m) (Takishita et al. 2007); Eastern Pacific, anoxic white bacterial mat in the Gulf of California at |

(continued)

**Table 12.2** (continued)

| Environmental Sequence Group | Deep-sea water and sediments | Hydrothermal vents | Anoxic waters |
|---|---|---|---|
| | | | a depth of 1575 m (Bass et al. 2007); Cold methane seep, Gulf of Mexico (2400 m) (Thaler et al. 2012); Bacterial mats and methane seeps near Japan (640 m; Takishita et al. 2007) |
| PCG: branches close to well-known fungal cultures *Penicillium* sp., *Eupenicillium* sp., and *Aspergillus* sp | Drake passage of the Antarctic polar front (250–3000 m) (López-García et al. 2001); Off Japanese Islands (1200–10,000 m) (Nagano et al. 2010) | Mid-Atlantic Ridge hydrothermal area (2264 m) (López-García et al. 2003) | Bolinas tidal flat (coastal anoxic sediment) (Dawson and Pace 2002); Kagoshima bay (204 m) (Takishita et al., 2005); Cariaco basin (340 m) (Stoeck et al. 2006); L'Atalante basin (3499 m) (Alexander et al. 2009); Arabian Sea (25–200 m) (Jebaraj et al. 2010) |
| BCG1: Basal clone group most closely related to Chytridiomycota | Various locations (Bass et al. 2007) | | Methane cold seep sites Sagami Bay (1174–1178 m) (Takishita et al. 2007; Nagahama et al. 2011); Kagoshima bay (204 m) (Takishita et al. 2005) |
| MAST-9, most closely related to "Labyrinthula" sp. | Surface seawater, Blanes Bay (Catalan coast, NW Mediterranean, 41°40.0'N, 2°48.0'E) (Massana et al. 2004a, b) | Lost-city hydrothermal vent (750–900 m) (López-García et al. 2007); Guaymas Basin in the Gulf of California (Edgcomb et al. 2002) | |

– **Fungal hyphae have been detected in 5000 m deep-sea sediments of the Central Indian Ocean Basin using Calcofluor staining and epifluorescence microscopy** (Damare et al. 2006a). Immunofluorescence methods targeting a deep-sea isolate of *Aspergillus sydowii* confirmed that the hyphae belonged to this terrestrial species of fungus. These observations indicate that **mycetaen fungi are metabolically active in deep-sea sediments and produce mycelia. Fungi in deep-sea sediments may grow embedded within humic aggregates**. Prior treatment of sediment samples with ethylenediamine tetra-acetic acid (EDTA) enables the breakdown of aggregates and improves the chances of detecting fungal hyphae and estimating their biomass more correctly. Using this method, in combination with Calcofluor staining, the biomass contribution of fungi in 5000 m sediments of the Central Indian Basin was estimated to range from below detection to 215 μg C $g^{-1}$ sediment (Damare and Raghukumar 2008). Fungi grown under experimental conditions at 200 bar and 5 °C in a nutrient medium and in deep-sea sediment extract showed that the fungi themselves may be responsible for accretion of humic particles around them leading to aggregate formation in the sediments. This may be an important ecological role of deep-sea fungi.

– The rRNA gene is expressed when an organism is metabolically active. Amplification of a high number of fungal LSUrRNA copies, corresponding to $3.52 \times 10^6$–$5.23 \times 10^7$ gene copies $g^{-1}$ wet weight sediment from deep-sea sediments of the Pacific Ocean, provides evidence that fungi might be involved in important ecological functions in these environments (Xu et al. 2014).

**Yeasts belonging to Ascomycota and Basidiomycota are common in deep-sea sediments** (Fell 2012; Singh and Raghukumar 2014).

– The basidiomycetous yeast *Rhodotorula mucilaginosa* has been isolated from the deepest part of the ocean, the Mariana Trench at a depth of 11,000 m (Takami 1999), in addition to the ascomycetes *Candida parapsilosis* and three species of *Metschnikowia* (Nagano et al. 2010). *Candida parapsilosis* is an opportunistic human pathogen. In addition to the above, it has also been isolated from the Central Indian Basin (Singh et al. 2012a). The three *Metschnikowia* species are generally associated with terrestrial plants and beetles. The basidiomycetous yeasts *Rhodotorula glutinis* and *R. mucilaginosa* were isolated from northwest Pacific Ocean sediments (Nagahama et al. 2001). Ascomycetous yeasts were more prevalent at depths above 2000 m in sediments collected from these sediments than below that depth.
– **Yeasts are associated with deep-sea benthic animals. A number of basidiomycetous red yeasts belonging to** *Rhodotorula, Sporobolomyces Cryptococcus*, and *Pseudozyma* have been isolated from giant white clams, a tubeworm, a crab and a mussel from deep-sea Pacific sediments. *Rhodotorula glutinis*, was isolated from benthic animals. Animal-associated yeasts were more abundant than those found in the sediments. The reason for this could be that nutrients are more easily available when associated with animals than with sediments (Nagahama et al. 2001). Twenty-two strains identified as members of *R. glutinis*, which showed a wide distribution in the deep sea, and five isolates identified as *R. minuta* were isolated only from benthic animals.

**Terrestrial fungi might have made an adaptive transition to deep-sea habitats and live as facultative marine fungi, as is evident from isolation in culture, metagenomics, and experimental studies.**

- **Filamentous fungi and yeasts are capable of growing under deep-sea conditions of high hydrostatic pressure and low temperature, although their growth rate may be slow** (Lorenz and Molitoris 1997; Raghukmar et al. 2012). Their capability to grow at various combinations of temperature and pressure varies.
  - Most terrestrial species of fungi isolated from the deep sea grow best at a room temperature of 30 °C and 1 bar, compared to other combinations of temperature and high pressure conditions (Damare et al. 2006a; Singh et al. 2010). These fungi grow to some extent at 200 bar pressure, when the temperature is maintained at 30 °C. Many terrestrial isolates are also capable of such growth. Most fungi isolated from the deep sea grow poorly at 200 bars and 5 °C. Hence, pressure *per se* might not be as inimical to growth as low temperature. A combination of high hydrostatic pressure and low temperature imposes an extreme condition of growth for fungi.
  - Several fungi have been shown to grow equally well under conditions of low temperature and high pressure as at 1 bar and room temperature. Deep-sea isolates of the terrestrial species belonging to *Aspergillus*, *Sagenomella* and *Exophiala*, obtained from sediments of the Central Indian Basin at a depth of ~5000 m, grew equally well at 200 bar and 5 °C, as at 1 bar and 30 °C. A few fungi, including 5 isolates of *Aspergillus,* grew better at 200 bar/5 °C than at 200 bar/30 °C (Damare et al. 2006a). In another study, three fungi, two corresponding to *Tilletiopsis albescens* and one to *Sagenomella*, grew well under 200 bar pressure at 5 °C as at 30 °C (Singh et al. 2010). All these are terrestrial species isolated from the deep sea. Several yeasts also grew well at the combination of high pressure and low temperature.
- **Fungal hyphae detected directly from deep-sea sediments appear to be normal in their morphology, in contrast to those from experimental studies** (Fig. 12.2a–c). Immunofluorescence using FITC-tagged polyclonal antibodies raised against the fungus *Aspergillus terreus* (# A 4634) isolated in culture from deep-sea sediments helped in detection of hyphae both in the sediments and in particulate organic material in the sediments. The morphology of such hyphae was normal (Damare et al. 2006a). However, the hyphae of this fungus were abnormal when the fungus was grown in a diluted Malt Extract Broth under simulated deep-sea conditions of 200 bar/5 °C (Fig. 12.2 D). It produced hyphae in place of metulae and conidia-bearing phialides on the surface of the vesicle. There might be several reasons for this. (1) Culture media may differ from *in situ* nutrients available in the deep sea, which may be more conducive for normal growth. Thus, in contrast to nutrient media, use of deep sea sediment extract to culture fungi under deep-sea conditions led to normal hyphae (Damare and Raghukumar 2008a). (2) Abnormalities may also be induced as a response to deep-sea conditions. Thus, a deep-sea isolate belonging to *Aspergillus terreus* #

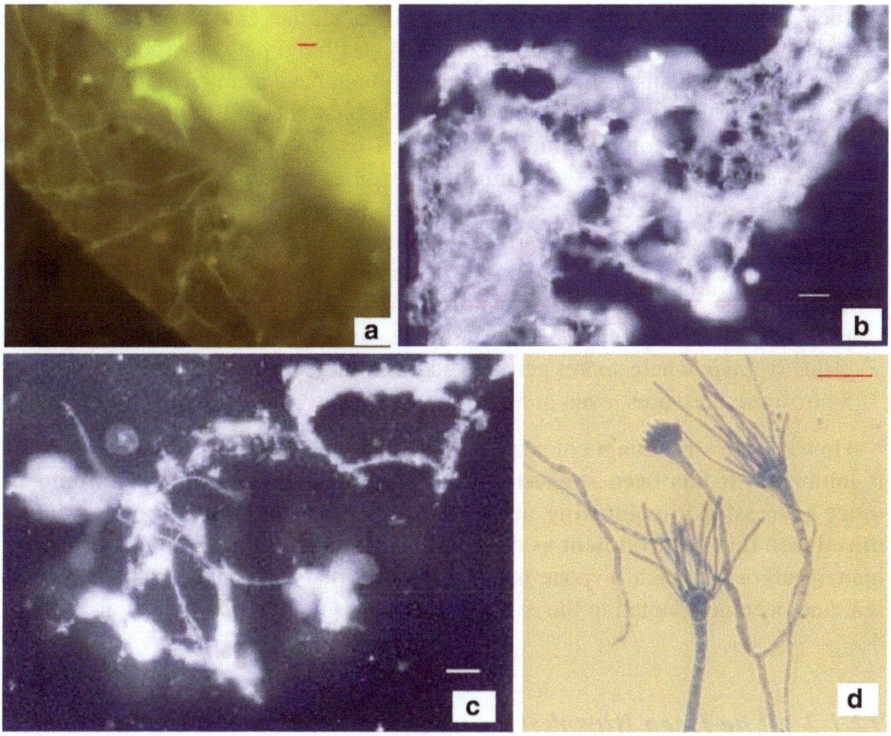


**Fig. 12.2** (**a**) Immunofluorescence detection of a dense cluster of hyphae of *Aspergillus terreus* in an organic particle from 15 to 20 cm core of a deep-sea sediment. Bar represents 10 μm. (**b**, **c**) The deep-sea fungal isolate *Aspergillus terreus* grown in CIB sediment extract at 200 bar 30 °C and viewed under an epifluorescence microscope after staining with Calcofluor. Bar represents 10 μm. (**b**) Hyphae concealed in aggregates. (**c**) Hyphae exposed after treatment with EDTA. (**d**) A deep-sea sediment isolate of *Aspergillus* sp. with abnormal morphology in culture grown at 200 bars and 5 °C. (**a**, **d**: Reprinted from Damare, S. et al. 2006. Fungi in deep-sea sediments of the Central Indian Basin Deep-Sea Research I 53. 14–27. ©2005 Elsevier Ltd. With permission from Elsevier; **b**, **c**: From Damare, S. and Raghukumar, C. 2008. Fungi and Macroaggregation in Deep-Sea Sediments. Microbial Ecology 56:168–177. ©Springer Science + Business Media, LLC 2007)

A 4634 showed normal mycelial growth at 1 bar pressure at 30° and 5 °C, while hyphal swellings and constrictions occurred at 30° and 5 °C/200 bar pressure.

- **An interesting role of soil aggregation has been suggested for fungi living in deep-sea sediments.** Consequent to the discovery that fungal hyphae could be detected more effectively by disrupting aggregates following EDTA treatment (see above), an experiment was carried out to see if the fungi themselves were responsible for the aggregation (Fig. 12.2b, c). Deep-sea fungi when grown in the laboratory at 200 bar and 5 °C in deep-sea sediment extract generated microaggregates, which stained for humic material (Damare and Raghukumar 2008a). Fungi may thus contribute to humus formation in deep-sea sediments. Such aggregates in terrestrial soils have a number of ecological implications, such as in preventing the leaching of extracellular enzymes and recycling organic matter. It remains to be seen if deep-sea fungi play any such major role.

- One of the mechanisms by which terrestrial fungi have adapted to the deep sea is alteration of the membrane composition (Simonato et al. 2006).
- **Germination of fungal spores is affected by low temperatures.** Spores of fungi derived from deep-sea sediments as well as those from terrestrial habitats germinate easily at 30 °C, even at a high hydrostatic pressure of 200 bars. However, they germinate poorly or not at all at 5 °C even at 1 bar pressure (Raghukumar and Raghukumar 1998; Damare et al. 2006a; Singh et al. 2010). Thus, low temperature is more inimical to germination than high hydrostatic pressures. An exception was noticed in the case of a deep-sea isolate of *Aspergillus sydowii*, where spores germinated when subjected to 100, 300, and 500 bar hydrostatic pressures, even at 5 °C.

How do terrestrial fungi colonize substrates in the deep sea if spore germination is inhibited? **It has been suggested that terrestrial species of fungi would be more successful in colonizing and growing in the deep sea if they are introduced into that environment as mycelia, rather than spores.** Particulate organic matter harboring fungal hyphae might be transported from land, sink to the deep sea, and become adapted to the extreme conditions therein (Damare et al. 2006a).

### *12.1.2 The Deep Biosphere*

The marine "deep biosphere," where microorganisms live in sediments deep below the seafloor, is an exciting habitat that has been discovered in the last 40 years (Wang et al. 2013). The discovery owes itself to the Integrated Ocean Drilling Program (IODP). The "deep biosphere" is found in the subsurface regions of marine sediments (Fig. 12.1). These sediments result from a constant sinking of particles from the overlying waters, which over a long period of geological time accumulate over the sea floor. Their thickness ranges from a few centimeters to nearly a kilometer in thickness depending on their proximity to the oceanic crust and cover almost the entire sea floor. A huge amount of carbon, equivalent of all plant life on Earth, is estimated to be stored in the deep biosphere. The deep subsurface comprises sediment layers with microbial communities that are distinct from those of the water column. The temperature increases along the sediment core at a rate of 30–50 °C per kilometer. The temperature at 1922 meters below the seafloor (mbsf)) was estimated to be in the range from 60 to 100 °C. Salinities in sediment samples near the seafloor are slightly lower than the salinity of normal seawater at 3.3%, and salinity rapidly declines to 3.0% at 28 mbsf.

**The deep biosphere is home to extremophilic Bacteria and Archaea.** The presence of bacterivorous protists that thrive on the bacteria or osmoheterotrophic microbial eukaryotes that utilize buried organic carbon is not well known. **Information on fungi in the "deep biosphere" has begun to appear in recent years.**

- **Fungi have been found to be the most dominant eukaryotes in samples taken 1.7 to 37.4 mbsf from sulfide-rich sediments of Peru Margin and the Peru Trench at 150 and 427 m water depth**, based on metagenomic studies using universal eukaryote primers (Edgcomb et al. 2011). Fungal sequences were recovered from both DNA- and RNA-based clone libraries. 18S rRNA clone libraries and cDNA-based clone libraries have shown fungi may be metabolically active and dominate the eukaryotic microbial community in the deep marine subsurface. Maximum eukaryotic diversity was present in shallowest sample at 1.75 mbsf. Basidiomycetes were the most frequent and diverse, particularly at the sediment surface. The basidiomycete yeast *Cryptococcus* appears to be abundant in this habitat. Two other yeasts, namely, *Malassezia* and *Trichosporon*, which are common opportunistic pathogens of deep-sea mammals, were also recovered.
- **The deep biosphere may be repositories of "paleobes" or ancient microorganisms.** Many microorganisms that are deposited on the seafloor may be gradually buried by the constant sedimentation. They may continue to be active or lie dormant in the subsurface sediments for thousands of years and may be amenable to retrieval by culturing. Such ancient organisms are called paleobes. Fungi have been cultured from subsurface ranging from 10 cm to a depth of 3.7 m of a core obtained from 5900 m depth in the Chagos Trench in the Indian Ocean (Raghukumar et al. 2004b). Radiolarian biostratigraphy indicated that the deepest samples from which fungi were cultured were 0.43 million years old. This is probably the oldest recorded age for recovery of culturable fungi. Fungal abundance ranged from 69 to 2493 CFU $g^{-1}$ dry weight sediment with a maximum abundance recorded at 160 cm depth of the core, corresponding to 0.18 Mya (million years ago). The fungus *Aspergillus sydowii* as well as a number of non-sporulating forms were isolated. These fungi might have been derived from past hyphae or spores that had been transported from land, buried, and preserved for long periods of time. Fungi in subsurface sediments are most likely the result of a continuous selection process that has eliminated all eukaryotes of water column or upper sediment origin that are not compatible with long-term survival under deep subsurface conditions. Study of paleobic fungi allows us a virtual time travel and helps us to bring fungal mammoths and dinosaurs to the present.
- **Fungi have been cultured from subsurface deep-sea sediments several hundred meters below the surface**. Subsurface cores from 4 mbsf to an astounding depth of 1884 mbsf of sediment lying at a water depth of 344 m from the Canterbury Basin in the eastern margin of the South Island of New Zealand yielded fungi in ten out of eleven samples (Fig. 12.3; Rédou et al. 2015). More than 200 isolates of mycetaen fungi belonging to 21 genera of terrestrial species of Ascomycota and Basidiomycota were obtained. These belonged mostly to mycelial forms (68%) and to a lesser extent to yeasts (32%). Maximum numbers occurred at 34 m below the surface. Treatment of samples under high hydrostatic pressure prior to culturing yielded more cultures than otherwise. Species of *Acremonium, Aspergillus*, *Cladosporium, Fusarium*, and

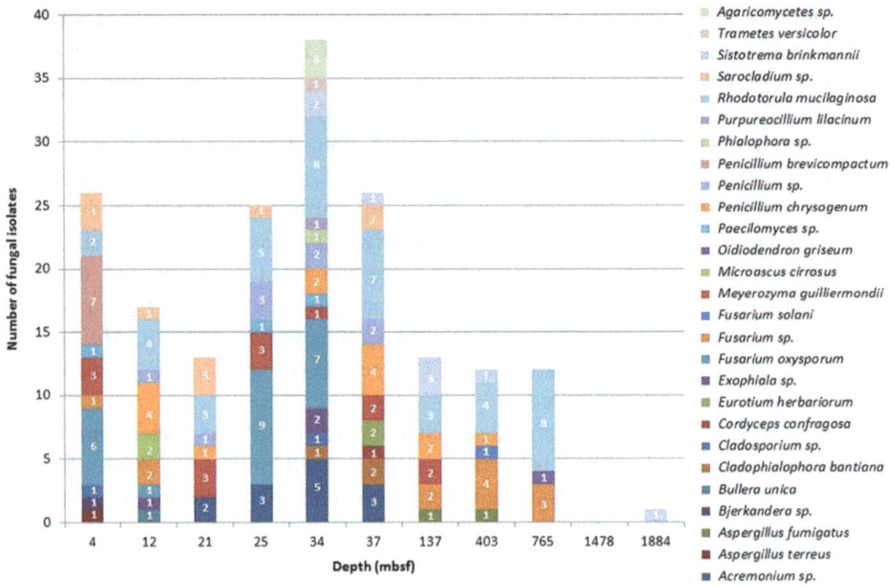


**Fig. 12.3** Diversity of fungi in the deep biosphere of eastern margin of the South Island of New Zealand, Canterbury Basin (Source: Rédou V, Navarri M, Meslet-Cladière L, Barbier G, Burgaud G. 2015. Species richness and adaptation of marine fungi from deep-subsea floor sediments. Appl Environ Microbiol 81:3571–3583. © 2015, American Society for Microbiology)

*Penicillium* were the most common. Among yeasts, the ascomycetous genus *Meyerozyma* and basidiomycetous genera *Rhodotorula* and *Bullera* have been found in the deep biosphere. The yeast *Rhodotorula mucilaginosa* was the most abundant. Fungi detected by culture-based and molecular approaches in the deep-subsea floor biosphere may vary from those that are buried alive and are simply dormant to those that adapt to subsurface conditions.

- The deep sea starts with the mesopelagic or the twilight zone at 200 m, from which depth onwards light is too feeble for photosynthetic production to take place.
- The deep sea is a region of high hydrostatic pressure and low temperature and darkness.
- Organisms in the deep sea depend almost entirely on the drizzle of food from above.
- Availability of oceanographic research vessels, sophisticated deployment and retrieval mechanisms, and appropriate samplers are important requirements that make deep-sea biology a highly specialized research topic.

(continued)

- Fungi appear to be the most dominant group of eukaryotes in deep-sea sediments.
- Yeasts and thraustochytrids are adapted to pelagic conditions because of their unicellular nature. They are also capable of growth on particulate organic matter.
- Labyrinthulomycetes are abundant in deep-sea waters of 200 to 3000 m in the Arabian Sea and the Equatorial Indian Ocean and the Atlantic, as revealed by culturing and molecular signatures. Deep-sea thraustochytrids apparently depend on POM generated in the mixed layer for their nutrition. The aplanochytrid, *Aplanochytrium kerguelense*, is the dominant species associated with the equatorial Indian Ocean zooplankton from surface, up to 1000 m deep-sea waters.
- Yeasts belonging to Ascomycota and Basidiomycota are common in deep-sea water and sediment samples. Basidiomycetous yeasts are more frequent in the deep sea, compared to ascomycetous yeasts. Yeasts are associated with deep-sea benthic animals.
- Fungal mycelium in deep-sea sediments confirms their active presence in these habitats.
- Fungi in deep-sea sediments may grow embedded within humic aggregates.
- Obligate marine, lignicolous fungi grow in wood deposited in the deep sea. Fungi occur within calcareous molluscan shells in the deep sea.
- Most filamentous fungi derived from culture or metagenomic methods belong to terrestrial species of Ascomycota and Basidiomycota. Zygomycota and Chytridiomycota are seldom encountered.
- Culture-independent methods have discovered a large diversity of hitherto uncultured and novel fungi in the deep sea, belonging to distinct environmental clone groups.
- Terrestrial fungi are capable of making the transition to deep-sea habitats and live as facultative marine fungi. Such fungi are capable of growth under deep-sea conditions of high hydrostatic pressure and low temperature. Low temperature conditions of the deep sea are generally inimical to germination of fungal spores.
- The unique, deep biosphere habitat of subsurface deep-sea marine sediments is inhabited by fungi, mostly belonging to terrestrial species. Subsurface cores of deep-sea sediments are potential repositories of paleobes.

**Future Directions**

1. Metagenomic studies have discovered a number of novel, uncultured deep-sea fungi. Novel culture methods should be developed to culture rare and unique fungi, which cannot be cultured using standard methods. Barophilic species are often lost using conventional methods for retrieval of samples.

2. Cells of thraustochytrids and aplanochytrids have been detected in the deep sea. However, few members of Labyrinthulomycetes have been cultured from the deep-sea sediments so far. Efforts should be made to bring deep-sea Labyrinthulomycetes into culture.
3. What is the role of fungi in calcareous animal shells in the deep sea?
4. What is the diversity of fungi in deep-sea sediments and what are their physiological adaptations to withstand high hydrostatic pressure and low temperatures?
5. It has been suggested that the deep sea may be a significant reservoir of saprobic fungi. What is the role of deep-sea fungi in degradation of organic matter in deep-sea sediments and what is their role in biogeochemical cycles?
6. How common are fungi in humic aggregates of deep-sea sediments?
7. Enormous quantities of wood become deposited in the deep sea. Mycetaen fungi are adept at lignocelluloses decomposition. Diversity of both obligate and facultative marine fungi and their role in deep-sea wood needs to be understood.
8. What is the diversity and role of fungi in the deep biosphere?

## 12.2 Hydrothermal Vents

In 1977, scientists aboard the submersible Alvin of Woods Hole Oceanographic Institution stumbled upon hydrothermal vents at a depth of 2500 m near the Galapagos Islands in the Pacific when they were exploring the deep-sea bed. This turned out to be one of the most stunning discoveries in deep-sea oceanography and unveiled an incredibly diverse and unique ecosystem.

Hydrothermal vents occur along mid-ocean ridges at the edges of spreading tectonic plates, where new crust is being formed (Fig. 12.2). Hydrothermal vents are formed when seawater penetrates through fissures and permeates into the deeper underlying rocks, close to the magma chamber. The water becomes heated, pressurized, and chemically transformed. It then becomes buoyant and gushes out of the ocean floor as a plume of super-heated, mineral-rich hydrothermal fluid. The water is typically bereft of dissolved oxygen, is strongly acidic with a pH of 2–3, has a high concentration of electron donors in the form of reduced compounds such as methane and hydrogen sulfide, and is rich in heavy metals. Some plumes reach a temperature of up to 350 °C and are black in color because of the high content of metal sulfides and sulfate particles. These are the "black smokers" (Fig. 12.4). The water does not boil even at these extreme temperatures because of the high hydrostatic pressure. The metals precipitate when the water cools down and form a chimney-like structure around the plume. Other vents, where the water permeates shallower depths of the crust, form plumes of lower temperatures, being around 6 to 23 °C. A gradient of temperature exists around the vents, ranging from the plume temperature to the ambient, deep-sea water temperatures of around 2 °C. Following the first discovery, numerous hydrothermal vents have been discovered in many oceans.

**Fig. 12.4** A black smoker vent with jets of particle-laden with fine-grained sulfide minerals. (Source: National Ocean Service, National Oceanic and Atmospheric Administration, US Department of Commerce. http://oceanservice.noaa.gov/facts/vents.html)

Chemolithotrophic bacteria, which oxidize sulfides in the vent fluids and fix carbon dioxide, support the ecosystem around the vents (Dick et al. 2013). Many of these are hyperthermophilic bacteria and archaea, which can grow at temperatures up to 121 °C, while others further from the fluid emissions grow at lower temperatures. Molecular studies are revealing an astonishing diversity of such organisms, many of which have biotechnological applications. An amazing diversity of animals, such as tubeworms, clams, mussels, shrimps, anemones, crabs and others are found around the vents, most of these harboring mutualistic, chemolithotrophic Bacteria and Archaea. The organic matter generated supports other chemoheterotrophic microorganisms. **Hydrothermal vents are hot spots of biodiversity. Fungi have been found to be inhabitants of this ecosystem.**

Deep-sea sampling at vent sites is generally carried out using Deep Submergence Vehicles (DSV) and Remotely Operated Vehicles (ROV).

**Fungi have been discovered in several shallow and deep-sea hydrothermal vent regions during oceanographic cruises** (Bass et al. 2007; Le Calvez et al. 2009; Burgaud et al. 2009; Colaço et al. 2006). Some of these regions are as follows.

1. Northeast Pacific hydrothermal vent environments of Guaymas Basin in the Gulf of California
2. East Pacific Rise at the Elsa site (12°48N, 103°57W; depth, 2630 m)
3. Lau Basin, South-west Pacific (20°3.0′S, 176°7.8′W; 2620 m); TAG (26°02′N, 44°54′W, 3630 m)

4. Mid-Atlantic Ridge at: The Menez Gwen site (37°51N, 31°31W; depth, 860 m); Lucky Strike site, adjacent to a fluid emission at the Tour Eiffel chimney (37°17N, 32°16W; 1700 m); Rainbow hydrothermal sediment (metal-rich, 2264 m deep; 36°08′N, 34°00′W); Lost City, located approximately 15 km off the Mid-Atlantic Ridge axis 30°N and at a depth of 750–900 m.
5. Shallow water hydrothermal vents at D. João de Castro Seamount in the North Atlantic.

### *12.2.1 Presence and Diversity of Fungi in Hydrothermal Vents*

**Actively growing hyphae of fungi have been found in shallow water hydrothermal vents and a large diversity of fungi cultured therefrom.** D. João de Castro Seamount in the North Atlantic contains active shallow hydrothermal vents. The yellow zone of these vents is characterized by high concentrations of Fe, Ba, and Mn and has a temperature of around 60 °C. Melanized fungal hyphae were detected in organic particles in this zone (Colaço et al. 2006; Raghukumar et al. 2008a).

- **Many terrestrial species of fungi such as species of *Aspergillus* and *Cladosporium* have been isolated from the yellow zone of shallow water hydrothermal vents of D. João de Castro in North Atlantic.** The white zone which is characterized by high content of $H_2S$, $CH_4$, $Hg_2$, Pb, and Co and temperature of ~37 °C yielded comparatively less fungi. In addition, non-sporulating fungi with dark brown septate hyphae were also isolated from this vent.
- **Thraustochytrids are prevalent in these shallow hydrothermal vents.** A thraustochytrid isolate from the vent region showed better growth in the presence of metals, and its protease was not inhibited in the presence of S, Mn, Fe and Pb. Tolerance to high NaCl concentration and optimum activity of the protease in one of the thraustochytrid isolates at 45 °C further indicated its strategy to participate in the degradation of detrital material around the hydrothermal vent site.

**Yeasts are prevalent in hydrothermal vent habitats** (Fell 2012).

- Yeast densities ranged from 0 to 5940 CFU per liter in water samples collected immediately above sediments around hydrothermal vents of the Mid-Atlantic Rifts southwest of the Azores archipelago (Gadanho and Sampaio 2005). Both marine- and terrestrial species of yeasts were cultured. Pigmented yeasts are generally more common in marine regions. However, these hydrothermal vent habitats unusually harbored more nonpigmented yeasts. It has been surmised that this might be because of the rich animal and microbial diversity and a greater availability of utilizable organic compound in these habitats. These non-pigmented yeasts consisted of *Candida atlantica, C. atmosphaerica, C. lodderae, C. parapsilosis* and *Pichia guilliermondii*. Pigmented yeasts include *Rhodosporidium diobovatum, R. sphaerocarpum, R. toruloides*, and *Rhodotorula mucilaginosa*. Eight of the yeasts appeared to be novel phylotypes, five belonging to the genus *Candida* of Ascomycota and three to *Rhodotorula* of Basidiomycota.

- *Cryptococcus* sp. and *Debaryomyces hansenii* have been detected using molecular sequences. A number of Basidiomycota phylotypes belonged to the yeast genus *Cryptococcus*, a phylotype previously detected in another hydrothermal vent and the *Filobasidium* anamorph (Le Calvez et al. 2009; Bass et al. 2007).
- Fe-oxide mats and rock surfaces from the crater of the volcanically active Vailulu'u seamount (Samoan chain) were sampled within six months of its most recent eruption in November 2004 (Connell et al. 2009). A higher number of pigmented yeasts such as *Dioszegia antarcticum, Rhodotorula graminis, Sporidiobolus salmonicolor, and Rhodosporidium toruloides* were isolated compared to non-pigmented ones such as *P. guilliermondii, Cryptococcus saitoi*, and *Clavispora lusitaniae*. All seven species of yeasts in this study showed strong siderophore production, a class of molecules used to acquire and utilize Fe(III). One of the isolates, *Rhodotorula graminis*, oxidized Mn(II) to Mn(IV), suggesting a functional role of fungi in sea-floor alteration and cycling of metals under such extreme conditions.

**Metagenomics are providing insights to our understanding of fungal diversity in hydrothermal vents. Sequences of a number of unique and novel fungal phylotypes have been retrieved from hydrothermal vents** (Table 12.2).

- **The "Hy-An" or "Hydrothermal-Anoxic" environmental sequence group encompassing a number of novel phylotypes is typical of hydrothermal vents.** This is most closely related to the basidiomycetous yeast "*Malassezia*." For example, sequences of many phylotypes closely related to *Malassezia furfur* were retrieved from Lost City, which is a unique hydrothermal vent field in the mid-Atlantic ridge, characterized by fluids that are slightly less hot (generally 200–212 °C, up to 333 °C in black smokers) and hydrogen sulfide-rich (López-García et al. 2007). Uncultured clones related to *Malassezia furfur* were also the most frequent at the Rainbow hydrothermal vent site in the mid-Atlantic ridge (Bass et al. 2007). Sequences relating to this group have also been found to occur in the Atlantic and Pacific ridges by others (Edgcomb et al. 2002; Le Calvez et al. 2009).
- **Labyrinthulomycetes are represented by the environmental sequence group "MAST-9" comprising novel phylotypes belonging to these straminipilan fungi.** Clones belonging to MAST-9 were retrieved from the mid-Atlantic ridges of Lost City (López-García et al. 2007). This group occurs mostly in hydrothermal vents, probably being adapted to anoxic or microoxic habitats (Massana et al. 2004a, b). A number of novel phylotypes related to thraustochytrids and aplanochytrids have also been reported from **hydrothermal vent environments of Guaymas Basin in the Gulf of California** (Edgcomb et al. 2002).

- **Hydrothermal vent habitats are possible sources of novel mycetaen fungi.** Out of 20 distinct phylotypes detected from hydrothermal vents at the East Pacific Rise at the Elsa site and the Mid-Atlantic Ridge at the Menez Gwen sites, 4 ascomycetes, 3 basidiomycetes, and 2 chytridiomycetes that appeared to be hitherto undescribed taxa were detected using small-subunit (SSU) rRNA gene sequences (Le Calvez et al. 2009).
- The Cryptomycota environmental group, comprising *Rozella* and LKM 11 cluster, is an enigmatic group at the base of the fungal lineage (Sect. 1.2). A novel phylotype placed in deep branches within the phylum Chytridiomycota with *Rozella* spp. as the closest related organism has been recovered from The East Pacific Rise at the Elsa site and the Mid-Atlantic Ridge at the Menez Gwen sites (Le Calvez et al. 2009) (Table 12.2).

## *12.2.2 Fungi in Hydrothermal Vent Animals*

**Many hydrothermal vent animals harbor fungi.**

- **Mycetaen fungi have been detected by culturing from hydrothermal vent animals** such as shrimps, mussels, alvinellids, and tubeworms, in addition to sediments and smoker rock scrapings, collected from a number of sites at Pacific and Atlantic (see above) (Burgaud et al. 2009). Most fungi were recovered from deep-sea hydrothermal shrimps and the deep-sea mussel *Bathymodiolus azoricus*, particularly from inside animal (from flesh) and outside (from shell and byssi). These belonged to terrestrial Ascomycota and Basidiomycota, including *Aspergillus sydowii* and other *Aspergillus* species, *Eurotium herbariorum*, *Exophiala*, *Aureobasidium pullulans*, *Lecythophora hofmannii*, *Paecilomyces lilacinus*, *Lecanicillium lecanii*, *Geomyces pannorum*, *Hyphodiscus hymeniophilus*, and *Helicodendron paradoxum*. Most of these were psychotrophs which could grow at low temperatures but also above 20 °C. Molecular detection using FISH (Fluorescent *in situ* hybridization) showed that some of these yeasts were resident inside the vent animals. These yeasts may have a role in the decomposition of organic matter entrapped in the filaments of mussel byssi (Burgaud et al. 2010).
- **There is evidence that some fungi are pathogens of hydrothermal vent animals** (Moreira and López-García 2003). **Metagenomic studies have revealed several novel phylotypes close to known pathogens.** Two unusual species, namely, the clinically important *Exophiala dermatitidis* and *Trichosporon dermatis*, were also cultured from hydrothermal vents of the Mid-Atlantic Rifts southwest of the Azores archipelago (Gadanho and Sampaio 2005). **The environmental sequence group 'Hy-An" or "Hydrothermal-Anoxic" environmental clones are most closely related to the yeast *Malassezia furfur* that has been found in many vent areas** (Table 12.2). These sequences were retrieved from Lost City hydrothermal vents in the

mid-Atlantic ridge (López-García et al. 2007). Uncultured clones related to *Malassezia furfur* were also the most frequent at the Rainbow hydrothermal vent site in the mid-Atlantic ridge (Bass et al. 2007). It remains to be seen if the presence of *Malassezia* spp., which are well-known infectious agents of skin diseases in marine mammals (see Sect. 9.3.1), has any implication of diseases in hydrothermal vent animals. ***Candida parapsilosis*** is an opportunistic human pathogen. This yeast had been isolated previously from water samples collected in hydrothermal fields as well (Gadanho and Sampaio 2005).

**An infection of the hydrothermal vent mussel *Bathymodiolus brevior*, probably caused by the black yeast *Capronia***, has been reported from Hill hydrothermal vent in Fiji Basin at a depth of 1990 m (Van Dover et al. 2007). The infection appears as scattered spots of brownish discoloration of mantle tissues that progresses to more discolored bodies. In extreme cases, the tissues turn black and emit a distinctive odor. Tissue deterioration is accompanied by a pronounced host immune response. The association of *Capronia* with the infection has been confirmed by culturing and identified by the SSU rDNA sequence. The yeast was also detected in mussel tissue by fluorescent *in situ* hybridization. Severe infection of the mussel is feared to increase to an epizootic proportion in these vents. The same mussel species in another vent region was not infected nor were other invertebrates in the same region, such as gastropods, limpets, and polychaetes.

**A gill infection of an endemic, hydrothermal vent gastropod** by an unidentified ascomycete has been reported from vents along the upper continental slope of the Gulf of Mexico.

## 12.3 Hypoxic and Anoxic Habitats

Fungi are mostly aerobic and are therefore seldom talked about when discussing microorganisms in hypoxic or anoxic environments. However, the anaerobic mode of life is not absent in fungi. Yeasts are well known for their anaerobic, fermentative activities. Fungi are capable of dissimilatory nitrate reduction under suboxic conditions. Fungi have been shown to account for nearly 80% of the nitrous oxide production in a grassland ecosystem (see Jebaraj and Raghukumar 2009). The importance of the obligate anaerobic fungi belonging to Neocallimastigomycota in the anaerobic rumen microbiology is well recognized (Webster and Weber 2007). There is now increasing evidence for occurrence and activity of fungi also in low oxygen extreme environments in marine ecosystems (Jebraj et al. 2012).

**Hypoxia and anoxia are extreme conditions found in several parts of the oceans.** Most parts of the oceans are well oxygenated, the average oxygen contents being 4–6 mg $L^{-1}$. However, hypoxic and anoxic conditions occur in many parts of the ocean, where the oxygen levels may drop much below 2 mg $L^{-1}$ (Fig. 12.5a, b).

a

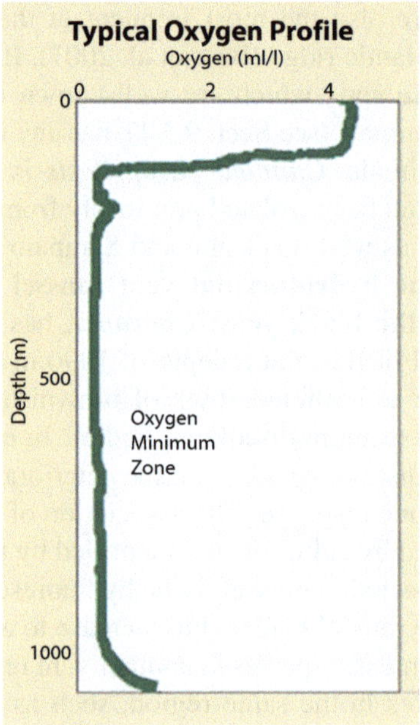


b

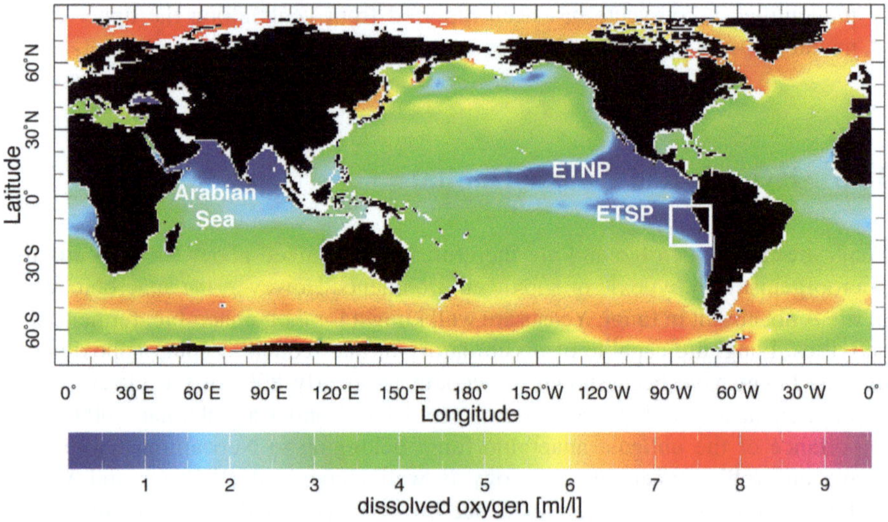


**Fig. 12.5** Hypoxic and anoxic oceanic environments. (**a**) An oxygen profile at an oceanic location with an oxygen minimum zone (OMZ) (Source: Wishner, K. 2000. Zooplankton in the deep sea. Maritimes 42:9–13). (**b**) Major oceanic oxygen minimum zones (OMZs) in the world (Source: http://iridl.ldeo.columbia.edu/, IRI/LDEO Climate Data Library, Columbia University; original raw data from World Ocean Atlas 2005, NODC, NOAA)

(1) Hypoxic and anoxic conditions are caused in regions of high organic matter production, where heterotrophic microbial degradative activities result in a rapid utilization of oxygen and where strong water column stratification restricts water exchange and prevents oxygen replenishment. The Oxgen Minimum Zone (OMZ) occurs usually at 200 to 1000 m water depth. Although OMZ waters constitute only about 0.1% of the total ocean volume in the world, 20–40% of total oceanic nitrogen loss is estimated to occur therein. Some examples of anoxic zones are in the eastern Pacific Ocean and Arabian Sea (Naqvi 1994).
(2) **Anoxic to hypoxic conditions prevail in many "methane cold seeps"** (Levin 2005; Singh et al. 2011). As with hydrothermal vents, seawater in cold seeps percolates into the crust through fissures on the seafloor caused by plate tectonics. Subsequently, fluid with high concentrations of methane, sulfide, and hydrocarbons seep out of these fissures, get diffused by sediment, and spread over an area several hundred meters wide. Seepage of this fluid in different areas of the deep sea often results in the formation of brine pools. Most of these habitats are located in the Atlantic, Eastern and Western Pacific Ocean, and Mediterranean Sea. Cold seep habitats are characterized by carbonate rocks and reefs. Seep sediments possess a shallow, anoxic layer that is less than 10 cm thick. Chemosynthetic bacteria support rich animal communities.

The Arabian Sea contains one of the world's largest perennial OMZs in the intermediate waters between 200 and 1000 m depth in the open ocean. This OMZ is the result of upwelling of bottom water nutrients and high productivity during the south-west and north-east monsoons (Warren 1994; Naqvi 1994). It is also the region of a seasonal hypoxic zone along the western coast of India at the end of the southwest monsoon during September and October, when a strong water column stratification along the continental shelf down to 200 m depth takes place (Naqvi et al. 2006). Bacterial dentrification plays a major role in such hypoxic environments.

### *12.3.1 Diversity of Fungi in Hypoxic and Anoxic Marine Environments*

**A number of terrestrial species of Ascomycota and Basidiomycota have been cultured or detected using metagenomic methods from anoxic habitats.**

- Twenty three of the 35 phylotypes detected at 850–1200 m depth in methane cold-seep area in Sagami-Bay using **fungal-specific PCR-based analysis of environmental DNA were similar to known species of Ascomycota and Basidiomycota and which had also been reported from other deep-sea studies** (Nagahama et al. 2011).

- Fungal diversity from five different sediment DNA samples in methane hydrate-bearing deep-sea marine sediments in the South China Sea, based on internal transcribed spacer (ITS) regions of rRNA genes were similar to those of *Phoma glomerata, Hortaea werneckii* and *Pichia ohmeri* (Lai et al. 2007).
- Likewise, most of the fungi cultured from seasonally anoxic coastal as well as the perennially anoxic offshore sediments of the Arabian Sea, under aerobic as well as anaerobic conditions, belonged to terrestrial species, particularly *Aspergillus* and *Humicola* (Jebaraj and Raghukumar 2009; Jebaraj et al. 2010). Metagenomics also revealed the presence of the terrestrial fungi *Penicillium namyslowskii, Fusarium oxysporum*, and the ascomycetous yeast *Kodamaea* sp.
- Most phylotypes obtained from sediment cores from the Alaminos Canyon 601 methane seep in the Gulf of Mexico at 2400 m belonged to known species (Thaler et al. 2012).

**Anoxic environments may be home to novel and unique fungi. Standard culture methods may not be suitable to isolate many fungi that grow under such extreme conditions. Metagenomic methods have begun to unravel the unknown fungal diversity of anoxic habitats** (Table 12.2; Jebraj et al. 2012).

- **The environmental sequence group "Hy-An" or "hydrothermal and/or anaerobic fungal group," related to the basidiomycetous yeast *Malassezia*, is one of the most prevalent novel sequences in anoxic habitats** (López-García et al. 2007). The Hy-An group has been found in many anoxic and hypoxic habitats as well as in hydrothermal vent habitats. Signatures of this group of fungi have been found in the permanent offshore OMZ as well as seasonal anoxic regions of the Arabian Sea (Jebaraj et al. 2010; Manohar and Raghukumar 2013), such as anoxic intertidal sediments (Dawson and Pace 2002), a deep-sea hypersaline anoxic basin (DHAB) in the Mediterranean Sea (Alexander et al. 2009), anoxic salt marsh water (Stoeck and Epstein 2003), anoxic Norwegian Fjord, anoxic aquifer sediment (Brad et al. 2008), a hypersaline anoxic Mediterranean deep-sea basin (Alexander et al. 2009), anoxic coastal sediments (Dawson and Pace 2002) and other deep-sea sites (Bass et al. 2007).
- **The DSF Group 1, most closely related to the ascomycetous yeast *Metschnikowia bicuspidata*, is an environmental sequence group typical of the deep-sea, anoxic habitat. It may represent a novel fungus.** Sequences of this group have been exclusively reported from different deep-sea environments (Bass et al. 2007; Nagano et al. 2010; Nagahama et al. 2011; Thaler et al. 2012). DSF-Group1 was frequently observed at 1200 to 10,000 m sediments from five different sites off Japanese islands. DSF Group 1 sequences have been detected from the deep-sea, methane cold-seep of Sagami Bay, Japan, at a depth of 1170 m as well as from anoxic white bacterial mat in the Gulf of California at a depth of 1575 m (Takishita et al. 2007; Bass et al. 2007). This was also the dominant fungal group in sediment cores from the Alaminos Canyon 601 methane seep in the Gulf of Mexico at 2400 m (Thaler et al. 2012). The constant association of this group with anoxic habitat suggests that it is either a facultative

anaerobe or lives in association with other organisms generally found in this environment. *Metschnikowia bicuspidata* is a parasite in freshwater *Daphnia* species. It is probably also parasitic in planktonic animals living in deep-sea environments (Nagano et al. 2010).

- **Sequences of phylotypes belonging to the LKM 11 cluster, most closely related to the basal fungal clade Cryptomycota, are some of the most common ones recovered from anoxic, marine habitats**. As mentioned earlier, the Cryptomycota along with Microsporidia share the root of the earliest divergent fungi. Cryptomycota comprises the **endoparasitic** genus *Rozella* (Lara et al. 2010). Members of the LKM11 clade have appeared from aquatic (mostly freshwater) environments and marine environments (Savin et al. 2004; Takishita et al. 2007). The LKM11 clade appears to be associated predominantly with anoxic environments and may participate actively in this ecosystem (van Hannen et al. 1999; Takishita et al. 2005; Takishita et al. 2007).
- The KD14 clade consists of a clone group that is phylogenetically isolated from known lineages and is associated with the anaerobic environment. These clones have been detected by various researchers from sediments of anaerobic bacterial mat (Bass et al. 2007), methane cold seep (Takishita et al. 2007; Nagano et al. 2010), Svalbard in Arctic, and the Sea of Marmara (see Nagahama and Nagano 2012).
- The SCG1 group or the Soil Clone Group represents a distinct monophyletic group comprising only the environmental DNA clones. This was reported for the first time from the Pacific methane cold-seep area in Sagami-Bay samples (Nagahama et al. 2011).
- A new fungus, *Tritirachium candoliense*, was described from coastal anoxic waters of the Arabian Sea (Manohar et al. 2014). This species grows slowly under anoxic conditions and produces mainly hyphae with only few blastoconidia. Electron microscopy revealed differences when the culture was exposed to anoxic stress. Notable ultrastructural changes occur in the structure of mitochondrial cristae, irregularly shaped fat globules, and the presence of intracellular membrane invaginations.
- **A number of novel Ascomycota and Basidiomycota inhabit anoxic habitats**. Highly novel lineages related to *Entorrhiza* sp. *Penicillium chrysogenum*, *Phoma* sp., *Cladosporium* sp., and *Geomyces* sp. and agaricomycetous fungi were recorded from the Pacific methane cold-seep area in Sagami-Bay samples (Nagahama et al. 2011). A number of such novel sequences were detected in 25 m depth sediments during the seasonally anoxic period of October from the Arabian Sea as well as 200 m sediment and overlying water from the permanent OMZ using a multiple primer approach (Jebaraj et al. 2010). A study based on 18S rRNA sequences on eukaryotes present in black, reducing sediments of intertidal mud in California revealed that 20% of the novel phylotypes belonged to mycetaen fungi, being next only to alveolates (Dawson and Pace 2002). Fungal diversity based on internal transcribed spacer (ITS) regions of rRNA genes from five different sediment samples in methane hydrate-bearing deep-sea marine sediments in the South China Sea at about 3000 m depth revealed a high

prevalence of a novel yeast with <97% similarity to *Lodderomyces*. Other yeasts were *Malassezia*, *Cryptococcus*, and *Candida*, besides the filamentous fungi *Cylindrocarpon* and *Aspergillus* (Lai et al. 2007).

- **A number of novel straminipilan fungi have been detected in anoxic habitats.** Novel labyrinthulomycetes related to *Aplanochytrium* as well as others that form an ancestral lineage to these fungi were detected in the Cariaco Basin off the Venezuelan coast, the world's largest, truly marine permanently anoxic basin, using an 18S rRNA approach (Stoeck et al. 2003). Novel sequences of oomycetes, hyphochytriomycetes, and labyrinthulomycetes have also been detected in intertidal mud in California coast (Dawson and Pace 2002). These were most closely related to *Aplanochytrium minutum* (*Labyrinthuloides minuta*) and probably belonged to the environmental sequence group MAST-9.

### *12.3.2 Activity of Fungi in Anoxic Environments*

**Metabolically active mycetaen and straminipilan fungi are present in marine, hypoxic and anoxic habitats.**

- **Actively growing fungal hyphae have been detected in coastal, seasonally anoxic sediments of the Arabian Sea along the west coast of India, using the Calcofluor direct detection method** (Jebaraj and Raghukumar 2009). Fungal biomass in four samples obtained from different depths of sediment cores at the two stations varied widely and was lower than that of bacteria. The highest fungal biomass attained nearly half the maximum bacterial biomass. Bacterial biomass ranged from 1.2 to 500 × $10^3$ pg C $g^{-1}$. Fungal biomass varied between 0.01 and 0.206 × $10^3$ pg C $g^{-1}$ sediment.
- **Labyrinthulomycetes are common in the permanent OMZ of the Arabian Sea** at depths of 200–1000 m (Raghukumar et al. 2001). Thraustochytrids were regularly found below 150 m (Fig. 11.5). Their existence in the intermediate layers assumes special importance in the context of the Arabian Sea. Dense populations of thraustochytrids were detected even at depths up to 2000 m. Similar to the upper 150 m, thraustochytrids below 200 m were least frequent in the summer pre-monsoon (present in 49% of samples), although high numbers, up to 266 × $10^3$ cells $L^{-1}$, were found at 1000 m in one station. Their numbers in the OMZ ranged from 0 to 38 × $10^3$ $L^{-1}$ during the end of the summer and winter monsoons. Both thraustochytrids and bacteria showed a distinct peak at 250–500 m in the OMZ during these 2 seasons. Dense populations of thraustochytrids were detected even at depths up to 2000 m. Culturable thraustochytrids are also abundant in coastal sediments, even during the peak anoxic post-monsoon period.

**A variety of anaerobic respiratory modes exist in marine, anoxic, and hypoxic habitats.** Nitrate is sequentially reduced to nitrite and further to dinitrogen by a consortium of microorganisms in the case of denitrification. Dissimilatory nitrate reduction (DNRA) is a process whereby nitrate is sequentially reduced to ammonium.

Anaerobic ammonium oxidation (anammox) is a process in which ammonium is oxidized by nitrite to form dinitrogen. Facultatively anaerobic bacteria capable of these modes play an important role in the biogeochemical cycle of anoxic waters and sediments in oceans. Denitrification and DNRA have been discovered in a limited set of eukaryotic microorganisms, including marine foraminifers and diatoms in recent years. The role of fungi in such anoxic habitats is not yet clear.

**Mycetaen fungi may play an active role in denitrification in marine anoxic habitats** (Manohar and Raghukumar 2013).

- **Five terrestrial species of fungi isolated during seasonal anoxic conditions from coastal Arabian Sea grew under experimental anaerobic conditions** in a nutrient medium (Jebaraj and Raghukumar 2009). Of these, an isolate of *Fusarium, Aspergillus terreus*-# An-2, and *Tritirachium*-# 11 (later described as a new species, *Tritirachium candoliensis*) grew equally well under both aerobic and anaerobic conditions. *Penicillium* sp. and *Aspergillus* sp. are ubiquitous fungi from anoxic zones that are capable of anaerobic denitrification (Takasaki et al. 2004; Jebaraj and Raghukumar 2009).
- **Two fungi accumulated significant amounts of nitrite** in the culture medium when grown under anaerobic conditions.
- **Ammonia formation has been observed in cultures maintained under anaerobic conditions** (Jebraj and Raghukumar 2009; Stief et al. 2014). Four fungi generated ammonia under anaerobic conditions. *Aspergillus terreus* # An-4 was capable of dissimilatory nitrate reduction to ammonium under anoxic conditions (Fig. 12.6). A $N_{15}$-labeling experiment proved that An-4 produced and excreted ammonium through nitrate reduction. The products of dissimilatory nitrate reduction were ammonium (83%), nitrous oxide (15.5%), and nitrite (1.5%), while dinitrogen production was not observed. This process in fungi appears to be widespread as 15 of 17 fungi tested by Zhou et al. (2002) showed ammonia formation under anaerobic condition.
- **Many fungi store nitrate intracellularly and carry out dissimilatory nitrate reduction when oxygen is absent.** *Aspergillus* # An-4 used intracellular nitrate stores (up to 6–8 μmol $NO^3$ $g^{-1}$ protein) for dissimilatory nitrate reduction.

In the words of Stief et al. (2014), "In the currently spreading oxygen-deficient zones in the ocean, an as yet unexplored diversity of fungi may recycle nitrate to ammonium and nitrite, the substrates of the major nitrogen loss process anaerobic ammonium oxidation, and the potent greenhouse gas nitrous oxide." Future research will confirm how important fungi are in this process.

**The role of fungi in relation to other organisms in anoxic habitats remains to be explored.**

- The terrestrial basidiomycetous yeast *Cryptococcus curvatus*, which is an opportunistic pathogen of humans and animals, was detected based on SSU rDNA from 0 to 9 cm sediment samples at the Kuroshima Knoll at a depth of 642 m (Takishita et al. 2006). This knoll is characterized by emission of methane and

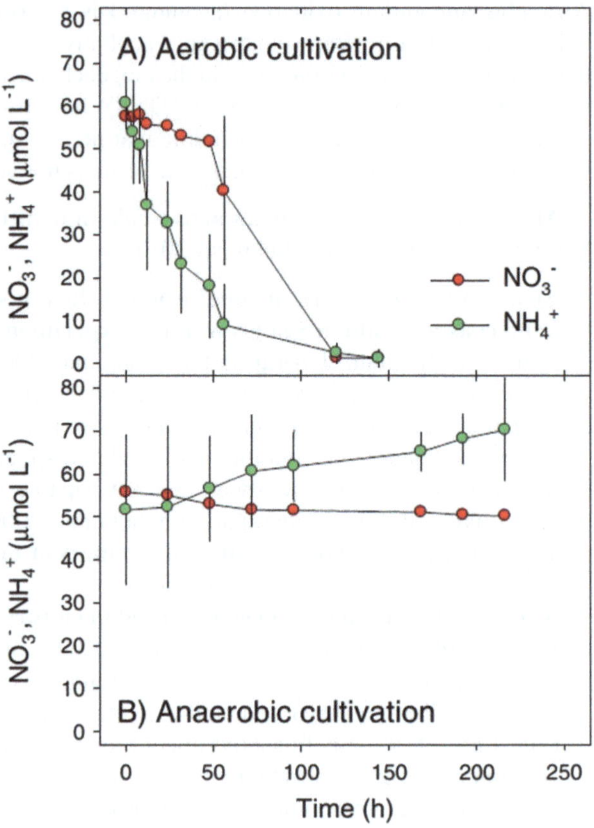


**Fig. 12.6** Time course of nitrate and ammonium concentrations during axenic cultivation of *Aspergillus terreus* isolate #An-4 (Source: Stief, P. et al. 2014. Dissimilatory nitrate reduction by *Aspergillus terreus* isolated from the seasonal oxygen minimum zone in the Arabian Sea. BMC Microbiology 2014, 14:35. Courtesy: Creative Commons Attribution)

hydrogen sulfide through fissures in the large carbonate pavement of this knoll. It is possible that this yeast also infects bivalves found in the same sediment.

- Another possible parasite in animals of anoxic habitats is the yeast *Malassezia,* a common parasite in marine animals (Sect. 11.2). *Malassezia* phylotypes are ubiquitous in anoxic environments, but have not been cultured. The reason for this might be that these yeasts require anaerobic conditions for growth (López-García et al. 2007). Species of this genus are potential parasites of small marine invertebrates, such as nematodes or polychaetes in anoxic zone sediments (Edgcomb et al. 2011).

## 12.4 Polar Environments

Life in the polar oceans of the Arctic and Antarctic, symbolized by penguins, seals, walruses, and polar bears, has always fascinated human beings. These waters, characterized by extreme low temperatures and intense seasonal variations in primary production, impose extreme conditions on organisms living in them or depending on them. Despite these common features, the Arctic and Antarctic differ in some key aspects. The Arctic Ocean is a land-locked basin and receives a large amount of fresh water influx. The Antarctic or Southern Ocean surrounds an ice-covered continent and is distinctly separated from lower latitude waters by a circumpolar front. "Sea ice" and "polynyas" in polar oceans harbor unique life forms.

Sea ice is a characteristic of polar oceans (Hollibaugh et al. 2007). Sea ice cuts off light to the water column below it and provides a matrix for special life forms. Freezing water excludes salts which form channels of brine within the sea ice, where both the salt content and dissolved organic matter are high. Sea ice provides an important matrix for prokaryotes, as well as a number of eukaryotes, which are termed the "Sea ice microbial communities" (SIMCO). Polynyas are persistent open water habitats, which harbor characteristic organisms (Fig. 12.7).

Fungi in the Antarctic continent have been studied by a few (see Ruisi et al. 2007; Newsham et al. 2015). Lichens, endoliths within rocks, and soil fungi have been well documented. However, the polar environment is probably the least studied with respect to marine fungi. This is in contrast to Bacteria, Archaea, and

**Fig. 12.7** Antarctic sea ice. A trail of open water left by a research ship and an iceberg in the background are seen (Credit: Ted Scambos, NSIDC https://nsidc.org/cryosphere/icelights/2012/01/sea-ice-down-under-antarctic-ice-and-climate)

phytoplankton that have been intensively studied in polar oceans. Hence, our knowledge on fungi in polar waters is very limited. One of the early significant findings was the discovery of thraustochytrids inhabiting polar waters. Günther Bahnweg, working with the eminent mycologist Prof. J.F.K. Sparrow, studied the taxonomy and ecology of these fungi in Antarctic waters and discovered a number of new species.

### *12.4.1 Fungi in Diverse Polar Habitats*

Arctic and Antarctic polar waters, sea ice, and allochthonous wood are some common habitats of polar fungi. Both obligate and facultative marine fungi occur in these. Many marine-derived fungi have also been reported.

**Thraustochytrids and yeasts are found in polar waters.**

- Four new species of thraustochytrids were discovered from subantarctic and Antarctic waters of the southeastern Indian Ocean, the southwestern Pacific Ocean, and the Antarctic Ross Sea during two cruises to these regions in the early 1970s (Bahnweg and Sparrow 1974a, b). These were *Thraustochytrium antarcticum* Bahnweg & Sparrow, *T. kerguelensis* Bahnweg & Sparrow, *T. amoeboideum* Bahnweg & Sparrow, *and T. rossii* Bahnweg & Sparrow. They were also cultured from subtropical, subantarctic, and Antarctic regions of the Southern Ocean, by baiting samples with pine pollen or *Artemia* larvae (brine shrimp). The latter bait was found to be more effective. Culturable numbers ranged from 1 to 100 cells $L^{-1}$ water. Highest numbers were found near the Antarctic convergence zone. Although mycetaen fungi were recovered in culture in high numbers with proximity to continents or islands, thraustochytrid fungi did not show a similar increase. Thraustochytrids were more abundant below the photic zone at 25–250 m. These results suggested that they were saprotrophic and grew on detritus derived from the photic zone.
- **High abundance of thraustochytrids has been reported in Arctic surface water samples from the Greenland and Norwegian Seas.** Thraustochytrid populations ranged from $<8.1 \times 10^2$ to $2.3 \times 10^5$ cells $L^{-1}$, with an overall average of $3.1 \times 10^4$ cells $L^{-1}$ (Naganuma et al. 2006). In comparison, bacterial numbers varied from $2.2 \times 10^7$ to $6.0 \times 10^8$ cells $L^{-1}$, with an overall average of $2.1 \times 10^8$ cells $L^{-1}$. Particulate ATP concentration ranged from 79 pmol to 676 pmol at the mid-transect (68°N) and southernmost (62°N) sites, respectively, with an average of 222 pmol. Particulate ATP was correlated with the abundance of bacterioplankton, but not with that of planktonic thraustochytrids. This is probably related to differences in nutrient and substrate preferences of the two groups of organisms.
- **Basidiomycetous yeasts appear to be prevalent in polar waters.** Yeasts belonging to the genera *Rhodotorula*, *Cryptococcus*, *Debaryomyces*, and *Candida*, which are common in deep-sea waters also seem to be the most common in polar waters

(Zalar and Gunde-Cimerman 2014). The basidiomycetous yeast *Leucosporidium antarcticum* has been found in concentrations of 1 to 22 CFU $L^{-1}$ in Antarctic waters adjacent to ice packs. The yeast could be isolated in waters up to a depth of 3000 m (Jones and Fell 2012).

**Sea ice is an interesting habitat for fungi.**

- Sea ice form seasonally and have temperatures ranging from −1 to −50 °C in winter. Sea ice are permeated by a network of channels and pores, which are filled with brine. Salt concentrations of sea ice brines can reach up to 20% NaCl, owing to which the brine remains liquid even at −35 °C. Metagenomic biodiversity studies of Arctic sea ice samples have shown that fungi comprised many of the eukaryotic clones. Culturable fungal abundance in sea ice samples from Svalbard in the Arctic reached up to 7000 CFU fungi $L^{-1}$ (Zalar and Gunde-Cimerman 2014). Nearly 85% of the cultured fungi belonged to basidiomycetous yeasts, particularly *Cryptococcus adeliensis* and *Rhodotorula mucilaginosa*. Adjacent sea water yielded cultures of *Debaryomyces hansenii*, *D. maramus*, *Meyerozyma guilliermondii* (formerly *Pichia guilliermondii*), and a novel species resembling *Candida galli* and *Metschnikowia bicuspidata*. All of these isolated strains were characterized as psychrotolerant and xero/halotolerant. When sea ice melts, the entrapped yeasts and other microbes are released into the oceans.
- **Thraustochytrids colonize mucilage tubes of the sea ice pennate diatom *Berkeleya adeliensis* Medlin** (Riemann and Schaumann 1993). This diatom was found to grow in the lower 218–228 cm depth section of a fast ice core drilled close to the southern shelf ice margin of the Weddell Sea. Cell densities of the thraustochytrid in mucilage of this diatom reached values of $84 \times 10^3$ cells $ml^{-1}$ molten ice.
- **Mycetaen fungi have been cultured from sea ice** as well as seawater from Konigsfjorden in western coast of Spitsbergen of the Svalbard Archipelago, in the Arctic Ocean (Gunde-Cimerman et al. 2003). In order to isolate fungi from salt brines trapped in sea ice, where the water potential is low, media with low water activity were employed, in combination with incubation at low temperatures. This treatment yielded more fungi than mesophilic media. CFUs ranging from 6000 to 7000 $L^{-1}$ were obtained. The dominant taxa were ascomycetous and basidiomycetous yeasts, melanized fungi, mainly represented by the genera *Cladosporium* and *Aureobasidium*, and different species of *Penicillium*. Preliminary taxonomic analyses revealed several new species and varieties. Nonmelanized yeasts dominated in all samples, although the ratio of melanized to non-melanized fungi increased considerably in the glacier and sea-ice samples, where they occasionally represented up to 90% of all mycobiota.

**Fungi may play a role as diatom parasites**. Probably, the earliest report of a fungus in polar waters is that of chytrid parasites of the diatoms *Nitzschia cylindrus* and *Navicula specula* from ice cores in Beaufort Sea in the Arctic (Horner and Schrader 1962). No further research on this aspect has been carried out in recent years.

**Allochthonous wood in polar environments is an excellent habitat for obligate and facultative marine fungi.** A considerable amount of driftwood originating from Siberian and North American-boreal forests wash up along the coasts of the Arctic ocean. These host a large diversity of lignicolous marine fungi that are adapted to the harsh and stressful cold environment. Mycologists in 1960s to 1980s first discovered cold tolerant marine, lignicolous fungi in polar waters (Tubaki and Asano 1965; Pugh and Allsopp 1982; Pugh and Jones 1986).

- **Two of the unique fungi on driftwood in polar habitats are the ascomycetes *Havispora longyearbayensis* and *Remispora spitsbergensis*.** These were discovered in a study on fungi in driftwood collected at Tromsø, mainland Norway, and Longyearbyen, Spitsbergen, Svalbard, Norway. This study has also suggested the presence of a distinctive lignicolous fungal community (Pang et al. 2010). These appear to be psychrotolerant fungi adapted to Arctic conditions. A total of 9 species were detected on these wood samples.
- **Experimental wood samples placed in polar habitats become colonized by lignicolous fungi.** Meyers and Reynolds (1960a) submerged basswood and yellow pine panels at far northern marine sites of Alaska and Canada during periods of water temperature ranging from 0 °C to 15 °C. Many lignicolous marine ascomycetes, such as *Lulworthia, Ceriosporopsis, Peritrichospora*, and *Halosphaeriopsis*, as well as the anamorphic fungi *Piricauda, Humicola*, and others colonized the wood. Wooden panels of balsa, deal, and a tropical hard wood placed in Antarctic waters of about 0.5 °C for 42–84 days were colonized predominantly by the anamorphic fungus *Monodictys pelagica* and the ascomycetes *Ceriosporopsis tubulifera, C. halima,* and *Remispora maritima* (Pugh and Jones 1986). Those placed at Grytviken, South Georgia, were colonized more frequently than those at Signy Island off South Orkney Islands. The former had slightly higher sea surface temperature and stranded wooden material from erstwhile constructions.
- **Culture and metagenomic studies of fungi in driftwood and sea floor logs along the North Norwegian coast have revealed a large diversity of obligate as well as facultative marine fungi** (Rämä et al. 2014, 2016). The metagenomic studies were carried out based on pyrosequencing of ITS2 amplicons. Nearly 50 intertidal and sea-floor logs were examined. The cultures were identified based on 97% ITS sequence similarity to those known. Nearly half of the Operational Taxonomic Units (OTUs) belonged to non-marine fungal taxa, suggesting that facultative marine fungi may be important in wood decomposition in polar waters. Of the remaining marine mycetaen fungi, most belonged to Ascomycota. Other OTUs belonged to Basidiomycota, Mucoromycota, and Chytridiomycota. *Emericellopsis maritima, A. cruciatus, Halosphaeriaceae* sp., *Lulworthia* sp., and *Amylocarpus encephaloides* were the common, obligate marine ascomycetes. *Tolypocladium cylindrosporum* and *Cadophora* sp. were common terrestrial species. Many OTUs could not be assigned to any known fungi, suggesting the presence of novel fungi. Different fungal communities were detected in coniferous and deciduous logs. Geography, substratum, and site level variables contributed to shaping these communities.

- **Marine lichens are an interesting component of Antarctic coasts.** *Verrucaria serpuloides* has been found on dredged-up stones from 30 m in Antarctica (Kristiansen 2014).

## 12.5 Hypersaline Environments

Aquatic environments with salinities higher than that of seawater, which is about 34 ppt, are hypersaline. Such environments impose extreme conditions of life (Ventosa and Arahal 2002). Hypersaline environments that originate from seawater are thalassohaline, while those that originate from other sources are termed athalassohaline.

The high osmotic concentration in hypersaline environments results in low water potential or low water activity ($a_w$). Hence, cells cannot easily transport water into the interior for biochemical functions. This is a major constraint to life in hypersaline environments.

Salterns and hypersaline oceanic basins comprise two different thalassogenic hypersaline environments. Examples of athalassohaline environments are the Dead Sea and alkaline or soda lakes.

Despite the extreme conditions, many sturdy organisms still manage to subsist in hypersaline conditions or even prefer such environments, as with any extreme environment on earth. Organisms with an optimal salt requirement above 3% are halophiles. Those with an optimum between 3 and 15% are moderate halophiles and those that have a salt requirement above 15% and up to halite saturation of 34% are extreme halophiles. Many prokaryotes as well as eukaryotes such as protozoa, algae, diatoms, fungi, and larvae grow at a salinity between 3 and 10%. Diversity reduces sharply above this, and only a few eukaryotes, such as the green alga *Dunaliella salina* and the brine shrimp *Artemia salina,* can tolerate conditions up to ~16%. Salinities above this are the realm of halobacteria, which belong to Euryarchaeota (Archaea).

**Hypersaline conditions occur in marine ponds, salt marshes, coastal lagoons, and marine soils that are subjected to evaporation, salt or soda lakes, and man-made salterns.**

- **Solar salterns** are found mostly in tropical and subtropical parts of the world (Fig. 12.8). Salt production in salterns takes place by evaporation of sea water first to brine and subsequently to NaCl and other salts. The process leads to a continuous gradient of salinity and ends in hypersaline conditions. Their halite or NaCl concentrations exceed 10% (Oren 2002).
- **Deep Hypersaline Anoxic Basins** or DHABs are some of the most extreme environments on earth and are another relatively exciting modern discoveries in the oceans, having been found only in the 1980s. They combine conditions of deep-sea high hydrostatic pressure, hypersaline conditions, and anoxic environments. The salinity in DHABs is often 10 times more than that of normal sea

**Fig. 12.8** Salt pans near Chennai, India (Credit: Deepa Krishnan, Mumbai Magic)

water (350 ppt, or 35% salts). The high salinity water in the basin is capped off by seawater of normal salinity. The anoxic DHABs contain methane and hydrogen sulfide. Life in DHABs depends on chemosynthetic microorganisms (van der Wielen et al. 2005). They are now known to occur widely in world oceans and have been discovered in the eastern Mediterranean Sea, the Gulf of Mexico, and the Red Sea.

- **The Great Salt Lake in Utah** and the Dead Sea are two of the saltiest bodies of water in all of the Earth. Salinity in the Great Salt Lake ranges from 50 to 70 ppt. The organisms inhabiting the Great Salt Lake range from the primary producers or microbes to the consumers. The high salt accumulation results from water runoff with a high amount of salt and minerals. Salts are left behind when the water evaporates.
- **The Dead Sea,** located on the border between Israel, Jordan, and the Palestinian Authority, has a maximum depth of 300 m. It is an extremely harsh hypersaline environment, where the salt concentration is ~348 ppt, being about 10 times that of seawater. Magnesium and calcium are the major cations and not sodium. The former attains almost toxic levels of 2 M. The pH is ~6.0. Halophilic Archaea have been frequently isolated from the Dead Sea. Microbial communities reach high densities after exceptionally high rains that cause dilution of the water (Bodaker et al. 2010). Blooms of the green alga *Dunaliella* and red Archaea are triggered during these times.

Fungi had never been considered capable of living in extreme, hypersaline environments. However, numerous recent studies have shown that this is not true. Many fungi are indeed extremely halotolerant (Butinar et al. 2005a, b, c; Zajc et al. 2012; Nazareth 2014). Halotolerant or extremely halotolerant fungi can grow across a range of different salt concentrations. Those that prefer reduced $a_w$ have commonly been called xerophilic and osmophilic. Such fungi can grow at $a_w$ below 0.85 that corresponds to 17% NaCl or 50% glucose added to a growth medium.

### *12.5.1 Diversity of Fungi in Hypersaline Environments*

**Different fungi inhabit hypersaline environments** (Gunde-Cimerman et al. 2000; Butinar et al. 2005a, b, c; DasSarma et al. 2010; Zalar et al. 2005a; Zajc et al. 2012).

Only a single obligate marine, hypersaline fungus has been discovered so far. All other fungi described from this extreme environment belong to terrestrial genera or terrestrial species that are xerophilic or osmophilic. Fungi living in hypersaline environments, therefore, seem to be a group apart from the regular marine fungi. Melanized yeasts, non-pigmented yeasts, filamentous fungi, and straminipilan fungi are found in hypersaline environments. Characteristics of many of these fungi are given in Table 12.3.

- Many of the hypersaline fungi are "black yeasts." These are melanized fungi that possess both a filamentous as well as yeast-like morphology. The black yeast *Phaeotheca triangularis* de Hoog & Beguin is considered an oligotroph. It can grow on nutritionally poor agar media and has been most frequently isolated from low-nutrient storage ponds. This fungus can form biofilms on solid and liquid saline media and has often been isolated from microbial biofilms. The ascomycete *Hortaea werneckii* (Horta) Nishim. & Miyaji is an extremely halotolerant black yeast. *Hortaea werneckii* has been isolated from seawater, marine fish, salted freshwater fish, and beach soil (Fig. 12.9a, b).
- The basidiomycetous black yeast *Wallemia* is a model fungus to understand the mechanisms of extreme salinity tolerance (Fig. 12.9c). The genus comprises three species, namely, *W. ichthyophaga* Johan-Olsen, *W. sebi* (Fr.) Arx and *W. muriae* (J.J. Kickx) Zalar & de Hoog. *Wallemia ichthyophaga* is a true halophile. This xerophilic and extremely halotolerant fungus is often isolated from sea salt and hypersaline water. The fungus can thrive in media with NaCl above 1.7 M and up to saturation (5.3 M NaCl); this makes it the most halophilic fungi known to date.
- Non-melanized, true yeasts belonging to Saccharomycetales of Ascomycota, such as *Pichia guilliermondii*, *Debaryomyces hansenii*, *Yarrowia lipolytica*, and *Candida parapsilosis*, are also common in salterns. These yeasts are known to be osmophilic.
- Two new species, provisionally named as *Candida-atmosphaerica*-like and *Pichia philogaea*-like, have been reported in low numbers of 0–1000 cells $L^{-1}$

**Table 12.3** Common fungi from hypersaline environments

| Fungus | Salinity tolerance and occurrence |
|---|---|
| *Hortaea werneckii*<br>Black yeast; Capnodiales, Ascomycota | 0–32%; with a broad optimum from 6 to 14% NaCl. In intermittently drying salty pools or highly saline water of the crystallization ponds in the solar salterns. The natural ecological niche may be the highly saline water of the crystallization pond. |
| *Phaeotheca triangularis*<br>Black yeast; Capnodiales; Ascomycota | 0–26%; optimal concentration as 6–12% NaCl. Most abundant in the range of 18–25% NaCl; often found in storage ponds with relatively constant salinity. A halophilic species with a narrow ecological amplitude |
| *Trimmatostroma salinum*<br>Black Yeast; Helotiales; Ascomycota | 0–24%; with optimal values at 2–8% NaCl. Peaks at 25% salinity |
| *Aureobasidium pullulans*<br>Black yeast; Dothideales, Ascomycota | 0–18%; Halotolerant species. Grows optimally on medium without NaCl; at environmental salinities below 8% NaCl. The highest at 5% salinity, before or after the salt production season. Also in surface sediments and detritus in salt marshes and in the hypersaline waters of the Adriatic salterns. |
| *Cladosporium* spp.<br>Filamentous fungus; Capnodiales; Ascomycota | 0–20%. In habitats characterized by low $a_w$, like sugary and salty foods, salt marshes, the rhizosphere of halophytic plants, Adriatic salterns, hypersaline. Detected as well in high CFU numbers at 28% NaCl concentration, they might nevertheless be metabolically active in water. |
| *Candida parapsilosis*; *Debaryomyces hansenii*; *Pichia guilliermondii*<br>Saccharomycetales, Ascomycota | 0–17% |
| *Aspergillus* spp.<br>Eurotiales; Ascomycota | 0–27.5% |
| *Penicillium* spp.; Eurotiales; Ascomycota | 0–17% |
| *Wallemia:* Wallemiomycetes, Basidiomycota | 0–27% for *W. sebi* and *W. muria;* 9–32% for *W. ichthyophaga*; one of the most xerophilic and halophilic fungal taxa known to date. Found in hypersaline waters of man-made salterns on different continents |
| *Rhodosporidium* spp.; Sporidiobolales, Ascomycota | 0–17% |

in bittern (water rich in magnesium chloride) from the La Trinidad salterns (Spain). These are found in a season outside salt production or in waters with NaCl concentrations below 20% (Butinar et al. 2005a).

- Many species of Eurotiales, Ascomycota, particularly species belonging to *Aspergillus* and *Penicillium*, such as *Aspergillus niger, A. sydowii, Penicillium chrysogenum, P. brevicompactum, P. citrinum, P. oxalicum, P. nordicum,* and *P. steckii*, are examples of filamentous fungi found in hypersaline waters (Pitt and Hocking 1997).

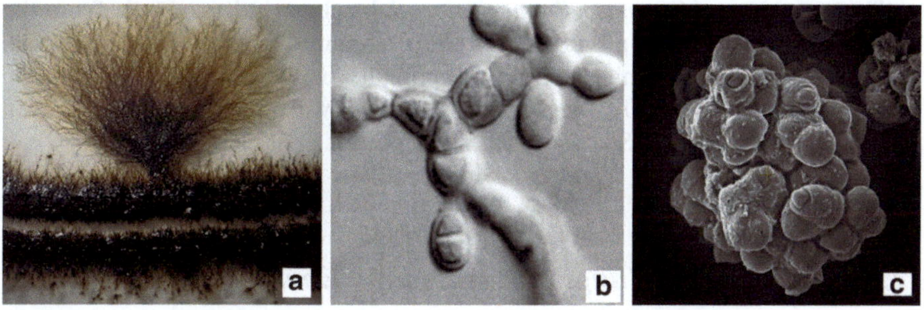


**Fig. 12.9** (**a**, **b**) The black yeast *Hortaea werneckii*. (**a**) Growth morphology in culture. (**b**) Cells. (**c**) The basidiomycete yeast *Wallemia ichthyophaga* (Courtesy: Dr. Nina Gunde-Cimerman. Photo by Marjetka Kralj Kunčič)

- Many species of *Cladosporium, such as C. halotolerans, C. dominicanum, C. velox, C. psychrotolerans, C. spinulosum, C. salinae* and *C. fusiforme. C. psychrotolerans, C. spinulosum, C. tenellum, C. subinflatum*, and *C. herbaroides* are known to be part of the hypersaline mycobiota (Schubert et al. 2007).
- Fungi are found in DHABs. Molecular signature of a fungus related to the basidiomycete yeast *Malassezia*, belonging to the Hy-An environmental sequence group, was detected in the L'Atalante deep-sea basin located in the eastern Mediterranean Sea (Alexander et al. 2009). This is also an anoxic basin.
- Interestingly, halophilic fungi have also been isolated from non-hypersaline environments. A number of halophilic fungi, particularly belonging to species of *Aspergillus* and *Penicillium* have been isolated from mangroves of Goa on the west coast of India (Nazareth 2014).
- An interesting discovery is that of the presence of typically marine, straminipilan fungi belonging to Labyrinthulomycetes in the Salt Lake, Utah. *Aplanochytrium minutum, Schizochytrium* sp., and *Thraustochytrium* sp. were recovered from sandy beach water of 120 ppt salinity, grass fragments, and insect exuviae (Amon 1978).
- **A number of fungi, including Oomycota, Ascomycota, and Basidiomycota, have been cultured from near-shore localities, as well as off-shore samples of the Dead Sea**. Nearly 70 species of fungi have been reported from the Dead Sea (Wasser et al. 2003). Terrestrial species of Ascomycota, such as *Aspergillus sydowii, A. versicolor, Eurotium herbariorum, Cladosporium cladosporoides*, and *C. sphaerospermum* are most frequent. Surprisingly, halotolerant black yeasts, the dominant representatives of fungi in hypersaline waters of salterns worldwide, have not been isolated from the lake (Butinar et al. 2005c; Gunde-Cimerman et al. 2000). Butinar et al. (2005a) isolated the yeasts *Candida glabrata*-like isolate, *Candida parapsilosis, Cryptococcus albidus, Meyerozyma guillermondii, Rhodotorula laryngis*, and *Trichosporon mucoides*. More than 40% of the diversity was isolated >0.5 km offshore. Most species recovered were isolated during one season only and from a limited number of sampling sites.

- **The ascomycete, *Gymnascella marismortui*, has been cultured only from the Dead Sea and may be a truly hypersaline, endemic species of the Dead Sea.** It is an obligate halophile and grows optimally between 0.5–2 M NaCl or in media containing 10–30% by volume of Dead Sea water. Some growth was even obtained in 50% Dead Sea water media (Molitoris et al. 2000).

## *12.5.2 Ecology of Fungi in Hypersaline Environments*

Few studies have addressed ecological activity by fungi in brine environments. **Hypersaline brine conditions do not support a large diversity, and the role of microbes may be expected to be confined to the utilization of dissolved organic nutrients.** Particulate organic matter is negligible.

- Salinity of seawater gradually increases to NaCl saturation levels during the evaporation process leading to salt formation. **Different fungi may appear at different times depending on the salt concentration.**

  The non-melanized yeasts *Pichia guilliermondii, Debaryomyces hansenii*, and *Candida parapsilosis* were isolated primarily outside the salt production season (Adriatic salterns) or in waters with NaCl concentrations below 20% (Butinar et al. 2005a). The black yeasts *Hortaea werneckii, Phaeotheca triangularis, Trimmatostroma salinum, Aureobasidium pullulans* and another extreme halotolerant filamentous, melanized fungus *Cladosporium* spp. appear during the crystallization in the solar salterns (Gunde-Cimerman et al. 2000; Butinar et al. 2005b). Members of Eurotiales displayed two peaks of occurrence, the first at the 10–15% NaCl range and later with a more prominent peak in the 18–25% NaCl range.

- **Wood samples located in hypersaline conditions support growth of lignicolous fungi** (Zalar et al. 2005b). Melanized hyphae belonging to *Hortaea werneckii* and *Trimmatostroma salinum* have been detected in the black-stained parts of wooden boards that support the walls of the crystallization ponds of salterns during high saline conditions. These fungi are capable of xylanolytic and lignolytic activities. *Trimmatostroma salinum also showed* cellulolytic activity.
- The yeast *Metschnikowia bicuspidata var. bicuspidata* may be a parasite in brine schrimps (*Artemia salina*). The yeast has been found in diseased brine shrimp in salt lakes and ponds with 10% NaCl from Great Salt Lake brine. The yeast requires 2% NaCl to grow in vitro (Zajc et al. 2012; Samson et al. 2002).
- **The exact substrates on which filamentous, mycetaen fungi grow in Dead Sea are not known, although they have been isolated from there.**

Experimental studies have provided **circumstantial evidence** that fungi are capable of growing in this environment (Oren and Gunde-Cimerman 2012).

– **Fungi are capable of producing various enzymes when grown in media made up with Dead Sea water.** *Penicillium westlingii* and *Ulocladium chlamydosporum* were excellent urease producers, while the endemic *Gymnascella marismortui* produced cellulase. Enzyme production generally decreased with increasing temperature and salinity, but some enzymes such as amylase, cellulase, and urease in the case of *G. marismortui* and amylase and caseinase in the case of *U. chlamydosporum* were produced at intermediate salinities and temperatures. Some fungi decolorized dyes indicating phenol oxidase activities (Buchalo et al. 1998; Molitoris et al. 2000; Oren and Gunde-Cimerman 2012).
– **Mycelia and spores of many fungi are capable of surviving in 100% Dead Sea water for many weeks.** Survival seems to be better when the water is diluted to 50–80%. *G. marismortui* mycelia retained their viability for 4 weeks in undiluted Dead Sea water and 12 weeks in Dead Sea water diluted to 80% (Kis-Papo et al. 2003). Also its spores survived prolonged suspension in Dead Sea water better than the spores of other species.

However, since such fungi grow in particulate matter and cannot live planktonically, it is not clear as to what substrates in the Dead Sea support their growth. Development of mycelium may be possible at times when salinity in upper layers of the Dead Sea becomes diluted to massive freshwater inflow. However, such events are rare (Oren 1999). These fungi may actually thrive in waters of low salinity at the mouth of the Jordan River and freshwater springs around the lake where they might germinate, grow, develop mycelia, and mature.

### *12.5.3 Physiological Adaptations*

**Halophilic fungi possess various structural and physiological adaptations that help them withstand the extreme conditions and enhance their survival in stressful environments.**

- **The black yeasts** are all melanized and belong to a single order of the Ascomycetes, the Dothideales. **They all have thick, melanized cell walls, slow, often meristematic growth, and proliferation with endoconidiation.** This has been suggested as an extremophilic ecotype morphology.
- **The basidiomycetous yeast *Wallemia* spp. show distinct cell morphology at high concentrations of NaCl** (Zajc et al. 2012). Hyphal compartments of *W. sebi* and *W. muriae* are thicker and shorter, and their mycelial pellets are larger at high salinity. Furthermore, an increase in the thickness of the multilayered cell wall occurs at higher salinities in all the three *Wallemia* spp., although it is especially pronounced in *W. ichthyophaga*. *Wallemia ichthyophaga* forms sarcina-like multicellular clumps that are composed of compactly packed spherical cells. Clustered growth allows the sheltering of the cells in the interior and minimizes the number of cells that are directly in contact with the hostile environment. A pronounced extracellular polysaccharide layer has been

observed in all the three *Wallemia* spp. and for *T. salinum*. Extracellular polysaccharides are part of the protection against desiccation in rock-inhabiting fungi, and it might also have a protective function at high salinities.

- **An important adaptation of halotolerant and halophilic fungi is to lower their intracellular water potential below that of the ambient environment in order to facilitate transport of water**. This is commonly achieved by synthesizing and accumulating a mixture of polyols such as glycerol, erythritol, arabitol, and mannitol as compatible solutes. Glycerol is the most significant of these and is accumulated in high amounts in the extremely halotolerant black yeast *H. werneckii* and the halophilic *W. ichthyophaga*. *Hortaea werneckii* also accumulates mycosporine-glutaminol-glucoside. *Hortae werneckii* and *Aureobasidium pullulans* are $Na^+$ excluders when growing in hypersaline conditions and maintain low intracellular concentrations of sodium and potassium. Interesting exceptions are the halotolerant yeasts *Debaryomyces hansenii* and *Pichia guilliermondii*. These yeasts are capable of maintaining normal cellular functions despite relatively high internal concentrations of sodium when coping with salt stress, while accumulating glycerol and other solutes to maintain suitable osmotic potential. Escape of the low molecular glycerol through the plasma membrane is prevented by altering the sterol and fatty acid membrane composition in a way that membrane fluidity is maintained suitably.
- Glycerol synthesis is determined by the high-osmolarity glycerol (HOG) signaling pathway in the yeast *Saccharomyces cerevisiae*. The existence of a similar signaling pathway has been demonstrated and extensively studied in *H. werneckii*. Unlike the salt-sensitive *S. cerevisiae*, a different set of genes is expressed in *H. werneckii* at high salt concentrations and interacts with HwHog1 MAP kinase (Vaupotič and Plemenitaš 2007).
- The HOG gene has also been found in *Eurotium herbarum* (EhHOG gene) from the Dead Sea. The EhHOG was expressed in a *S. cerevisiae* hog1D mutant. This restored growth and rectified the aberrant morphology of the mutant under conditions of high osmotic stress. Intracellular glycerol content also increased in the transformant, implying an important role for the EhHOG gene in that fungus (Oren and Gunde-Cimerman 2012).

- Hydrothermal vents are hot spots of biodiversity.
- Fungi have been discovered in several shallow and deep-sea hydrothermal vent regions. Actively growing hyphae of fungi have been found in shallow water hydrothermal vents. A large diversity of terrestrial species of fungi has been cultured from hydrothermal vent habitats. Thraustochytrids are prevalent in the shallow hydrothermal vents of the North Atlantic. Yeasts are present in hydrothermal vent habitats.
- Sequences of a number of unique and novel fungal phylotypes belonging to Mycetae have been detected in vent habitats.

(continued)

- Many hydrothermal vent animals harbor fungi. Mycetaen fungi have been cultured from hydrothermal vent animals, of which some are pathogens. Metagenomic studies have revealed many novel phylotypes close to known pathogens. The environmental sequence group "Hy-An" or most closely related to the yeast *Malassezia furfur* has been found in many vent animals. An infection of the hydrothermal vent mussel *Bathymodiolus brevior*, is probably caused by the black yeast *Capronia*. An unidentified fungus causes gill infection of an endemic, hydrothermal vent gastropod.
- Hypoxia and anoxia are extreme conditions found in several parts of the oceans. Anoxic to hypoxic conditions prevail in many "methane cold seeps."
- Anoxic environments may be home to novel and unique fungi. Standard culture methods may not be suitable to isolate many fungi that grow under such extreme conditions.
- The environmental sequence group "Hy-An" related to the basidiomycetous yeast *Malassezia*, the DSF Group 1, most closely related to the ascomycetous yeast *Metschnikowia bicuspidate*, and the LKM 11 cluster, most closely related to the basal fungal clade Cryptomycota are some of the most common ones recovered from anoxic, marine habitats.
- A number of novel Ascomycota and Basidiomycota and straminipilan fungi inhabit anoxic habitats.
- Actively growing fungal hyphae have been detected in coastal, seasonally anoxic sediments of the Arabian Sea along the west coast of India, using the Calcofluor direct detection method.
- Labyrinthulomycetes are common in the permanent OMZ of the Arabian Sea at depths of 200 to 1000 m.
- A variety of anaerobic respiratory modes are found in marine, anoxic, and hypoxic habitats. Mycetaen fungi may play an active role in denitrification in marine anoxic habitats
- Terrestrial species of fungi occur during seasonal anoxic conditions from coastal Arabian Sea.
- Arctic and Antarctic polar waters, sea ice, and allochthonous wood are some common habitats of polar fungi. Both obligate and facultative marine fungi occur in these. Many marine-derived fungi have also been reported.
- Thraustochytrids and yeasts are found in polar waters. High abundance of thraustochytrids has been reported in Arctic surface water samples from the Greenland and Norwegian Seas.
- Many fungi inhabit sea ice. Thraustochytrids have been found in mucilage of diatoms growing in fast ice.
- Mycetaen fungi have been cultured from sea ice.
- Fungi may occur as diatom parasites in polar environments.

(continued)

- Allochthonous wood in polar environments is an excellent habitat for obligate and facultative marine fungi such as the ascomycetes *Havispora longyearbayensis* and *Remispora spitsbergensis*. Culture and metagenomic studies of fungi in driftwood and sea floor logs along the North Norwegian coast have revealed a large diversity of obligate as well as facultative marine fungi.
- Marine lichens are an interesting component of Antarctic coasts.
- Hypersaline conditions occur in solar salterns, Deep Hypersaline Anoxic Basins (DHABs), the Great Salt Lake in Utah and the Dead Sea.
- All the fungi described from this extreme environment are xerophilic or osmophilic terrestrial species.
- Many of the hypersaline fungi are "black yeasts." The ascomycete *Hortaea werneckii* and the basidiomycetous black yeast *Wallemia* are model fungi to understand the mechanisms of extreme salinity tolerance.
- Non-melanized, true yeasts, members of Eurotiales and many species *of Cladosporium* have been found in hypersaline habitats.
- Fungi are found in DHABs. Molecular signature of a fungus belonging to the Hy-An environmental sequence group, was detected in the L'Atalante deep-sea basin located in the eastern Mediterranean Sea.
- Labyrinthulomycetes have been isolated from the Salt Lake, Utah.
- A number of fungi, including Oomycota, Ascomycota, and Basidiomycota have been cultured from near-shore localities, as well as off-shore samples of the Dead Sea. The ascomycete, *Gymnascella marismortui*, has been cultured only from the Dead Sea and may be a truly hypersaline, endemic species of the Dead Sea.
- Different fungi may appear at different times depending on the salt concentration.
- Wood samples located in hypersaline conditions support growth of lignicolous fungi.
- The yeast *Metschnikowia bicuspidata var. bicuspidata* may be a parasite in brine shrimps (*Artemia salina*).
- Fungi can produce various enzymes when grown in media made up with Dead Sea water. Mycelia and spores of many fungi are capable of surviving in 100% Dead Sea water for many weeks.
- Halophilic fungi possess various structural and physiological adaptations that help them withstand the extreme conditions and enhance their survival in stressful environments.
- An important adaptation of halotolerant and halophilic fungi is to lower their intracellular water potential below that of the ambient environment in order to facilitate transport of water. This is commonly achieved by synthesizing and accumulating a mixture of polyols, particularly glycerol.
- The high-osmolarity glycerol (HOG) signaling pathway gene has important roles in osmoregulation.

### Future Directions

1. Fungi are apparently common in hydrothermal vent habitats, but our information is very rudimentary. Further studies are required to understand the diversity, biomass, activity, and animal association of fungi in hydrothermal vents.
2. Mycelial fungi have been cultured from anoxic waters and sediments. What are the substrates in which they are found?
3. How important are thraustochytrids in anoxic waters and do they have an ecological role therein?
4. Anaerobic respiration has been found to occur in fungi from anoxic marine waters and sediments. Further studies are required to understand their role in the nitrogen cycle.
5. Fungi in the Antarctic continent have been studied by many but we know very little of the diversity, ecology, and physiology of fungi in Arctic and Antarctic waters and sediments. Are there psychrotolerant or psychrophilic fungi? What are the habitats of filamentous mycetaen fungi and single-celled thraustochytrids?
6. The physiological mechanism of high salt concentrations by fungi is a fascinating area that has important implications in biotechnology. One important aspect is related to the HOG gene.

# Chapter 13
# Physiology, Biochemistry, and Biotechnology

*There are no such things as applied sciences, only applications of sciences.*

Louis Pasteur

During the 1980s, it was realized that the marine environment was a source of a vast, untapped biodiversity which could be utilized for the welfare of the human society. The substantial information on the diversity and ecological role of fungi that had accrued by this time provided an impetus to studies on the biotechnological potential of marine fungi.

Applications of marine fungi in biotechnology are facilitated by knowledge on their physiological adaptations. In terms of their physiology and biochemistry, **marine fungi are adapted to in situ conditions of salinity, pH, temperature, hydrostatic pressure, oxygen concentrations, and chemical nature of organic material available for their growth. Many adaptations are determined by the evolutionary and taxonomic diversity of obligate or facultative marine fungi.** Obligate marine lignicolous and algicolous fungi and straminipilan fungi belonging to oomycetes and labyrinthulomycetes may possess unique properties not found in terrestrial fungi. Even terrestrial species of fungi living in the marine environment, the facultative marine fungi may be genetically different from their counterparts on land. Marine fungal biotechnology has also been approached from the perspective of habitats from which they are isolated, such as endophytes, deep-sea environments, and others (Raghukumar 2008).

**The unique nature of the marine environment as well as extreme conditions of many marine habitats are of biotechnological interest.** Several pharmaceutical and biotechnology companies are engaged in bioprospecting extreme marine environments. They are focussing on extremophilic bacteria for their biodiversity, thermostable and cold-tolerant enzymes, novel secondary metabolites, metal-tolerant enzymes, stress proteins, and bioremediation potentials to develop technologies. The famous thermostable enzyme Taq polymerase, vent polymerase, detergents supplemented with enzymes, surfactants, antioxidants, industrial enzymes, and colorants are some of the examples of the commercial products obtained from extremophilic microorganisms (Bull et al. 2000). Highly solvent-



S. Raghukumar, *Fungi in Coastal and Oceanic Marine Ecosystems*,
DOI 10.1007/978-3-319-54304-8_13

tolerant bacteria, useful in degradation of xenobiotic compounds, have been isolated from deep-sea sediments and waters (Inoue et al. 1999). Fungi from extreme marine environment have not yet been tapped for any commercial applications.

## 13.1 Growth Conditions

### *13.1.1 Salinity*

**Capability to grow at seawater salinities is an important criterion of success for marine fungi. Understanding the mechanisms behind such growth is also relevant to biotechnology** (Agarwal et al. 2013).

**Marine fungi respond in different ways to salinity for growth.** Numerous earlier studies have shed much light on salinity requirements of marine fungi for growth (Goldstein 1973; Jones and Harrison 1976; Jones and Byrne 1976; Jennings 1986).

- **Obligate marine mycetaen fungi are capable of growth in culture media with seawater as well as fresh water. However, they grow better in seawater media with salinities of at least 5–34 ppt, depending on the species. Terrestrial species of fungi also grow well in media with seawater. However, spore germination and reproduction of terrestrial isolates of fungi are affected by salinity, while those of obligate marine species is not** (Fig. 13.1). Optimal salinity of obligate marine, lignicolous fungi varies from 3 to 17 ppt in *Dendryphiella salina*, *Asteromyces cruciatus*, *Corollospora maritima*, *Zalerion maritimum (Helicoma maritima)*, *Halosphaeriopsis (Halosphaeria) mediosetigera*, and *Sporidesmium salinum* and from 20 to 34 ppt in *Trichocladium achrasporum* (*Culcitalna achraspora*). *Orbimyces spectabilis*,

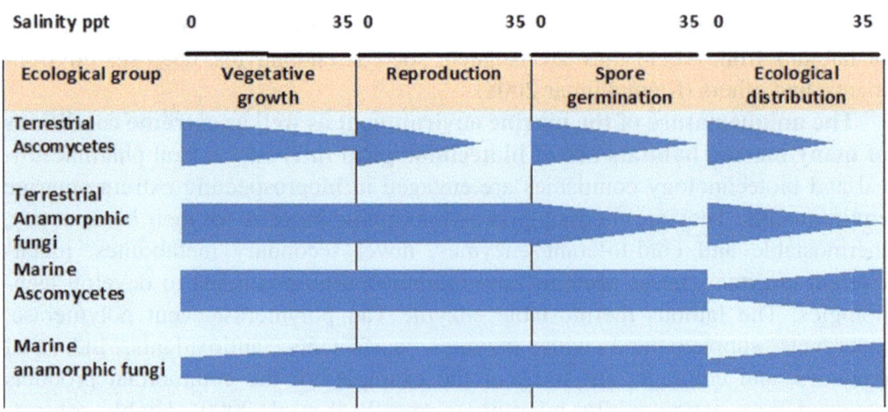


**Fig. 13.1** Responses to salinity of mycetaen fungi (Adapted from Jones and Byrne 1976)

*Corollospora maritima*, and *Lindra thalassiae* grow best at full strength seawater of about 34 ppt (Jones and Jennings 1965; Sguros and Simms 1964; Jones and Byrne 1976). Salinity optima may often depend on the environment in which the fungi grew. Isolates of *Lulworthia* sp. from brackish water grew best at 10 ppt salinity. Members of Basidiomycota may be more sensitive to salinity than those of Ascomycota (Jennings 1983). Although lignicolous marine fungi do not have an obligate requirement for $Na^+$, they are not found in freshwater conditions. The reasons for this are not clear. It is important to consider the effect of individual cations on the growth of marine fungi.

**Individual cations play an important role in determining growth of obligate marine fungi at various salinities.** Sodium, except at low levels, is inhibitory to growth. It inhibits glucose uptake, increases efflux of potassium from cells, and makes the plasma membrane permeable, resulting in leaching out of polyol osmotic solutes. Growth becomes inhibited through a reduction of the mycelial potassium content. Calcium reduces this inhibition. Magnesium and strontium are also effective in some species. The cells need to maintain optimum potassium level to be able to grow under saline conditions. Calcium appears to be necessary for the retention of potassium and organic solutes within the hyphae. It has been suggested that the plasma membrane ATPase in cells of these fungi has a higher pH optimum compared to that of terrestrial fungi. This may help to maintain a higher ratio of potassium to sodium in cells, compared to the ratio in seawater. Contrary to sodium, potassium promoted growth. Best growth of *Dendryphiella salina* was achieved in potassium concentrations approximating that of seawater. $K^+$ and $Ca^+$ appeared nontoxic at seawater concentrations.

- **The "*Phoma* pattern" is a phenomenon in which several obligate as well as facultative mycetaen marine fungi grow better at higher salinities when growth temperature is higher** (Ritchie 1957). The phenomenon was discovered first in the facultative marine fungus *Phoma* isolated from sea water. The optimal salinity for this fungus was 19 ppt at 16 °C, 23 ppt at 25 °C, and 34 ppt at 30 °C. A similar pattern was found in the fungi *Pestalotia* and *Curvularia*. It has also been found in the obligate marine, lignicolous fungus *Dendryphiella salina*, but not in *Lulworthia* (Lorenz and Molitoris 1992). The "*Phoma* pattern" might be genetically controlled.
- **Straminipilan fungi belonging to Labyrinthulomycetes and Oomycetes are obligately marine and require seawater for growth** (Jones and Harrison 1976; Raghukumar 2009). Most are euryhaline and grow at salinities of 5–35 ppt. Optimal salinity for growth, however, is narrower and ranges from 20 to 35 ppt, depending on the species. Thraustochytrid strains growing even at just 2 ppt are known (Burja et al. 2006). *Schizochytrium* species with low salinity requirements have been cultured from mangrove habitats. An exceptional strain that requires no sodium and grows at freshwater salinity has been reported from mangroves (Yokochi et al. 1998).

**The oomycetan fungus *Halophytophthora* exhibits a wide tolerance of 0–30 ppt** to salinity in nature and under laboratory conditions (Nakagiri et al.

1996; Leaño et al. 1998). Optimum salinity of *Halophytophthora* species varies from 10 to 30 ppt. Exceptionally, the mycoparasitic species, *H. mycoparasitica*, grows and reproduces better at higher salinities of 30–40 ppt (Nakagiri et al. 1996). The holocarpic oomycetes *Haliphthoros philippinensis* and *H. milfordensis* parasitic in crab eggs do not grow at freshwater salinities, but only in salinities of 10–40 ppt (Leaño 2002).

- **Terrestrial fungi isolated from marine habitats have been known to grow better in media with freshwater than those with seawater** (Jones and Jennings 1965). However, they are still capable of growing well in sea water media.
- **Fungi from hypersaline environments tolerate much higher levels of salinity compared to other marine fungi.** Thus, their tolerance to low water activity is remarkable. Some of these are generalists and can tolerate salinities ranging from 0 to those above 15% NaCl (150 ppt, which is nearly 5 times that of seawater). The "black yeast" *Aureobasidium pullulans* grows best at 0% NaCl, but can tolerate up to 18% NaCl. *Hortaea werneckii* can grow from 0% NaCl concentration to saturated salt concentrations. The three species of the basidomycetous fungus *Wallemia* can grow in media of very low water activity. *Wallemia muriae, W. ichthyophaga*, and *W. sebi*, as well as *H. werneckii*, are halophilic with an obligate sodium requirement and require at least 10% NaCl for growth. They have growth optima at extreme salinities (Zalar et al. 2005a; Gostinčar et al. 2010; Kunčič et al. 2010; Zajc et al. 2012).

Seawater has a lower water potential ($\psi$) or water activity ($a_w$) compared to freshwater. Intracellular water potential of marine fungi as with marine organisms has to be lower than that of seawater in order to maintain a flow of water into the cells.

- **Fungi lower intracellular water potential by synthesizing a mixture of solutes including polyols and amino compounds or by accumulating ions** (Oren 1999).
- Mechanisms of salt tolerance have been studied in detail in the marine halotolerant yeast *Debaryomyces hansenii* (Jennings 1983). This yeast as well as *P. guilliermondii* maintain relatively high internal concentrations of sodium when coping with salt stress, together with the production and intracellular retention of compatible solutes. **Ions may contribute about 60% and polyols about 30% to the solute potential in *D. hansenii* growing in sea water**. The polyols are mainly glycerol, mannitol, and inositol. Polyols also protect enzymes from the deleterious effects of ions in the cytoplasm. Glycerol is the most abundant and common osmolyte. Higher fungi produce more glycerol as the salinity of the external medium is increased. The highest amounts of glycerol have been measured in the extremely halotolerant black yeast *Hortaea werneckii* and the halophilic *Wallemia ichthyophaga*. *Hortaea werneckii* also accumulates erythritol, arabitol, and mannitol, as well as mycosporine-glutaminol-glucoside, as compatible solutes (Zajc et al. 2012).

- **Labyrinthulomycetes and many members of Oomycetes are obligately marine and may use a high concentration of inorganic ions to maintain their osmotic concentrations** (Jennings 1986).
- **Labyrinthulomycetes have an obligate requirement for $Na^+$ions** (Jennings 1986). Though the ion is required at high concentration for growth, sodium cannot be replaced by potassium. About 91–94% of the solute potential in cells of the thraustochytrid *Thraustochytrium roseum* seems to be made of inorganic ions, while the rest is made of proline. Proline is presumably synthesized from glutamate which has been found in significant concentrations in both *T. roseum* and *T. aureum*. Evidence indicates that sodium is involved in the transport of solutes across the plasma membrane. Phosphate uptake in thraustochytrids requires sodium, since they are cotransported.

**High concentrations of intracellular sodium can be toxic to cellular mechanisms in mycetaen fungi and negatively affect the activity of enzymes. Therefore, an efficient sodium extrusion or sequestration mechanism is essential.**

- Sodium may be sequestered into vacuoles in *Debaryhomyces hansenii*. Generally, excess sodium in the cytoplasm is effectively extruded by cells growing in saline environments. *Hortaea werneckii* and *Aureobasidium pullulans* are $Na^+$-excluders and keep their intracellular concentrations of sodium and potassium cations low when growing in hypersaline environments with high concentrations of salt.
- Since sodium is required for phosphate transport in thraustochytrids, these fungi may also possess an efficient sodium extrusion pump.

**Several genes that confer resistance to salinities have been found in various marine fungi. These have the potential to improve salinity tolerance in plants leading to biosaline agriculture in coastal environments** (Agarwal et al. 2013).

- The extremely halotolerant *H. werneckii* is a promising source of salt-tolerant transgenes for agriculture. 3′-phosphoadenosine-5′-phosphate is a toxic compound produced during sulfur assimilation in eukaryotes. This compound is removed by 3′-phosphoadenosine-5′-phosphatase coded by the HAL2 gene. This gene is inhibited by sodium or lithium ions, resulting in toxicity because of accumulation of this compound. However, two novel *HAL2*-like genes from the halotolerant black yeast *Hortaea werneckii* conferred tolerance to a high level of 0.8 M NaCl or 0.8 M LiCl when expressed in a salt-sensitive *Saccharomyces cerevisiae* (Vaupotič et al. 2007).
- Modulation of cation transport and maintenance of a high internal $K^+$ concentration and decreasing intracellular $Na^+$ are important during salt stress. Two genes of *Saccharomyces cerevisiae*, HAL1 and HAL3, are involved and modulate such cation transport systems. Overexpression of the HAL1 gene in yeast considerably enhanced salt tolerance in yeast. The HAL1 gene has been successfully introduced from *Saccharomyces cerevisiae* into tomato (Gisbert et al. 2000). Higher level of salt tolerance was recorded in the progeny (see Damare et al. 2012).

- The high-osmolarity glycerol (HOG) pathway in many fungi such as *Saccharomyces cerevisiae* and *Hortaea werneckii* is responsible for sensing of osmolarity changes in the environment. Such genes from Dead Sea fungi may be a promising resource to improve salt tolerance of plants. Preliminary results showed that HOG gene from *Eurotium herbarum* isolated from the Dead Sea may provide increased stress tolerance in transgenic *Arabidopsis* (Jin et al. 2005; Vaupotič and Plemenitaš 2007; Oren and Gunde-Cimerman 2012). In an article on the halotolerant *Debaryomyces hansenii*, which the authors call the "Cinderella of nonconventional yeasts," the authors discuss the enormous biotechnological potential of understanding the salinity tolerance of this fungus (Prista et al. 2016).

The mechanisms of salinity tolerance in fungi are still inadequately understood, 30 years after Jennings (1986) reviewed this aspect.

### 13.1.2 pH

Seawater has a pH ranging from 8.0 to 8.4. **Fungi growing in seawater must be capable of coping with its alkaline pH** (Jennings 1986).

- **Marine fungi generally have a near-neutral pH optimum for growth.** The earliest study on marine lignicolous fungi showed that 5 of the 6 obligate, lignicolous marine fungi that they studied grew best at pH above 7.4 (Barghoorn and Linder 1944). The lignicolous fungi *Asteromyces cruciatus*, *Dendryphiella salina*, *Corollospora cristata*, and *Lulworthia* sp. possessed two pH optima for growth, namely, 6.0–6.6 and 7.0–8.0 (see Kohlmeyer and Kohlmeyer 1979). Conidia were produced in *Varicosporina ramulosa* only at a pH of 6.9–7.5. Contrary to the above, several marine-derived terrestrial species as well as obligate marine mycetaen fungi such as *Zalerion maritimum*, *Dendryphiella salina*, and *Monodictys pelagica* appear to prefer acidic pH to alkaline ones (Curran 1980). Thraustochytrids have a broad pH tolerance of 5–8 for growth.
- **Many fungi cause a change in the pH of the growth medium.** *Halosphaeria mediosetigera* grown on a medium with pH 4.1 raised the pH to 8.1 by 6 days, but caused a decrease to 4.0 later by 21 days. Thraustochytrids often cause an increase in the pH of the medium.
- Presence of ammonium salts causes acidification of the medium in the marine lignicolous fungi, such as *Halosphaeria mediosetigera*, *Culcitalna achraspora*, and *Humicola alopallonella*, where the pH drops down to 3.0. The same is also known to occur in thraustochytrids.

### 13.1.3 Temperature

Temperature optima for growth of marine fungi depend on the environmental temperatures where they grow.

- **Most marine fungi are mesophilic** and grow well between temperatures of 15 and 30 °C. Strains of *Lulworthia* species from West African as well as North Sea coasts showed a temperature optimum of 25–30 °C. Temperatures of 25–30 °C generally favor optimal growth in thraustochytrids and labyrinthulids (Raghukumar 2009).
- **Psychrotolerant fungi are common in the Arctic and Antarctic environments.** The marine lignicolous fungus *Havispora longyearbyenensis* from an Arctic environment is a psychrotolerant fungus and grows at temperatures of 4–20 °C, the optimal growth being at 20 °C. The fungus does not grow at 25 °C (Pang et al. 2010).
- **Some thraustochytrids are psychrophilic and are ideal candidates to understand growth at low temperatures.** Ulken (1968) reported a strain of *Schizochytrium aggregatum* that grew best between 5 and 10 °C. *Thraustochytrium antarcticum* Bahnweg and Sparrow grew well at 0 °C, had an optimum at 4 °C, and did not grow beyond 17 °C (Bahnweg 1979b; personal observations). This extremely interesting species does not seem to have been isolated again. Interestingly, several marine lignicolous fungi, such as *Neptunella (Gnomonia) longirostris, Byssothecium obiones (Leptosphaeria discors), Lindra thalassiae, Remispora (Halosphaeria) maritima, R quadriremis, Torpedospora radiata, Clavatospora (Clavariopsis) bulbosa, Dendryphiella arenaria, Nia vibrissa*, and *Orbimyces spectabilis*, prefer a temperature below 10 °C for growth (Jones and Byrne 1976).
- *Halophytophthora spinosa* var. *lobata* grows even at 37 °C, a unique characteristic among the *Halophytophthora* species.
- Cold-adapted microorganisms have the potential to produce a number of compounds that are useful in biotechnology. Some of these are cold-active enzymes, polyunsaturated fatty acids, ice-nucleation, and antifreeze proteins. These have various applications in chemical and food industries, cryoprotectants, cold active catalysts, bioremediation, and frost protection in plants (Cavicchioli et al. 2002).
- **Deep-sea fungi are interesting candidates to study low temperature tolerance.** A deep-sea yeast most closely related to *Sporidiobolus johnsonii* produced substantial biomass when grown at 5 °C. Deep-sea isolates of terrestrial fungi such as *Sagenomella* sp. and *Aspergillus* sp. grew equally well at 200 bar hydrostatic pressure and 5 °C, as they did at room temperature and pressure (Singh et al. 2010).

### *13.1.4 Carbon and Nitrogen Nutrition*

Marine mycetaen fungi are not significantly different from their terrestrial counterparts in their carbon, nitrogen and vitamin nutrition (Jennings 1983). An early study on the marine lignicolous fungi *Halosphaeria mediosetigera*, *Culcitalna achraspora*, and *Humicola alopallonella*, based on 79 organic carbon and 38 inorganic and organic nitrogen sources showed that glucose, fructose, mannose, cellobiose, and xylose were good C sources. Urea, valine, leucine, alanine, arginine, glutamate, glutamine, aspartate, asparagine, hypoxanthine, and xanthine as nitrogen sources supported excellent growth (Sguros et al. 1973).

The most detailed studies on the carbon and nutrition of Labyrinthulomycetes are those of Bahnweg in 1979, wherein he studied nearly 30 isolates of thraustochytrids and aplanochytrids (Bahnweg 1979a, b). **These studies have shown that both groups are well equipped to decompose diverse compounds of terrestrial and marine origin.** Among the various pentoses, hexoses and derivatives, di- and oligosaccharides, polysaccharides, sugar alcohols and glycosides, alcohols, and organic acids tested, D-glucose, cellobiose, maltose, soluble starch, dextran, laminarin, and glycerol were best utilized. Many species are versatile in their nitrogen requirements, while some utilize only a few. Only a few strains can utilize nitrate and urea. Most species require an organic nitrogen source. Among various amino acids, L-glutamate is best utilized. L-proline, L-arginine, L-leucine, L-isoleucine, L-glutamine, and L-aspartate are also well utilized. There is a general requirement for B12 or B1 vitamins, some species also requiring biotin and riboflavin.

## 13.2 Enzymes and Cellular Proteins

Global market for industrial enzymes is expected to reach US$ 7.1 billion by 2018 from the estimated market of nearly US$ 4.8 billion in 2013 (Dalmaso et al. 2015). Marine bacteria and fungi are now being extensively screened for better and novel industrially useful enzymes, such as proteases, lipases, and a wide range of glycoside hydrolases, such as cellulase, xylanase, amylase, galactosidase.

### *13.2.1 Enzymes and Proteins Tolerant to Extreme Conditions*

**An important approach of marine biotechnology is to explore unique enzymes and proteins that serve as survival tools in marine microorganisms that face challenges in the ecosystem in which they live**. Some of these challenges are as follows:

- Extreme living conditions of pressure, temperature, pH, salinity, oxidative stress, radiation, chemicals (oxygen, $H_2S$, $CH_4$), and metals (Fe, Cu, Mo, Zn, Cd, Pb, and others) (Dalmaso et al. 2015)
- A symbiotic mode of life with marine plants and animals that requires the production of enzymes and proteins that help achieve the symbiosis
- Defense mechanisms for survival against competition from other antagonistic macro- and microorganisms
- The need to survive saprobically by utilizing unique polysaccharides, proteins, and lipids generated by other dead marine organisms. For example, some of the unique marine algal polysaccharides are carrageenan and agar from red algae, alginate, laminaran, and fucan from brown algae, and ulvan from green algae.

Research on extremophilic bacteria is aiding the discovery of enzymes and proteins that enable organisms to survive extreme conditions (Dalmaso et al. 2015; Zhang and Kim 2010). For example, enzymes of halophilic bacteria do not denature in high salt concentrations and alkaline conditions and are able to maintain their high activity. Psychrophilic bacteria produce cold-adaptive cold shock proteins (CSP) that bind to RNA to preserve its single-stranded conformation. Unique proteins from hyper/thermophiles possess sufficient structural rigidity to resist unfolding. Several marine bacteria produce peptidases which are useful catalysts for inorganic synthesis.

**Marine fungi from different habitats hold a lot of promise for discovery of novel enzymes and proteins.** Both straminipilan and mycetaen fungi grow on a wide variety of substrates, produce a diverse range of enzymes, and are potential candidates for discovery of novel enzymes and proteins. A number of studies have been carried out on obligate and marine-derived mycetaen fungi (Burtseva et al. 2010; Velmurugan and Lee 2012; Bonugli-Santos et al. 2015; Wang et al. 2016). Thraustochytrids exhibit a wide spectrum of enzymes including lipases, proteases, and carbohydrate-degrading enzymes that hydrolyze a variety of organic compounds (Bongiorni et al. 2005b; Taoka et al. 2009; Leaño and Damare 2012). Many promising leads of interesting degradative enzymes from marine fungi are now emerging. Of particular interest are enzymes that act at extremes of temperature, salinity, or pH. A few examples are given in Table 13.1.

- **Proteases:** A deep-sea fungus, *Aspergillus ustus* (Damare et al. 2006b), a shallow water hydrothermal vent thraustochytrid (Colaço et al. 2006), and a yeast from a salt pan (Zhang and Kim 2010) have been shown to produce proteases that can operate at alkaline pH, high salt concentrations, and in the presence of heavy metals. Protease from *Aspergillus ustus* was produced when grown at 100 bar pressure, corresponding to a depth of 1000 m.
- **Lipases**: Two strains of thraustochytrids that efficiently produce lipase have been reported (Kanchana et al. 2011). Several yeasts belonging to *Candida intermedia, Pichia guilliermondii, Candida parapsilosis, Lodderomyces elongisporus, Candida quercitrusa*, *Candia rugosa*, *Yarrowia lipolytica, Rhodotorula mucilaginosa*, and *Aureobasidium pullulans* produce high amounts of lipases (Wang et al. 2007a).

**Table 13.1** Examples of enzymes produced by some marine fungi

| Enzyme | Fungus | Habitat | pH | Temperature | Salinity | Heavy metals | Pressure |
|---|---|---|---|---|---|---|---|
| Proteases | *Aspergillus ustus* (Damare et al. 2006a) | Deep-sea sediments, 5100 m; Central Indian Basin | Broad pH range of 6–10, with an optimum at pH 9 | Optimum at 45 °C; 10% of the activity was retained at 2 °C | Unaffected in the presence of 0.5 M NaCl, equivalent to seawater salinity of 29 ppt | | The protease was produced when the fungus was grown at 100 bar pressure |
| | Thraustochytrid *Ulkenia* sp. #2a (Raghukumar et al. 2008a) | Shallow water hydrothermal vents of Azores Island; brown alga *Sargassum vulgare* | Maximum activity between pH 9 and 10 | Activity between temperature 45 and 50 °C | 100% active in a wide range of NaCl concentration ranging from 90–900 mM (0.5–5.0%) | Not inhibited in the presence of S, Mn, Fe, and Pb | |
| | *Aureobasidium pullulans* (yeast) (Bonugli-Santos et al. 2015) | Sea saltern of the China Yellow Sea | 9.0 | 45 °C | | | |
| Lipases | *Thraustochytrid isolates* (Kanchana et al. 2011) | Mangroves sediments | Maximum production at 6.0 | Maximum production at 30 °C | Maximum production at 34 ppt | | |
| Laccase | *Cerena unicolor* NIOCC #2a (Raghukumar et al. 2008a) | Decaying mangrove wood | Optimum pH at 3.0 | Optimum activity at 70 °C, with half-life for 90 min at 70 °C | Not inhibited at 1mM NaCl; retained 75 % activity in 50 % seawater | Not inhibited by Pb, Fe, Ni, Li, Co, and Cd at 1 mmol | |
| Chitinase | *Beauveria bassiana* (see Velmurugan and Lee 2012). | Marine sediment | Maximum yield at pH 9.5 | Maximum yield at 27 °C | 1% NaCl and 2.5% $KH_2PO_4$ w/w in solid substrate fermentation | | |

- **Chitinase**: Many marine-derived fungi such as *Aspergillus, Penicillium, Rhizopus*, and *Beaueria assiana* are known to produce chitinases.
- **Endopolygalacturonase**: A deep-sea yeast isolated from the Japan Trench at a depth of 4500–6500 m produced endopolygalacturonase active at 0–10 °C with no loss in activity up to 100 MPa at 24 °C (Abe et al. 2006).
- **Nucleases**: Utilization of nucleic acids in the DOM pool is known for many marine microorganisms. A number of marine-derived fungi isolated from marine animals and sediments have been shown to produce nucleases, particularly with a preference to poly-U substrate and resulting in the formation of 5′-phosphate mononucleotides. Some of the fungi were capable of degrading RNA (Balabanova et al. 2012).
- **The methylotrophic yeast** *Malassezia* sp., occurs in methane hydrate-bearing deep-sea sediments (Lai et al. 2007) and marine subsurface (Edgcomb et al. 2011). The fungus is capable of converting methane into more accessible carbon and energy substrates which is of great interest in physiological studies and industrial applications.
- **Novel deep-sea genes:** It may be possible to discover novel genes by studying gene expression under specific conditions. Thus, deep-sea conditions may result in expression of novel genes. For example, a study using the subtractive hybridization technique showed that genes were differentially expressed in the psychrotolerant, deep-sea yeast *Cryptococcus* sp. when grown at 50 Mpa hydrostatic pressure (corresponding to 500 m) at room temperature or 5 °C (Singh et al. 2012b). Elevated hydrostatic pressure alone resulted in the upregulation and expression of certain genes that were different from those expressed when subjected to a combination of both hydrostatic pressure and low temperature. At least 50% of the nearly 20 expressed sequence tags (ESTs) corresponding to different genes were those not found in existing databases, suggesting novelty. Even the known, expressed genes such as proteins coding for arachidonic acid metabolism, amino acid transport, and unsaturation of membrane fatty acids, which have been previously demonstrated to assist in the survival of microorganisms under stress conditions, might be those with novel properties.
- The extremely halotolerant black yeasts *Trimmatostroma salinum* and *H. werneckii* have been shown to produce extracellular hydrolytic enzymes that are active at high salt concentrations and that could therefore have important roles in different industries (Zalar et al. 2005b).

### *13.2.2 Lignocellulolytic Enzymes*

A large diversity of obligate and facultative marine lignicolous fungi inhabit lignocellulosic substrates in the marine ecosystem present in the form of wood and leafy material of marine plants and allochthonous material (Chaps 4–7).

Understandably, therefore, marine mycologists turned their attention to studies on production of lignocellulolytic enzymes by such marine fungi even about 50 years ago. **It is likely that marine fungi produce unique lignocellulolytic enzymes.**

Lignocellulose comprises cellulose, hemicellulose, and lignin (Chap. 4). **Ascomycetes and Basidiomycetes of the Kingdom Mycetae are generally excellent degraders of the lignocellulose complex** in vascular plants by virtue of production of cellulases, hemicellulases, and lignin-degrading enzymes or LDEs (Fig. 13.2; Motta et al. 2011; Sajith et al. 2016).

- **The cellulose component of wood is degraded by three cellulase enzymes, namely, endocellulase, cellobiohydrolase, and beta-glucosidase** (Fig. 13.2). Endocellulase (EC 3.2.1.4) randomly cleaves internal bonds at amorphous sites and create new chain ends. Exocellulase or cellobiohydrolase (EC 3.2.1.91) is an exoglucanase that cleaves two to four units from the ends of the exposed chains produced by endocellulase and produces tetrasaccharides or the disaccharide cellobiose. Beta-glucosidase or cellobiase (EC 3.2.1.21) hydrolyses the tetra- and disaccharides into individual glucose units.
- **Hemicellulases are glycoside hydrolases or carbohydrate esterases represented by xylanases, β-mannanases, and arabinofuranosidases** (Fig. 13.2). **Xylanases are the most important among these** and include several hydrolytic enzymes that together break down xylan to its constituent sugars (Motta et al. 2011). These include endoxylanase (endo-1,4-β-xylanase, E.

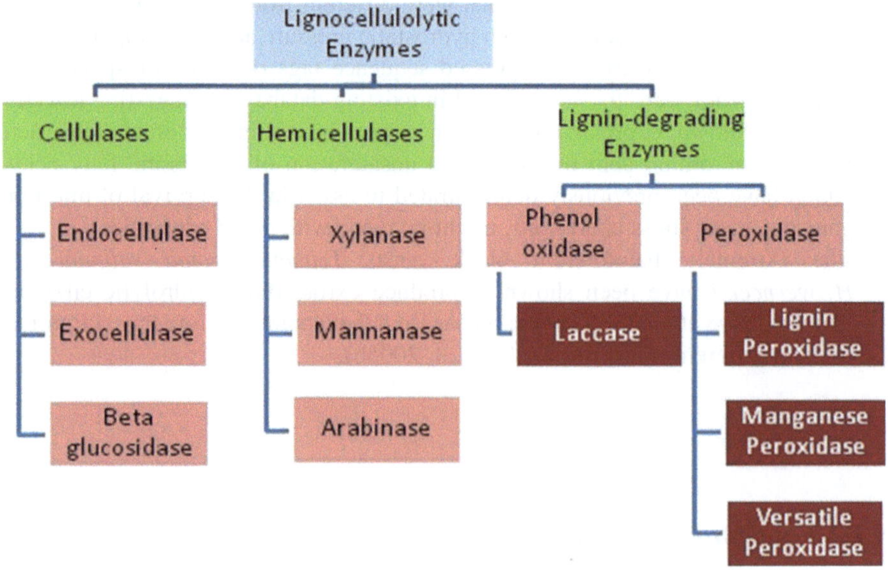


**Fig. 13.2** Lignocellulolytic enzymes of fungi (From: Sajith S, Priji P, Sreedevi S, Benjamin S. 2016. An Overview on Fungal Cellulases with an Industrial Perspective. J Nutr Food Sci 6: 461. doi:10.4172/21559600.1000461, distributed under the terms of the Creative Commons Attribution License)

C.3.2.1.8), β-xylosidase (xylan-1,4-β-xylosidase, E.C.3.2.1.37), α-glucuronidase (α-glucosiduronase, E.C.3.2.1.139), α-arabinofuranosidase (arabinosidase E. C.3.2.1.55), and acetylxylan esterase (E.C.3.1.1.72). Endoxylanases are the most important among these because they cleave the glycosidic bonds and liberate short xylooligosaccharides.

- **Lignin-degrading enzymes** (Fig. 13.2): Lignin is a complex, heterogeneous, phenolic polymer, characteristic of the cell walls of vascular plants (Chap. 4). Fungi belonging to Mycetae are excellent degraders of lignin by virtue of their capability to produce lignin-degrading enzymes (LDEs) (Pollegioni et al. 2015). **LDEs belong to two classes, namely, the heme-containing peroxidisaes and the copper-containing diphenol oxidases or laccases. Lignin peroxidase (LiP), manganese peroxidase (MnP), and versatile preroxidase (VP) are heme-containing peroxidises** (Raghukumar et al. 2008b). MnP requires Mn as a cofactor and LiP requires veratryl alcohol and association with a living cell. LDEs comprise a group of polyphenoloxidases that degrade lignin through a series of free radical-mediated redox reactions that results in the oxidation of phenolic and non-phenolic components until the ring structure is cleaved. This is followed by further degradation with other enzymes. Many marine fungi can decolorize lignin model compounds, such as the aromatic polymeric dyes PolyR-478 and Azure B (Raghukumar et al. 1994a; Pointing et al. 1998). This demonstrates their capability to degrade lignin. The capability to mineralize lignin to $CO_2$ has been demonstrated in several marine fungi (Sutherland et al. 1982).

**A number of obligate as well as facultatively marine fungi produce cellulases, xylanases, and LDEs** (Velmurugan and Lee 2012).

- **Cellulase production by marine fungi:** Extracellular endoglucanase, cellobiohydrolase, and β-glucosidase have been demonstrated from the lignicolous mangrove fungi *Halorosellinia oceanica, Julella avicenniae, Lignincola laevis, Savoryella lignincola*, and *Trematosphaeria mangrovei* based on quantitative enzyme assays. Of the three enzymes, endoglucanase is the most common among marine fungi. This has been demonstrated qualitatively by growth on carboxymethylcellulose, which is a treated, soluble form of cellulose (Pointing and Hyde 2000). *Corollospora maritima, Monodictys pelagica, Julella avicenniae, Lignincola laevis, Nia vibrissa*, and *Stagonospora* sp. can grow on crystalline cellulose throughout the salinity range 0–34 ppt, thus demonstrating their ability to produce all three cellulases (Pointing et al. 1998). Salinity adversely affects cellulase production in many marine fungi, while in others, such as *Julella avicenniae*, salinity has no such effect (Pointing and Hyde 2000).

  Thraustochytrids are another source of cellulases. Endocellulase has been detected in *Schizochytrium aggregatum* (Bremer and Talbot 1995). *Aurantiochytrium (Schizochytrium) mangrovei* has been shown to release glucose upon growth on microcrystalline cellulose, suggesting that some thraustochytrids may produce the entire complement of cellulases (Raghukumar et al. 1994b). Fourteen of nineteen thraustochytrids examined by Nagano et al. (2011) produced extracellular hydrolysed CMC, showing theprevalence of endocellulase in these fungi.

- **Hemicellulase production** by obligate marine fungi has been shown using qualitative as well as quantitative assays (Leightley 1980; Raghukumar et al. 1994a). The lignicolous fungi, *Aigialus mangrovei*, *Lophiostoma mangrovei*, and *Hypoxylon oceanicum*, as well as the facultative marine fungus *Gongronella* sp. are excellent xylanse producers (Raghukumar et al. 1994a). A cold-active thermo-labile xylanase from the psychrotrophic marine fungus *Penicillium chrysogenum* (F S010) isolated from the Yellow Sea has been cloned and expressed with fusion partner glutathione-*S*-transferase in *Escherichia coli* (BL21) (Hou et al. 2006). The recombinant marine fungal xylanase was highly active at 25 °C and pH 5.5 and showed 80% activity at 4 °C.
- **Many obligate and facultative marine fungi as well as marine-derived lignicolous fungi produce LDEs** (Pointing and Hyde 2000; Pointing et al. 1998; Raghukumar et al. 2008b; Bonugli-Santos et al. 2015).

  – Decolorization of the polymeric dyes PolyR-478 and Azure B has been used in growth studies to suggest lignin-degrading potential in marine fungi. Four fungi, *Nia vibrissa* (Basidiomycota), *Julella avicenniae* (Ascomycota), *Lignincola laevis* (Ascomycota), and *Stagonospora* sp. (anamorphic fungus) are notable in their ability to completely decolorize such polymeric dyes while utilizing either glucose or cellulose as primary carbon source (Pointing et al. 1998). *Ascocratera manglicola*, *Astrosphaeriella striatispora*, *Cryptovalsa halosarceicola*, *Linocarpon bipolaris*, and *Rhizophila marina* were shown to solubilize significant amounts of lignin, with indices of lignin solubilization comparable to those of terrestrial white-rot basidiomycetes (Bucher et al. 2004). Spectrophotometric enzyme assays have revealed manganese-dependent peroxidase activity in three isolates and laccase activity in 17 isolates of marine fungi (Raghukumar et al. 1994a). Basidiomycetes, particularly those classified as white-rot fungi, are the best producers of ligninolytic enzymes. Two remarkable facultative marine basidiomycete fungi have been reported to produce exceptionally high amounts of LDEs. A marine-derived strain of the basidiomycete fungus *Flavodon flavus* (#312), isolated from detritus of the sea grass *Thalassia hemprichii* produced all the three major lignin degradative enzymes, namely LiP, MnP, and laccase to varying extents, MnP being the predominant of these. The facultative marine fungus *Phlebia* sp. #MG-60 isolated from mangroves also produced high levels of MnP (Luo et al. 2005; Raghukumar et al. 2008b). *Cerena unicolor* NIOCC #2a isolated from decaying mangrove wood in the western coast of India produced high levels of laccase (Raghukumar et al. 2008b; Verma et al. 2010; Bonugli-Santos et al. 2015). The ascomycetous fungi *Sordaria fimicola* from mangrove sediment and *Halosarpheia ratnagiriensis* produced laccase and MnP. Many such fungi are useful candidates for biotechnological applications (see below).

- **Of the three LDEs, laccase is the most attractive because it is secreted in larger quantities, and it is not constrained by the need for a cofactor** or cell association as with MnP and LiP. A remarkable laccase-producing facultative marine fungus was isolated from mangrove detritus. This fungus, corresponding to *Cerena unicolor* (#NIOCC 2a), produced enhanced levels of laccase in the presence of several phenolics and lignin-derivatives (D'Souza et al. 2006; Raghukumar et al. 2008b). Many facultative marine fungi from mangrove habitats as well as obligate marine fungi produce laccases (Pointing et al. 1998; Raghukumar et al. 2008b). Many species of the obligate marine, lignicolous fungus *Lulworthia* are good laccase producers and can oxidize guaicol, naphthol, and benzidine (Schaumann et al. 1986). A high diversity of fungal laccase gene sequences have been recovered from the salt marsh grass ecosystem and from fungi in decaying blades of *Spartina alterniflora* (Lyons et al. 2003).
- **A complete degradation of any organic molecule results in mineralization leading to carbon dioxide. The capability of LDE-producing marine fungi to mineralize lignin to $CO_2$varies.** The benchmark lignin-degrading fungus, the terrestrial basidiomycete *Phanerochaete chrysosporium*, mineralizes about 21% of $^{14}C$-radiolabeled lignin to $^{14}CO_2$ within 21 days (Raghukumar et al. 1999). The facultative marine basidiomycete *Flavodon flavus NIOCC* #312 was equally efficient (Raghukumar et al. 1999). By comparison, the obligate marine fungus *Halosarpheia ratnagiriensis*, the facultative marine fungus *Sordaria fimicola*, a number of marine ascomycetes and mitosporic fungi, the mangrove basidiomycete *Nia vibrissa*, as well as *Phaeospheria spartinicola* from the salt marsh grass were shown to have low LDE activities and could mineralize up to 5–10% lignin to $CO_2$ in 21–60 days (Sutherland et al. 1982; Bergbauer and Newell 1992; Raghukumar et al. 2008b).
- **Many marine fungi secrete LDEs which are active in saline conditions.** Strong LDE producers such as *Nia vibrissa, Julella avicenniae, Lignincola laevis*, and *Stagonospora* sp. are highly active at all salinities. Several moderate or weak dye decolorizing isolates showed optimal activity at either low or high salinity (Pointing et al. 1998). *Flavodon flavus* produced MnP when grown in seawater. However, seawater inhibited enzyme activity when added during assay. On the other hand, MnP produced by *Phlebia* sp. could tolerate hypersaline conditions. *Cerena unicolor* (#NIOCC 2a) secreted laccase in media prepared with 50% diluted seawater. The enzyme showed optimum activity at 70 °C, with half-life for 90 min at 70 °C. It was active in the presence of 1 mmol NaCl and was not inhibited by Pb, Fe, Ni, Li, Co, and Cd at 1 mmol. Purified laccase from NIOCC #2a was not inhibited in the presence of NaCl roughly up to 0.3 M, above which it was reversibly inhibited (unpublished results).

**LDEs are of much biotechnological interest in treatment of highly colored and toxic industrial effluents because they degrade several aromatic,**

**recalcitrant compounds that cause environmental pollution** (Mtui and Nakamura 2004; Kiiskinen et al. 2004; Raghukumar et al. 2008b; Bonugli-Santos et al. 2015). **Many obligate and facultative marine fungi from lignocellulosic substrates are capable of breaking down industrial effluents by virtue of their LDE production.** Bioremediation of these pollutants can be achieved by using salt-tolerant fungi and their salt-tolerant enzymes (Passarini et al. 2011).

- **Paper and pulp industries** generate black liquor. These are dark in color because of toxic chlorolignin released from the raw material through chemical treatments and have a pH of 8–11. These are subsequently released as bleach plant effluents (BPE). Removal of lignin and biobleaching of paper pulp in paper industries using fungi is an alternative to chemical bleaching that results in toxic byproducts.
- **Molasses based-alcohol distilleries** release "molasses spent wash" (MSW) containing dark brown colored recalcitrant compounds, collectively termed as melanoidin. These are formed by the Maillard amino-carbonyl reaction and are toxic.
- **Textile and dye industries** release effluents containing residual dyes belonging to compounds with azo, anthraquinone, triphenylmethane, and heterocyclic polymeric structures. Azo dyes comprise most of the synthetically produced ones. Such dyes are of concern to human health, are often recalcitrant, and are extremely difficult and expensive to remove from the system.
- **Polycyclic Aromatic Hydrocarbons (PAHs)** are made up of two or more fused benzene rings and are widespread in natural environments. Anthropogenic activities, such as burning of fossil fuels, their byproducts, as well as incineration of municipal wastes, generate PAHs that cause environmental pollution. Certain PAHs are considered toxic, mutagenic, and carcinogenic.

**Facultative marine fungi that produce LDEs and hemicellulases may have a tremendous potential in treatment of industrial effluents** (Fig. 13.3; Raghukumar et al. 2008b; Verma et al. 2010; Velmurugan and Lee 2012; Bonugli-Santos et al. 2015).

- **Biobleaching is a process in paper industries whereby the lignin in the paper pulp is removed by enzymatic rather than chemical means**. Xylanase is a potential enzyme that can bring about this. Microbial xylanases that are thermostable, active at alkaline pH, and cellulase-free are generally preferred for biobleaching of paper pulp. A facultatively marine fungal isolate of *Aspergillus niger* # 3 from mangrove detritus produced a unique, cellulase-free xylanase that was thermostable at 55 °C and active at pH 8.5. The crude culture filtrate containing 580 U $L^{-1}$ of xylanase could bring about bleaching of sugarcane bagasse pulp in 60 min treatment at 55 °C, resulting in a reduction of 30% chlorine consumption during bleaching (Raghukumar et al. 2004a). Another marine-derived fungus *Phlebia* sp. #MG-60, belonging to basidiomycetes, has been shown to biodegrade sugarcane bagasse used in paper and pulp mills and to biobleach paper pulp (Li et al. 2003).

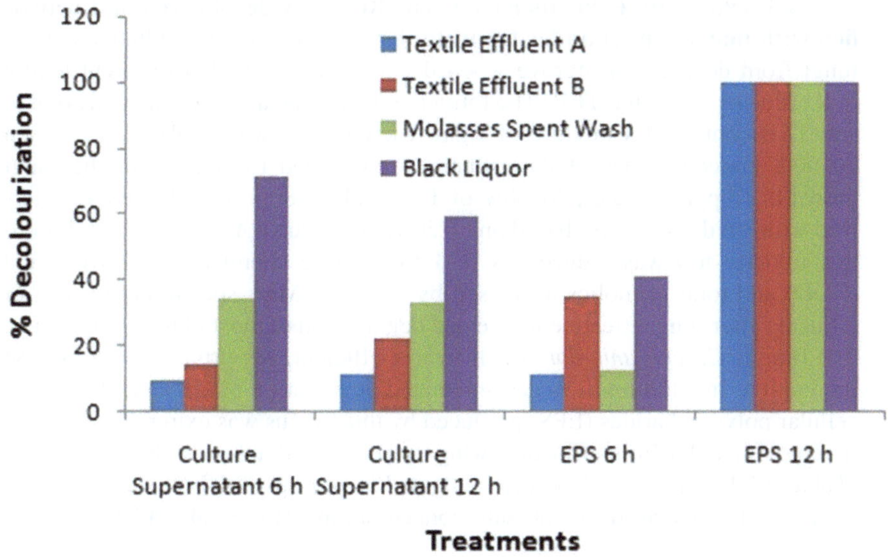


**Fig. 13.3** Decolorization of effluents by fungus-free extracellular culture supernatant and extracellular polymeric substance (EPS) produced by the facultative marine fungus *Cerena unicolo*r NIOCC#2a that produces high amounts of laccase (Data from D'Souza et al. 2006. Enzyme Microb. Technol. 38, 504–511)

- **Bioremediation of bleach plant effluents (BPE) from paper mills has been attempted using facultative marine fungi**. *Sordaria fimicola* (NIOCC #298) and *Halosarpheia ratnagiriensis* brought about 65–75% decolorization of bleach plant effluent within 8 days. These fungi produced MnP and laccase. The facultative marine fungus *Flavodon flavus*, an efficient MnP producer, decolorized bleach plant effluents remarkably at pH 11 (Raghukumar et al. 2008a).
- **Molasses spent wash (MSW) has been effectively treated using facultative marine fungi**. *Cerena unicolo*r NIOCC#2a, a remarkable laccase producer, removed 80% of color from MSW by 3 days. The fungus could be immobilized in polyurethane foam (PUF) in order to decolorize MSW and also reduce COD levels by 50% within 72 h. The MnP producing fungus *Flavodon flavus* also decolorized molasses spent wash (MSW) in addition to bleaching of paper pulp as given above. This did not seem to depend on its MnP production and the bleaching of MSW probably resulted from production by glucose oxidase. This fungus-treated MSW also detoxified it, as was shown by absence of liver damage in an estuarine fish exposed to it. The concentration of the PAH benzo(a)pyrene decreased by 68% in the treated MSW.

- **Colored textile mill effluents have been effectively decolorized and detoxified with marine fungi under laboratory conditions.** Four facultative marine fungi from decaying mangrove wood decolorized two highly different textile mill effluents, TEA and TEB. The former contained an azo dye with a pH of 8.9, and TEB contained a mixture of eight reactive dyes with a pH of 2.5. When 20–90% concentrations of the effluents were added to cultures of the fungi grown in 15 ppt seawater, 30–60% of TEA and 33–80% of TEB were decolorized within 6 days. Assays based on LC50 values against *Artemia larvae* showed that toxicity too was reduced by 2–3 fold, while chemical oxygen demand (COD) and total phenolics decreased by 70–80%. Mass spectrometric scan of effluents after fungal treatment revealed degradation of most of the components. The fungus *Cerena unicolor* was the most efficient, by virtue of high laccase production. In addition to laccase-mediated degradation of effluents, the extracellular polysaccharides (EPS) produced by this fungus was extremely effective in removing color from effluents owing to their remarkable adsorbent capacities (Table 13.2). The basidiomycete fungus *Phlebia* sp. #MG-60 decolorized a number of dyes at different salt concentrations (Li et al. 2003). Several

**Table 13.2** Known occurrence of various groups of fungi in marine habitats (common/poorly known) and the extent to which they have been studied for secondary metabolites (frequent/infrequent)

| Habitats | Obligate marine fungi: Kingdom Mycetae | Facultative marine fungi and marine-derived fungi: Kingdom Mycetae | Obligate marine fungi: Kingdom Straminpila |
|---|---|---|---|
| Saprotrophs in coastal lignocellulosic substrates (allochthonous wood; mangrove substrates; seagrassses; salt marsh grass) | Common[a] | Common[a] | Common[b] |
| Saprotrophs in macroalgae | Common[a] | Common[a] | Common[b] |
| Symbionts in seagrasses (endophytes); macroalgae (endophytes and parasites) | Common[b] | Common[a] | Common[b] |
| Symbionts in corals, invertebrates, sponges, tunicates, etc. | Poorly known[b] | Common[a] | Common[b] |
| Saprotrophs in animal substrates | Poorly known[b] | Common[b] | Common[b] |
| Coastal marine sediments | Poorly known[b] | Common[b] | Common[b] |
| Coastal and offshore pelagic | Poorly known[b] | Common[b] | Common[b] |
| Deep-sea sediments, including hydrothermal vents | Poorly known[b] | Common[b] | Poorly known[b] |

[a]Frequent
[b]Infrequent

marine-derived fungi isolated from sponges are known to decolorize textile dyes under both saline and nonsaline conditions (Bonugli-Santos et al. 2015). These include isolates belonging to the basidiomycete *Tinctoporellus* sp., and anamorphic fungi such as *Penicillium citrinum*, *Aspergillus sulphureus*, *Cladosporium cladosporioides*, and *Trichoderma* sp.

- **PAHs are efficiently degraded by many marine-derived fungi** (Wu et al. 2009; Raghukumar et al. 2008b; Passarini et al. 2011). PAHs may be degraded by fungal LDEs. *Aspergillus sclerotiorum* CBMAI 849 isolated from cnidarians degraded 99.7% of a 67 ppm pyrene by 8 days and 76.6% of 33 ppm benzo[a] pyrene by 16 days using the unique monooxygenase system of cytochrome P-450. The mechanism of the PAH metabolism by the fungus was elucidated by HPLC-DAD-MS data. *Aspergillus* sp. BAP14 isolated from Chinese coastal marine sediment degraded 30% and 60% of 10 ppm benzo[a]pyrene after 3 and 12 days, respectively. Phenanthrene added to cultures of the fungi *Flavodon flavus* NIOCC#312 and *Cerena unicolor* NIOCC#2a at 20 ppm concentration was adsorbed strongly by the fungal mycelium and removed from the culture medium (Raghukumar et al. 2008b). *Cerena unicolor* NIOCC#2a subsequently metabolized or transformed it into the more polar derivative phenanthrene (trans)-9,10-dihydrodiol. Protein expression may be altered in marine fungi as a response to PAH presence and enabling PAH degradation. A total of 304 individual protein spots were found to be differentially expressed in the fungus *Paecilomyces* sp. (SF-8). These were identified by using 2-DGE liquid chromatography-MS and other mass spectrometry techniques. Low molecular weight PAHs, phenanthrene and anthracene, and high molecular weight PAHs, benzo[a]anthracene and benzo[b]fluoranthene, were degraded after seven and 14 days, respectively (Velmurugan and Lee 2012).

### *13.2.3 Enzymes that Degrade Marine Polymers*

Most studies have focused on conventional enzymes such as proteases, lipases, lignocellulolytic enzymes, and amylases, particularly from marine-derived fungi. **Fewer studies have explored enzymes that degrade marine polymers such as polysaccharides from marine algae.**

- **Alginates**, namely, alginic acid and its salts are the major cell wall polysaccharides in brown algae. They are made up of D-mannuronic and L-glucuronic acids.
- **Laminarin (or laminaran)** is the major storage polysaccharide of brown algae and is made up of $\beta(1 \rightarrow 3)$-glucan with $\beta(1 \rightarrow 6)$-branches.
- **Fucoidan** is comprised of fucose-containing sulfated polysaccharide (FCSP), and is found in several brown algae.
- **Carrageenans** and agar are galactose containing linear sulfated polysaccharides found in cell walls of many red algae.

**Despite the well established presence and role of fungi in marine algae as endophytes, parasites, and saprobes, few attempts have been made to study production of enzymes that break down algal polymers by fungi.**

- The obligately marine, anamorphic fungi, *Dendryphiella salina*, *D. arenaria*, *Asteromyces cruciatus*, and *Corollospora intermedia* are known to break down **alginate** (Wainwright and Sherbrock-Cox 1981; Schaumann and Weide 1990). The former has been shown to degrade alginate from 10 to 40 °C and up to 3.0% NaCl. The optimal pH was 6.0. The enzyme system seems to be comprised of an endo-alginate hydrolase and alginate lyase. Marine-derived fungi belonging to *Calcarisporium* sp., *Tritirachium* sp., *Bartalinia robillardoides*, *Penicillium pinophilum*, *Scopulariopsis brevicaulis*, and *Pestalotiopsis* sp. have been shown to grow on all algal polysaccharides as carbon sources, thus demonstrating production of enzymes that can degrade them (Wang et al. 2016).
- A strain of the obligately marine fungus *Dendryphiella arenaria* produced a fucoidanase that hydrolyzed high molecular weight fucoidan to low molecular weight fucoidan rather than to fucose (Wu et al. 2011). Several marine-derived fungi also degrade fucoidan (Rodríguez-Jasso et al. 2010). A strain of *Aspergillus niger* was found to be the best among these, the enzyme activity being stimulated in the presence of sucrose.
- **Chitin** is the primary component of fungal cell walls as well as crustacean exoskeleton. Chitin degradation is first initiated by chitinases to produce soluble oligosaccharides, with further hydrolyzation to monosaccharides by *N*-acetyl-β-D-glucosaminidases. Endo-chitinase, exo-chitinase, and *N*-acetyl-D-glucosaminidase have been reported from a number of marine-derived fungi (Velmurugan and Lee 2012). The fungus *Beauveria bassiana* is an important source of this enzyme. It has also been demonstrated in *Trichoderma harzianum, Metarhizium anisopliae*, *Fusarium oxysporum*, *Penicillium janthinellum,* and an Antarctic strain of *Verticillium lecanii*. A novel chitinase of 67 kDa has been shown to occur in the marine fungus *Plectosphaerella* sp.

## 13.3 Metal Tolerance

Heavy metals include biologically essential ones such as copper (Cu), nickel (Ni), iron (Fe), and zinc (Zn), as well as nonessential ones such as lead (Pb), arsenic (As), mercury (Hg), cadmium (Cd), and tin (Sn). **Essential heavy metals are required in trace quantities, while the nonessential ones are tolerated in low levels. Both are toxic above a low threshold level.** Heavy metals are stable, are not broken down, and subsequently accumulate in the environment.

**Heavy metals in the marine environment arise through both natural and anthropogenic means.** Heavy metals enter the sea through riverine influx of weathered and eroded rocks, as well as atmospheric deposition in the form of particles. Deep-sea and shallow water hydrothermal vents emit fluids with heavy metals, such as Zn, Co, Cu, Fe, and Mn arising from the magma of the earth into the oceans. Anthropogenic inputs come primarily from industrial activities. **Bacteria and fungi overcome heavy metal toxicity by various mechanisms.** These include (1) intracellular interactions, particularly involving physical sequestration of metal by binding to proteins or other ligands to prevent it from damaging the metal-sensitive cellular targets, or methylation of heavy metals and their volatilization; (2) extracellular interactions, involving chelation and cell-wall binding, the production of acidic metabolites to leach out heavy metals from sediments, or production of extracellular polysaccharides that bind and transport them (Ford and Ryan 1995; Anahid et al. 2011). Studies on microorganisms that tolerate heavy metals are of interest to develop bioremediation technologies that can be applied to the natural environment. Several marine fungi are adept at overcoming heavy metal toxicity (Damare et al. 2012).

**Fungi from hydrothermal vents are ideal targets to study heavy metal tolerance.**

- For example, the brown alga *Sargassum vulgare* C. Agardh collected from the white and yellow zones of shallow-water hydrothermal vents of the Atlantic near Azores islands accumulated Mn and Fe. Average Mn concentrations ranged between 25.1 and 39.5 μg $g^{-1}$ and that of Fe between 6385 and 13,855 μg $g^{-1}$ at the two zones, respectively. **An isolate of the thraustochytrid Ulkenia sp. #2a, cultured from this alga, grew better when Fe, Mn, or Pb were included in the culture medium than without** (Fig. 13.4; Colaço et al. 2006;

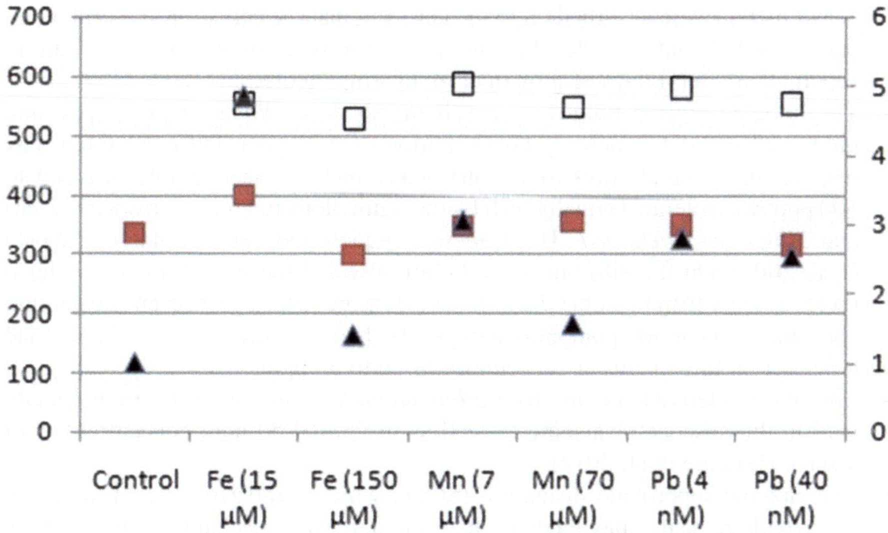


**Fig. 13.4** Heavy metal tolerance of a thraustochytrid from shallow water hydrothermal vent at Azores, Atlantic. Protease production (*open squares*) and protease activity (*filled squares*) are shown as Azocasein Units. Growth (triangles) is mg dry wt. per ml (From: Raghukumar, C. et al., 2008. Current Science 95: 1715–1723)

Raghukumar et al. 2008a). Growth as well as protease production were better at low levels of Fe, Mn, and Pb concentrations than in the absence of these heavy metals or at higher concentrations of them. The ability of this thraustochytrid to tolerate high concentrations of these metals is highly significant.

- Fluids in hydrothermal habitats contain reduced metals such as Fe (II) which oxidizes to insoluble Fe (III) on mixing with seawater, making it unavailable to organisms. **Yeasts isolated from Fe-oxide mats and rock surfaces from the crater of the volcanically active Vailulu'u seamount (Samoan chain) synthesize the organic chelators, siderophores, which increase the bioavailability of Fe(III)** (Connell et al. 2009). The yeast *Rhodotorula graminis* was found to be capable of oxidizing Mn (II) to Mn (IV).

**Many obligate as well as marine-derived fungi possess various levels of tolerance to different heavy metals.**

- In one of the earliest studies, it was shown that some fungi isolated from estuarine and seawater samples tolerated nickel better when grown in a medium containing seawater, rather than distilled water. This was attributed to the presence of magnesium in seawater (Babich and Stotzky 1983).
- A deep-sea, psychrolerant isolate of the yeast *Cryptococcus* sp. obtained from polymetallic nodule-bearing sediments of the Central Indian Basin showed considerable growth in the presence of 100 mg $L^{-1}$ of Zn, Cu, Pb, and Cd (Singh et al. 2013). This strain proved to be more efficient in tolerating these heavy metals, compared to terrestrial isolates of *Cryptococcus*, *Rhodotorula*, *Rhodosporidium*, and *Sporidiobolus*. It grew best at a pH of 6–8. Scanning electron microscopy together with FTIR and EDAX analysis revealed that the heavy metals were adsorbed on to the cell, and that the cell surface morphology was altered. About 30–90% of the heavy metals were removed from the culture supernatant after 4 days of growth at room temperature.
- The obligate marine fungus from salt marsh grass, *Phaeospharia typharum*, grew well even at a mercury concentration of 0.74 ppm. Different fungi may respond in different ways to various heavy metals. For example, lead up to 500 ppm was tolerated equally well by the lignicolous fungi *Corollospora lacera* and *Monodictys pelagica*. The lead was sequestered extracellularly (Taboski et al. 2005). On the other hand, cadmium affected the growth of *M. pelagica* more severely than that of *C. lacera*. *Corollospora lacera* grew even at cadmium concentrations of 500 ppm, although poorly than at lower levels. Cadmium had no effect on this fungus at concentrations up to 50 ppm.
- The marine-derived fungus *Aspergillus candidus* was shown to grow equally well in the absence of arsenic as well as in 25 and 50 ppm concentrations of arsenic (Damare et al. 2012).
- The enzyme superoxide dismutase (SOD), which catalyzes the scavenging of superoxide radicals, may have a role in the defensive mechanisms against high concentrations of $CuSO_4$. A nearly 4-fold increase of SOD was noticed in a strain of the yeast *Cryptococcus* sp. from 4500 to 6500 m deep-sea sediments of

the Japan Trench when grown with 10 mM $CuSO_4$ in comparison with 1 mM $CuSO_4$ (Abe et al. 2001).

## 13.4 Hydrocarbon Degradation

**Oil or hydrocarbon pollution is a serious environmental problem in the marine ecosystem** (GESAMP 2007; McGenity et al. 2012). Crude oil, which is the major cause, is a natural, heterogeneous mixture of numerous hydrocarbons, consisting mainly of alkanes with different chain lengths and branch points, cycloalkanes, and mono-aromatic and polycyclic aromatic hydrocarbons. Chronic oil pollution arises from drilling for oil as well as natural seeps. About 1.3 million tonnes of petroleum enters the marine environment each year this way. Acute pollution arising through accidental causes is much more serious and causes severe ecosystem damage. A number of these are well known. Chronic pollution is sublethal but causes physiological, genetic, and behavioral damage to organisms and affects the normal functioning of the ecosystem. **Consortia of natural microbial populations act in degradation of hydrocarbons in the environment.**

The aliphatic fraction of crude oils is the easiest to degrade microbiologically. Aromatic hydrocarbons are toxic, and only a few microorganisms are capable of degrading them. Asphaltenes are the most recalcitrant to microbial degradation.

A vast amount of literature on bacterial dynamics in hydrocarbon degradation is available. **Surprisingly, much less is known of fungi. Yeasts are commonplace in hydrocarbon polluted waters and are known to be capable of degrading hydrocarbons**. The earliest study on their hydrocarbon degradation dates back to that of Ahearn and Meyers in 1972 (Kutty and Philip 2008; Damare et al. 2012). Hydrocarbonoclastic yeasts, such as *Pichia ohmeri* and *Trichosporon* sp., were seen to increase in numbers when plots of *Spartina alterniflora* salt marsh in Louisiana were experimentally polluted with oil.

Yeasts belonging to *Candida parapsilosis, C. albicans, C. guilliermondii, Yarrowia lipolytica, C. tropicalis*, and *C. intermedia*, isolated from marine mud and water around Mumbai, degraded more than 10% of crude oil provided in culture (Zinjarde and Pant 2002). *Yarrowia lipolytica* utilized 78% of the aliphatic fraction of Bombay High crude oil. None of these isolates degraded the aromatic or ashphaltene fractions. All the isolates required aeration, nitrogen, and phosphate supplementation for optimal degradation.

**Obligately marine, lignicolous fungi, *Corollospora lacera*, *C. maritima,* and *Lulworthia* sp. are also capable of using hydrocarbons** as their sole carbon sources for growth. Cooney et al. (1993) have reported that these obligate marine fungi could grow in artificial sea water with single hydrocarbon as their sole source of carbon and energy. They observed that the unsaturated compound 1, -1,4-tetradecadiene and the methyl branched compound pristane were used by several fungi, while none of the fungi used aromatic hydrocarbons as their sole source of carbon. They found that four of the five fungi examined form microbodies

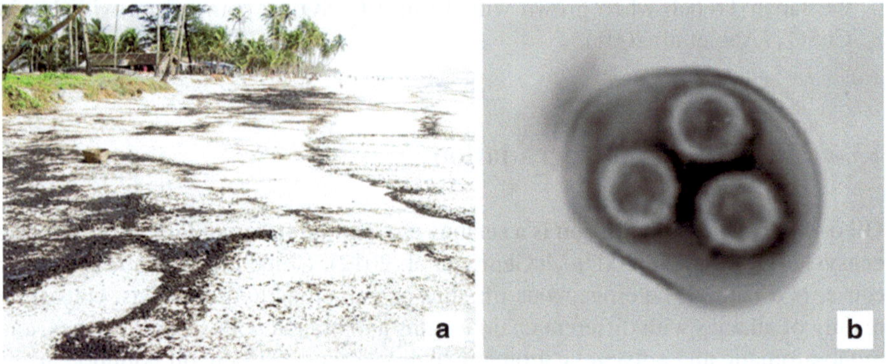


**Fig. 13.5** (**a**) Tar ball pollution on a beach (Photograph: X.N. Verlencar. Kind permission of CSIR-National Institute of Oceanography, Goa, India). (**b**) Adhesion of thraustochytrid cells to a drop of oil. Three cells are seen (Source: Raikar, M.T. et al. 2001. Indian Journal of Marine Sciences 30: 139–145. With permission from CSIR-NISCAIR, India)

when they were grown on hydrocarbons as sole C source but not when they were grown on glucose.

Polycyclic aromatic hydrocarbons (PAHs), which are highly toxic and difficult to break down, can be degraded by fungi that produce LDEs (see Sect. 13.2.2).

One of the most serious problems is that of tar balls that are often washed up on many beaches. Tar balls are made of asphaltenes, which are weathered crude oil fraction (Fig. 13.5a). These are not only difficult to degrade but also cause environmental damage to organisms in the coastal ecosystem. **Thraustochytrids are interesting fungi that are capable of degrading tar balls** (Raikar et al. 2001). Isolates from coastal waters and sediments of Goa coast, India, have been reported to grow in sediments enriched with tar balls and reduce 71% of the tar ball content in such sediments by 1 month. Thirty per cent of tar balls added to culture medium made of peptone broth was degraded by 7 days. Cells of several thraustochytrids were hydrophobic and adhere to hydrocarbons as tested using the MATH (Microbial Adhesion to Hydrocarbons) assay. This is an indication that such cells were capable of utilizing them as substrates (Fig. 13.5b).

## 13.5 Secondary Metabolites and Bioactive Compounds

The discovery of the first antibiotic penicillin in 1929 by Sir Alexander Fleming from the fungus *Penicillium notatum* and its subsequent use as an effective drug in the 1940s opened the doors for a vast research world on microbial metabolites. Exhaustive research in the last 60 years have contributed to the discovery of a number of microbial metabolites that have successfully helped treat human and animal diseases. **Over 20,000 microbial metabolites have been described so far,**

**most of which are from the terrestrial environment.** More than 60% of all drugs admitted worldwide between 1981 and 2006 are based directly or indirectly on structures from nature (Ebel 2012). Nearly 350 antimicrobials from natural products have entered the market (Silber et al. 2016). **The search for novel drugs from biological sources, particularly bacteria and fungi, continues because cures for a number of diseases are still not available.** A serious concern is also that many disease-causing microorganisms have developed resistance to existing antibiotics. Thus, MDRB (Multi-drug resistant bacteria) have become a serious concern.

**There are several reasons for exploring marine biodiversity for novel metabolites and drugs** (Bhatnagar and Kim 2010; Blunt et al. 2016). Firstly, the number of novel metabolites that have been discovered from terrestrial sources in the last thirty decades has declined. Besides, access to and research of unique marine habitats, such as the deep-sea, hydrothermal vents and coral reefs, has improved tremendously in the last three decades, revealing how unique the marine environment is compared to land. Our knowledge of the unique diversity of microorganisms, of prokaryotes and eukaryotes, and of bacteria and fungi with special adaptations inhabiting the marine environment has been coming to light in recent years. The search for novel marine metabolites has become an important pursuit of biotechnology in the last three decades. **Nearly 16,000 natural products have been discovered from marine sources. Of these only about a fourth have been explored for their biological activities** (Blunt et al. 2015; Rangel and Falkenberg 2015).

**The maximum number of marine natural products (MNPs) have come from sponges (Porifera), followed by cnidarians and fungi** (referred to as Ascomycota by the authors) (Fig. 13.6) (Blunt et al. 2015).

Marine fungi, therefore, deserve tremendous attention for developing drugs that can combat a variety of human illnesses.

The first natural product from a marine fungus, Cephalosporin C, was discovered as early as 1946, from *Acremonium chrysogenum*, isolated from seawater close to a sewage outlet off the Sardinian coast by Italian researchers and characterized in 1955 (Imhoff 2016). Synthetic analogues of Cephalosporin C such as Cefalotin are sold as cephalosporin antibiotics.

However, a systematic exploration for marine fungal metabolites started only in the late 1980s. Discovery of novel metabolites from marine fungi has been increasing steadily ever since (Bhadury et al. 2006; Imhoff 2016).

**Starting from a meager 15 compounds in 1992, over 2100 marine fungal metabolites have been discovered so far** (Ebel 2012; Gomes et al. 2015; Blunt et al. 2015, 2016). **Biological activities of marine fungal metabolites have been studied with reference to antibiotic and anticancer properties, cell cycle inhibition, antagonism of platelet activating factor, antiviral activity, neuritogenic activity, phosphatase and kinase inhibition, and radical scavenging activities.**

**Discovery of bioactive metabolites depends on a combination of the organism that is being investigated and the environment to which it is adapted.**

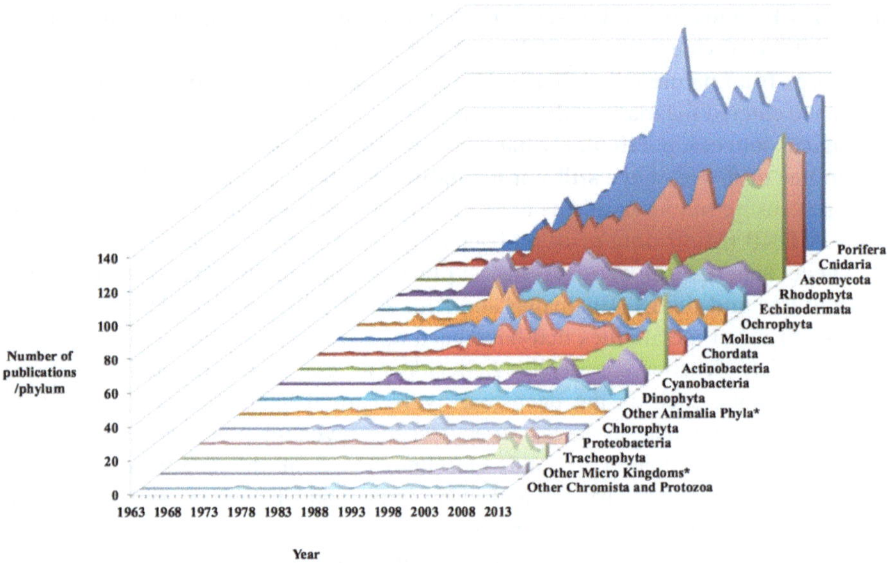


**Fig. 13.6** The phylum preferences of the marine natural product research community across a 50-year period from 1963. (Source: Blunt, J.W. et al. 2015. Nat. Prod. Rep. 2015, 32, 116–211. © The Royal Society of Chemistry)

Obligate and facultative marine fungi, as well as marine-derived fungi from a wide variety of habitats have been explored for bioactive metabolites. These include lignicolous and algicolous fungi, endophytes from seagrasses and algae, corals, invertebrates such as sponges, marine sediments, the deep-sea including hydrothermal vents and a number of other habitats (Table 13.2).

**Certain groups of fungi have been studied inadequately in some habitats; some common groups have not been studied in detail and several habitats are yet to be explored.** Members of the Straminipila have been least investigated for secondary metabolites. This is so possibly because zoosporic oomycetes and labyrinthulomycetes generally have a ruderal ecological strategy and are considered poor producers of secondary metabolites. Facultatively marine and marine-derived fungi have been the most popular targets for explorations of bioactive substances. They grow faster than obligately marine fungi and have been isolated from a large variety of marine habitats. Obligate marine fungi are not known from many marine habitats such as the deep-sea and marine sediments. Species that might occur in such habitats could be useful in marine natural product research. Many habitats have been poorly studied for fungi producing useful metabolites. These include coastal and offshore pelagic sediments as well as the deep-sea environments. There is much scope to further culture fungi and investigate them for useful natural products.

**Fungi that live endobiontically within animals and plants have been found to produce a number of novel and interesting metabolites.** Of particular interest are endophytic fungi that live within algae. Endophytes in terrestrial plants have long been considered a source of interesting metabolites (Suryanarayanan et al. 2009; Weber 2009). Most marine fungal metabolites have been discovered from endophytic fungi in algae (Fig. 13.7; Rateb and Ebel 2011; Ebel 2012; Suryanarayanan et al. 2010; Suryanarayanan 2012). Many fungi isolated from green, red, and brown algae also produce antialgal, antifungal, and anti-insect metabolites which may help in deterring colonization of algal thalli by other microbes thereby reducing competition and in warding off herbivores. Fungi inhabiting sponges are another prolific source of interesting metabolites. It is not clear whether fungi cultured from sponges occurred as dormant spores or as active mycelia. An important, emerging source of fungi for metabolite discovery is the deep-sea habitat and hydrothermal vents.

**Most fungal metabolites, terrestrial or marine, are polyketides. Other important structures belong to alkaloids, terpenoids, and peptides** (Fig. 13.8; Ebel 2012). Examples of the different types of these compounds are given in Table 13.3.

**Obligate marine mycetaen fungi have been shown to produce interesting bioactive compounds.** About 500 species of marine Ascomycota and Basidiomycota are known and these grow mostly in lignocellulosic substrates, macroalgae, or corals. Chytridiomycota are few in number in the marine ecosystem and hardly any research has been carried out on these. Experience from freshwater

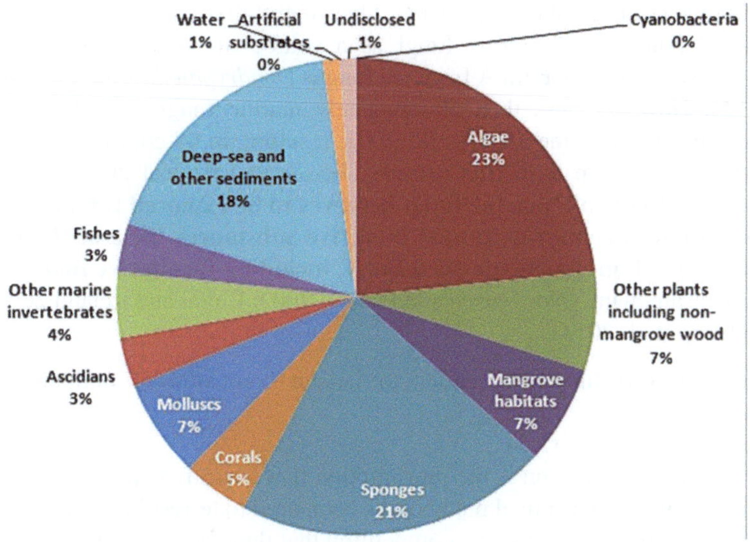


**Fig. 13.7** Habitats which have yielded novel natural products from marine-derived fungi (Source: Adapted from Rainer Ebel. 2012. Natural products from marine-derived fungi. Chap. 20. In: Marine Fungi: and Fungal-like organisms. (Eds.: Jones, E.B. Gareth; Pang, Ka-Lai). De Gruyter. Courtesy: De Gruyter. pp. 421, Fig. 20.3)

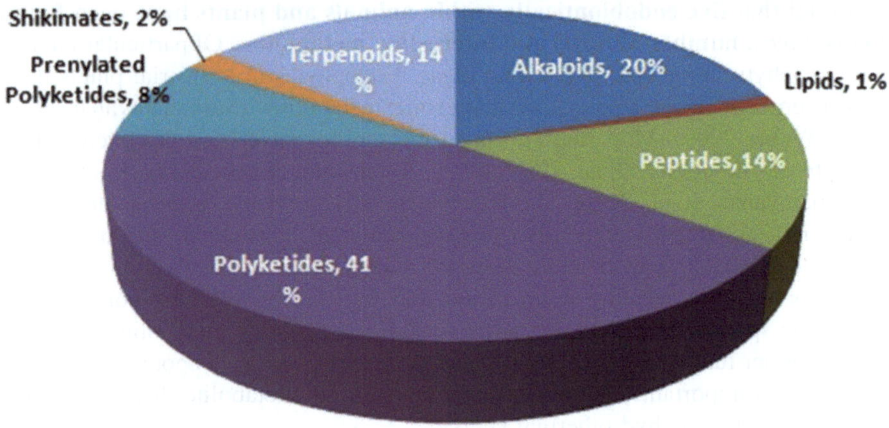


**Fig. 13.8** Different classes of novel natural compounds discovered from marine-derived fungi (Courtesy: De Gruyter. Source: Adapted from Rainer Ebel. 2012. Natural products from marine-derived fungi. Chap. 20. In: Marine Fungi: and Fungal-like organisms. (Eds.: Jones, E.B. Gareth; Pang, Ka-Lai). pp. 414, Fig. 20.2)

chytrids has shown that these are poor in the production of secondary metabolites. Unlike mycetaen fungi, the straminipilan fungi have rarely been studied for bioactive compounds. Most of the studies on obligate marine fungi have been carried out on ascomycetes and basidiomycetes. Several novel bioactive compounds have been recovered from obligate marine fungi (Table 13.4; Bugni and Ireland 2004; Raghukumar et al. 2008a; Ebel 2012). One of the first compounds that demonstrated the potential of marine fungi as a source of metabolites with unique chemistry was dendryphiellin A from the fungus *Dendryphiella salina* (Tables 13.4 and 13.6). However, less than 50 obligately marine fungi seem to have been searched for bioactive metabolites till 2002, resulting in 48 compounds. Seven of these are known from terrestrial sources (Jensen and Fenical 2002). **Secondary metabolites of obligate marine fungi deserves to be explored further**.

**Most studies on marine, fungal bioactive substances have in the last two decades focused on marine-derived fungi, including facultative marine fungi**. The reasons are many fold (Jensen and Fenical 2002; Bugni and Ireland 2004; Ebel 2012; Overy et al. 2014).

- Marine-derived fungi are easier to isolate compared to obligate marine, mycetaen fungi.
- They also grow faster.
- A very important reason is the presumption that terrestrial species from marine sources may produce novel metabolites compared to terrestrial counterparts as a chemical defense to overcome competition that they face in the marine environment, which they may not on land and in freshwater.

**Table 13.3** Examples of natural products discovered from marine-derived fungi

| | | | |
|---|---|---|---|
| Cephalosporin C originally isolated from *Acremonium* sp. 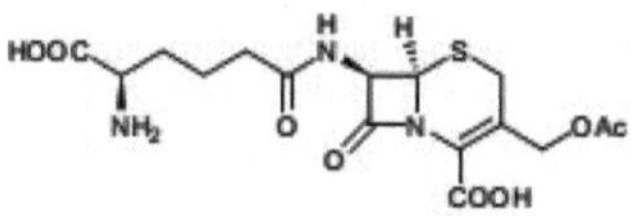 **1** | | Polyketide Monodictysin from *Monodictys putredinis* isolated from a marine green alga: inhibits cytochrome P450 1A activity which is involved in the metabolic conversion of procarcinogens into carcinogens | 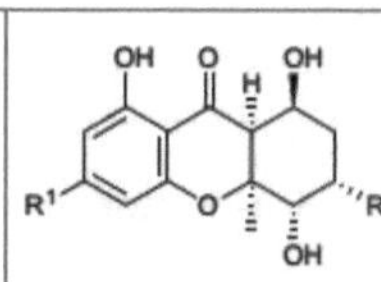 **5** ($R^1 = CH_3$, $R^2 = OH$) **6** ($R^1 = H$, $R^2 = CH_3$) **7** ($R^1 = OCH_3$, $R^2 = CH_3$) |
| Prenylated polyketide new macrocyclic trichothecenes, from *Myrothecium roridum*, isolated from a wood sample: has strong cytotoxic activity towards L1210 murine leukemia cells | 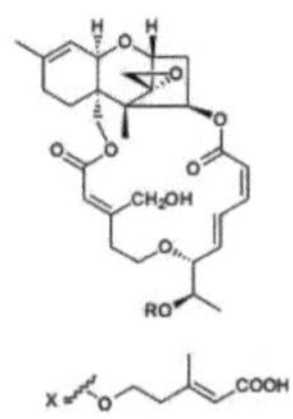 **12** (R = OH) **13** (R = X) | The diterpene phomactin (terpenoid) from *Phoma* sp. isolated from the shell of the crab *Chironoecetes opilio:* act as antagonists of platelet-activating factor | 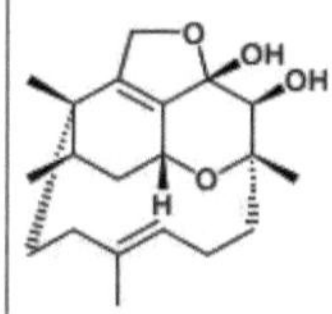 **16** |
| Lipodepsipeptides (Peptide) acremolides A from an estuarine isolate of *Acremonium* sp.: synergizes chaetoglobosin cytotoxicity | 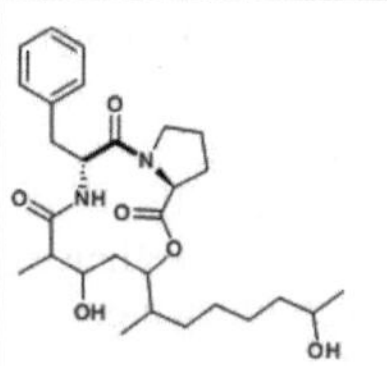 | Alkaloid Communesins from *Penicillium* spp.: possess antiproliferative activity against different leukemia cell lines | 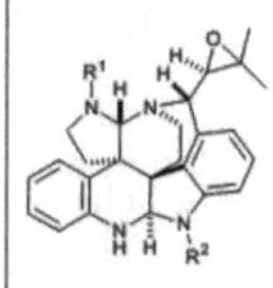 **35** ($R^1 = Ac$, $R^2 = H$) **36** ($R^1 = X$, $R^2 = CH_3$) **37** ($R^1 = X$, $R^2 = H$) **38** ($R^1 = X$, $R^2 = CHO$) |
| Terprenins—Shikimate-derived natural products (phenylpropanoids) from a marine isolate of *Aspergillus candidus:* moderately active against KB3—1 cells | 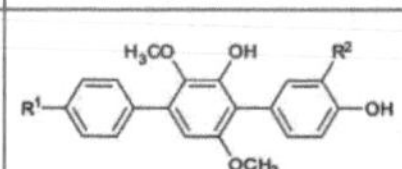 **39** ($R^1 = OH$, $R^2 = prenyl$) **40** ($R^1 = H$, $R^2 = prenyl$) **44** ($R^1 = OH$, $R^2 = H$) | Asperamides (Sphingolipids) from *Aspergillus niger,* isolated from the brown alga *Colpomenia sinuosa:* moderately antifungal 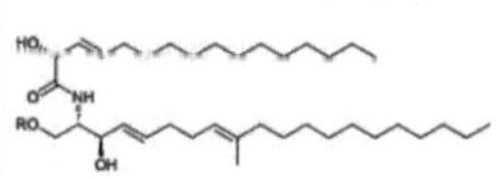 **45** (R = H) **46** (R = β-D-glucopyranosyl) | |

Source: Rainer Ebel 2012. Natural products from marine-derived fungi. Chap. 20. In: Marine Fungi: and Fungal-like organisms. (Eds.: Jones, E. B. Gareth; Pang, Ka-Lai). Göttingen. Walter de Gruyter. Courtesy: De Gruyter

- It has also been shown that such fungi do indeed produce a number of interesting and novel bioactive molecules not known in terrestrial isolates of the same species. These metabolites exhibit anticancer, antibacterial, antiplasmodial, anti-inflammatory, nematicidal, antiviral, and antiangiogenic activities.

**Table 13.4** Examples of secondary metabolites discovered from obligate marine lignicolous fungi

| Fungus | Habitat | Compound | Activity |
|---|---|---|---|
| *Aigialus parvus* S. Schatz & Kohlm | Decaying mangrove wood | Aigialomycin A–E; new resorcyclic macrolides | Aigialomycin D shows moderate antimalarial activity against*P. falciparum* and also cytotoxicity against human cancer cell lines |
| *Aschochyta salicorniae* Magnus | Associated with the marine green alga *Ulva* species | A novel polyketide ascosalipyrrolidinone-A | Antiplasmodial activity towards *Plasmodium falciparum* strain |
| *Corollospora maritima* Werderm. | Decaying wood | Corollosporine | Antibacterial |
| *Dendryphiella salina* (G.K. Sutherl.) G.J.F. Pugh & Nicot | Decaying algae | Dendryphiellins A1, B, C, and D, which are unusual terpenoid derivatives trinoreremophilanes. | – |
| *Halocyphina villosa* Kohlm. & E. Kohlm. | Decaying mangrove wood | Siccayne, a metabolite also known from terrestrial fungi | – |
| *Halorosellinia. oceanica* (S. Schatz) Whalley, E.B.G. Jones, K.D. Hyde & Lassøe | Decaying mangrove wood | Halorosellinic acid, halorosellins A and B, 17-dehydroxyhalorosellinic acid, the isochromanone, and the phenyl lactone | Phenyl lactone is against *Mycobacterium tuberculosis* |
| *Helicascus kanaloanus* Kohlm. | Decaying mangrove wood | Isomeric δ-lactones, helicascolides A and B | |
| *Kallichroma tethys* (Kohlm. & E. Kohlm.) Kohlm. & Volkm-Kohlm. | Decaying wood | Isoculmorin | |
| *Kirschsteiniothelia maritima* (Linder) D. Hawksw. | Decaying wood | Aromatic aldehyde, ascochital | Antibacterial activity against *B. subtilis* |
| *Phaeosphaeria oraemaris* (Linder) Khashn. & Shearer | Salt marsh grass | Oxidatively modified 2-aminohexose derivative—leptosphaerin, leptosphaerolide and leptosphaerodione, Obioninene. | |
| *Byssothecium. obiones* (P. Crouan & H. Crouan) M.E. Barr | Salt marsh grass | Obionin A | Inhibits binding of a dopamine D-1 selective ligand to bovine corpus striatum membrane |
| *Lignincola laevis* Höhnk | Salt marsh grass | 7-hydroxyergosterol and a very unusual phosphorohydrazide thioate | Cytotoxic against L1210 cells |

**Table 13.5** Examples of secondary metabolites discovered from facultative marine fungi and marine-derived fungi

| Source | Fungus | Compound | Activity |
|---|---|---|---|
| Sediment | *Aspergillus glaucus* | Anthraquinone derivative, aspergiolide | Cytotoxicity towards A-549 and HL-60 cells |
| | *Spicaria elegans* | New cytochalasin derivatives | Mildly cytotoxic to P388 and A-549 cells |
| Sediment sample from 5059 m | *Phialocephala* sp. | New bisorbicillinoids oxosorbiquinol | Weakly cytotoxic to five different cancer cell lines |
| Endophyte of the green alga *Ulva* sp. | *Ascochyta salicorniae* | New deoxytetramic acids, ascosalipyrrolidinones, and the new α-pyrone ascosalipyrone | Antibacterial and anti-plasmodial activities |
| Caribbean seagrass *Halodule wrightii* | *Scytalidium* sp., | Novel lipophilic, linear hexapeptides | Strong inhibition of *Herpes simplex* viruses 1 and 2 in vitro |
| Sponge *Halichondria japonica* | *Gymnascella dankaliensis* | Unusual nitrogen-containing polyketides, gymnastatins, as well as structurally unusual steroids | Pronounced cytotoxic activity against the P388 cancer cell line |
| Sea fan *Annella* sp. | *Nigrospora* sp. | New epoxydon esters, nigrospoxydons A–C, and the new pyrone, nigrosporapyrone | Nigrospoxydon A exhibited weak antibiotic activity against *Staphylococcus aureus* and methicillin-resistant *S. aureus* |
| Sea cucumber, *Eupentacta fraudatrix* | *Acremonium striatisporum* | New isopimaradiene diterpene glycosides virescenosides M | Cytotoxic on developing eggs of the sea urchin *Strongylocentrotus intermedius*, which is often associated with cytotoxic activity against Ehrlich carcinoma cells. |
| The common mussel, *Mytilus edulis* | *Aspergillus* sp. | Notoamides—structurally complex prenylated diketopiperazines | Moderate cytotoxicity towards HeLa cells |
| Japanese fish *Mugil cephalus* | *Chaetomium globosum* | Complex azaphilones named chaetomugilins | Significant cytotoxic activity |

- Some examples of compounds discovered from marine-derived fungi, as listed in Ebel (2012) from various habitats, are given in Table 13.5.

**A number of marine fungal metabolites are promising anticancer agents** (Fig. 13.9; Gomes et al. 2015).

- A diketopiperazine halimide (**phenylahistin**) was discovered by the research group of Bill Fenical at the Scripps Institute of Oceanography in the 1990s from

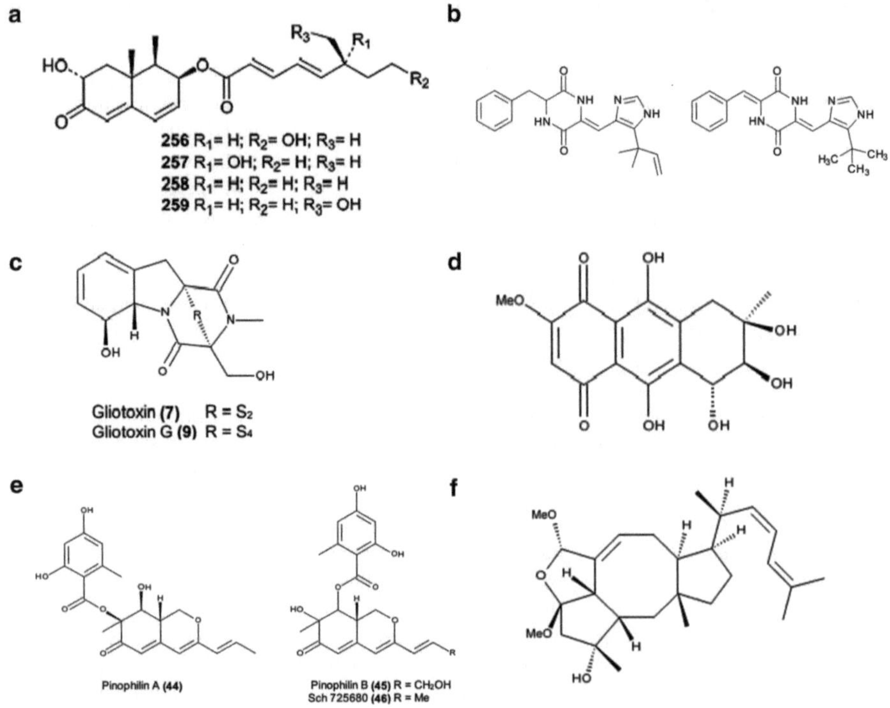


**Fig. 13.9** Some anticancer compounds obtained from marine-derived fungi. (**a**) Dendryphiellin from *Dendryphiella salina*. (Reproduced from Bugni, T.S. and C.M. Ireland. Marine-derived fungi: a chemically and biologically diverse group of microorganisms. Nat. Prod. Rep., 2004, 21:143–163 with permission of The Royal Society of Chemistry). (**b**) Phenylhistin from *Aspergillus* sp. and its synthetic analogue plinabulin. (Source: Rainer Ebel. 2012. Natural products from marine-derived fungi. Chap. 20. In: Marine Fungi: and Fungal-like organisms. (Eds.: Jones, E. B. Gareth; Pang, Ka-Lai). Göttingen. Walter de Gruyter. Courtesy: De Gruyter). (**c**) Gliotoxin from *Aspergillus* sp. (**d**) Bostrycin from *Nigrospora* sp. (**e**) Pinophilin from *Penicillium pinophilum*. (**f**) Ophiobolin O from *Aspergillus ustus* (Source, **c–f**: Gomes, M.G.M. et al. Can some marine-derived fungal metabolites become actual anticancer agents? Mar Drugs 2015, 13, 3950–3991. © 2015 by the authors; licensee MDPI, Basel, Switzerland. open access article of Creative Commons Attribution license (http://creativecommons.org/licenses/by/4.0/).

a species of *Aspergillus* collected in the waters off the Philippine Islands. This compound is capable of tubulin-depolymerizing activity. **A synthetic analogue of this compound, Plinabulin** (NPI-2358), is used to disrupt vascular endothelial architecture of tumors and promote apoptosis (Ebel 2012). Plinabulin is now undergoing Phase 1, Phase 2, and Phase 3 clinical trials against solid tumors and lymphomas by Nereus Pharmaceuticals and BeyondSpring Pharmaceuticals (Rangel and Falkenberg 2015).

- **Gliotoxin** was isolated from a strain of *Aspergillus* sp. cultured from a marine brown alga in Korean waters. This and some of its metabolites are actually potential anticancer drugs due to their activities against methylation regulation enzymes HMT G9a and HMT Set7/9.

- **Bostrycin** is a compound that was extracted from a culture of *Nigrospora* found as an endophyte in mangroves of South China Sea. Bostrycin and its analogues have various activities against non-small cell lung cancer (NSCLC), radioresistant nasopharyngeal cancer cells, and adriamycin-resistant breast cancer cells.
- The compounds **pinophilin A and B** are hydrogenated azaphilones obtained from a strain of *Penicillium pinophilum* Hedgcok. This fungus was isolated from seaweed collected along the coast of Kasai Marine Park in Tokyo, Japan. It is selectively cytotoxic against cancer but not against normal cells. Its activity is also unique because it specifically inhibits certain mammalian DNA polymerases.
- A marine-derived isolate of *Aspergillus ustus* yielded a compound called **Ophiobolin O**. It acts against several kinases. It induces apoptosis in human MCF-7 breast cancer cells, which are deficient in caspase-3 crucial for apoptosis. Ophiobolin A, its analog, kills glioblastoma cells that are apoptosis-resistant and also reverses the resistance of human breast cancer cells to the drug adriamycin. These compounds could be used against various types of cancers with different levels of resistance to pro-apoptotic stimuli.

**Novel approaches:** It is possible to induce the production of novel compounds or to enhance production of existing bioactive compounds in marine fungi by various ways (Jensen and Fenical 2002; Ebel 2012; Suryanarayanan 2012).

- **Cocultivation of a marine fungus with another organism may trigger the expression of silent gene clusters, resulting in the synthesis of novel metabolites** (Ebel 2012). For example, cultivation of the fungus *Pestalotiopsis* isolated from the brown alga *Rosenvingea* sp. and *Libertella* sp. *from* an ascidian together with a marine-proteobacterium resulted in the production of new compounds. Under such conditions, the former fungus produced a new chlorinated benzophenone antibiotic active against methicillin-resistant *Staphylococcus aureus* and vancomycin-resistant *Enterococcus faecium*. *Libertell*a sp. produced new pimarane diterpenoids, libertellenones A–D, that showed cytotoxicity towards HCT-116 human adenocarcinoma cells.

  Another example is that of *Emericella* sp., isolated from the Papua New Guinean green alga *Halimeda* sp. This fungus produced enhanced levels of the antibacterial cyclic lipodepsipeptides Emericellamides A and B upon cocultivation with the marine actinomycete *Salinispora*. This is the result of enhanced expression of emericellamide gene cluster. *Phomopsis asperagi* produces novel chaetoglobosins when the growth medium is amended with the sponge metabolite jasplakinolide.
- It has been suggested that culturing **marine-derived fungi in different salinities may alter the production of a given metabolite.** Miao et al. (2006) reported that the antibiotic activity of a marine-derived fungus increased with

salinity of the growth medium. Some marine-derived species exhibit increased growth with increasing seawater concentration in the medium while antimicrobial activity appeared to be maximum. Besides, altering salinity levels in culture may also change the metabolite profile (Bugni and Ireland 2004). For example, *Arthrinium c.f. saccharicola* exhibited higher antibacterial activity at 25 °C and 34 ppt seawater than under other conditions. Its activity also increased significantly when cell-free culture broth of the bacterium *Pseudoalteromonas piscicida* was added (Miao et al. 2006).

## 13.6 Omega-3 Fatty Acids, High Value Lipids, and Carotenoids from Thraustochytrids

**A major commercial biotechnology product of recent years is microbial oil rich in the omega-3 polyunaturated fatty acid, docosahexaenoic acid. The technology is based on the straminipilan fungi, the thraustochytrids** (Winwood 2013).

Polyunsaturated fatty acids (PUFAs) are characterized by two or more double bonds. PUFAs with the last double bond at the third position from the methyl end of the fatty acid chain are omega-3 fatty acids (ω3 or n3 PUFAs) and those having this at the sixth position are omega-6 PUFAs (Fig. 13.10).

Studies in the 1970s indicated that the **ω3 PUFAs, docosahexaenoic acid (DHA), and eicosapentaenoic acid (EPA) are essential fatty acids that provide cardiovascular health.** EPA and DHA may affect many aspects of cardiovascular function including inflammation, peripheral artery disease, major coronary events,

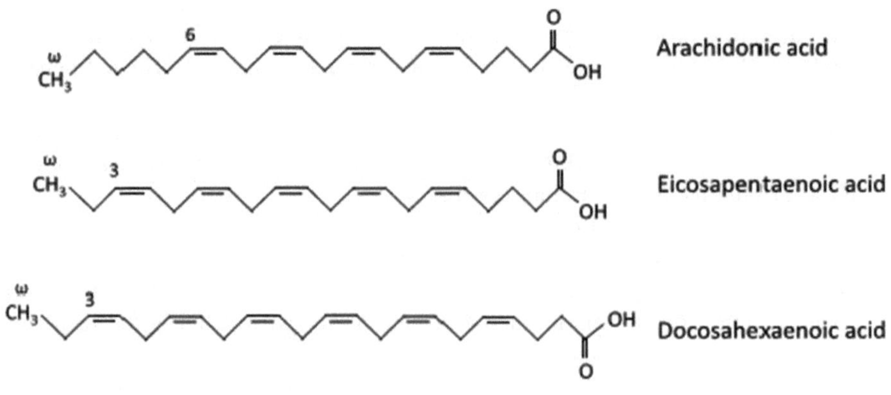


**Fig. 13.10** Structures of the omega-6 PUFA Arachidonic acid and the omega-3 PUFAs eicosapentaenoic acid and docosahexaenoic acid (Source: Maskrey, B.H. et al. 2013. Emerging importance of omega-3 fatty acids in the innate immune response: Molecular mechanisms and lipidomic strategies for their analysis. Mol Nutr Food Res 57: 57, 1390–1400. C 2013 Wiley-VCH Verlag GmbH & Co. KGaA, Weinheim with permission of John Wiley and Sons)

and anticoagulation. Numerous studies since then have confirmed this and also shown that **DHA and EPA are also important for proper fetal development, including neuronal, retinal, and immune function** (Simopoulos 2008; Swanson et al. 2012). DHA as well as the omega-6 fatty acid arachidonic acid (ARA) are important fatty acids in the brain. Oils containing PUFAs, therefore, are extremely useful for human health. They are also important in the diet and health of aquaculture fish that humans consume (Tocher 2015).

The major source of ω3 PUFA-rich oil has been fish oil. However, declining catches of the target fish species, increasing demand for the oil, fears of heavy metal contamination in fish oil, and a demand for vegetarian source of oil have resulted in a large market for microbial oil. There are two commercial sources for this, namely, the one from the heterotrophic dinoflagellate *Crypthecodinium cohnii* and the other from thraustochytrid fungi, particularly species of *Schizochytrium* and *Aurantiochytrium* (Ward and Singh 2005; Fan and Chen 2007; Fan et al. 2007; Raghukumar 2009; Barclay et al. 2010; Winwood 2013). Thraustochytrid oil is termed "microalgal oil" because the straminipilan fungi, labyrinthulomycetes, are related to algae of the Kingdom Straminipila than to the Kingdom Mycetae, which are commonly known as Kingdom Fungi.

**DHA is a signature fatty acid of the Labyrinthulomycetes.** Use of DHA-rich oil from thraustochytrids for human and aquaculture applications were patented in the early 1990s. Many thraustochytrids, particularly species of *Schizochytrium* and *Aurantiochytrium*, are extremely rich in storage triglyceride oils (Fig. 13.11). These often make up at least 50% of the dry weight biomass of these organisms, DHA frequently contributing 25% of the fatty acids or even more. ***Aurantiochytrium limacinum*****, a typical mangrove inhabitant, has been studied by numerous researchers for DHA production.**

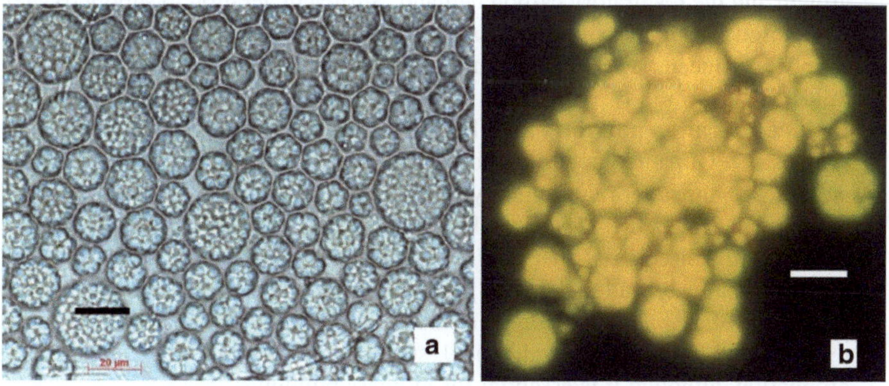


**Fig. 13.11** Cells of *Schizochytrium* sp. filled with lipid globules. (**b**) Epifluorescence microscopy of cells of *Schizochytrium* sp. stained with Nile blue and showing fluorescent lipids. Bar represents 20 μm (Source: Raghukumar, S. 2008. Thraustochytrid marine protists: production of PUFAs and other emerging technologies. Mar Biotechnol 10: 631–640. © Springer Science + Business Media, LLC 2008)

The major fatty acid in triglycerides of *Schizochytrium*, as well as in most other thraustochytrids, is the saturated fatty acid, palmitic acid. Other important fatty acids, such as the omega-3 docosapentaenoic acid (DPA), EPA, ARA, and mono-unsaturated fatty acids, may contribute low levels.

The path from discovery to commercialization achieved by Omega Tech and Martek Biosciences, pioneers in this field, is now well documented (Barclay et al. 2010).

The major steps in developing PUFA-rich microbial oil from *Schizochytrium* have been as follows.

1. **Screening and selection of the candidate strain**: The production strain should grow rapidly and have high lipid and DHA contents.
2. **Replacement of sodium chloride with sodium sulfate**: $Na^+$ ions are required by thraustochytrids. Seawater that is normally used for culturing these fungi has the required amount of sodium in the form of sodium chloride. Chloride ion causes corrosion of fermentors. A technology of replacing seawater with sodium sulfate was devised.
3. **Optimization of a fed batch culture mode**: Biomass yield depends upon the total amount of carbon fed to the organism. However, excess amount of glucose in culture medium causes severe osmotic stress. Hence, a steady dose of glucose is supplied in the form of fed batch cultures. Nitrogen is most easily supplied in the form of an organic source, such as peptone or sodium glutamate. Ammonium salts can be provided to bring down the costs. This, however, may cause rapid lowering of pH, and fermentation has to be regulated very carefully to avoid acidification.
4. **Enhancement of oil yield**: Thraustochytrids are grown in two stages during the fed batch mode. During the first stage, biomass is built up to a high level by using appropriate amounts of carbon and nitrogen that meet the requirements of cell growth and the respiratory carbon demand. Sufficient aeration to supply oxygen is required for biomass buildup. In the second, "fattening up" stage, nitrogen supply is reduced but the carbon supply is continued such that cells accumulate lipids.
5. **Reduced aeration and fatty acid buildup**: Conventional fatty acid pathway to synthesize PUFAs requires a number of steps involving various elongases and desaturases, the latter requiring oxygen. An interesting discovery was that DHA buildup in *Schizochytrium* can take place in the absence of oxygen. It was subsequently discovered that *Schizochytrium* possesses an unusual fatty acid synthesis pathway, called the PKS pathway, originally discovered in the marine bacterium *Shewanella* sp. (Hauvermale et al. 2006). Hence, supply of air is reduced or stopped during the final stages of fermentation to enhance oil production, although it is essential for biomass buildup.
6. **Downstream processing**: Whole cell or extracted oil may be used depending on the purpose. Animal feeds have been based on the former, while human nutraceuticals are developed from the oil. Biomass may be harvested by

centrifugation or preferably by drum drying. The oil is extracted using standard hexane extraction or supercritical drying, the former being more common. It is essential to prevent oxidation of PUFAs during this stage. Subsequently, the oil is refined to improve color, clarity, odor, and remove any particulate material and chemical contaminants.

**Using the above process, astounding values 170–210 g dry weight biomass $L^{-1}$ culture with at least 50% oil containing 35–45% of total PUFAs have been achieved.**

A representative fatty acid profile is given in Table 13.6. Such high levels are not easily achieved in commonly isolated strains under standard culture conditions.

**Aquaculture nutrition products and subsequently human nutraceuticals comprising *Schizochytrium* oil were marketed by Omega Tech and later by Martek Biosciences in the USA. Presently, the major manufacturer of *Schizochytrium* "microalgal" oil is DSM**, who make two different products. "Life's DHA™ oil" is standardized at 35% or 40% DHA and contain less than 2% of EPA. "Life's Omega™" resembles fish oil more closely and is made from a different *Schizochytrium* strain. It contains 40% total DHA and EPA and is specified individually as having minimum levels of 24% DHA and 12% EPA (2:1 ratio as in fish oil). These oils are now used in over 500 food, beverage, and supplement

**Table 13.6** A general fatty Acid Profile of omega-3 PUFA-rich microbial oil from *Schizochytrium* sp. with names of some common ones

| Fatty acid | % in total lipids | Name |
|---|---|---|
| 12:0 | 0.2 | Lauric acid |
| **14:0** | **7.6** | Myristic acid |
| 14:1 | 0.1 | Myristoleic acid |
| 15.0 | 0.4 | Pentadecylic acid |
| **16.0** | **22.1** | Palmitic acid |
| 16:1 | 0.4 | Palmitoleic acid |
| 18:0 | 0.5 | Stearic acid |
| 18:1 | 0.8 | Oleic acid |
| 18:2 | 0.5 | Linoleic acid (LA) |
| **18:3(n-6)** | **0.3** | Gamma linolenic acid |
| 18:4 | 0.3 | Stearidonic acid |
| 20:0 | 0.1 | Arachidic acid |
| **20:4(n-6)** | **1.0** | Arachidonic acid |
| **20:4(n-3)** | **1.0** | Eicosatetranoic acid |
| **20:5(n-3)** | **2.3** | Eicosapentaenoic acid |
| 22:0 | 0.1 | Behenic acid |
| **22:5(n-6)** | **16.6** | Docosapentaenoic acid |
| **22:6(n-3)** | 40.9 | Docosahexaenoic acid |
| 24:0 | 0.2 | Lignoceric acid |
| 24:1 | 0.2 | Nervonic acid |
| Others | 4.3 | |

Based on: Barclay et al. (2010)

products worldwide. *Schizochytrium* microalgal oil now has a large share of the world market for PUFA oils of several hundred million US$ per annum.

**Further improvement of the thraustochytrid microbial oil technology is being addressed by many researchers** (Barclay et al. 2010).

- Attempts are now being made to isolate **novel strains**, particularly those with specific fatty acid profiles. Aside from DHA, some strains of thraustochytrids also produce other PUFAs such as ARA, DPA, EPA, and docosatetraenoic acid (Burja et al. 2006). Such oils may find specific health applications.
- Methods to increase yield or modify fatty acid profiles by **improving or altering culture conditions and fermentation** are being attempted. For example, the use of Response Surface Methodology, a statistical method to optimize culture protocols, resulted in nearly 50% DHA in lipids (Manikan et al. 2015). Cheaper carbon sources such as glycerol will make the process even more economical. Continuous culture methods, instead of fed batch fermentation, may improve yield of total fatty acids and DHA productivity (Ganuza and Izquierdo 2007).
- **A better understanding of the PKS fatty acid synthesis pathway** of *Schizochytrium* is being pursued in order to be able to improve PUFA yields and also to enable expression of these genes in plants (Cheng et al. 2013).
- A genetic system for multiple transgene expression and targeted gene knockout in thraustochytrids is considered useful for production of large amounts of tailor-made oils. A versatile transformation system has been developed in *Aurantiochytrium, Parietichytrium, Schizochytrium*, and *Thraustochytrium aureum*. Heterozygous gene expression has also been demonstrated (Sakaguchi et al. 2012).
- **Attempts are being made to develop novel animal and feeds** to enhance health of crustaceans, fish, poultry, and livestock and also to provide PUFAs to human beings who depend on such products for their nutrition (Leaño and Damare 2012).

**The realization of the importance of thraustochytrids in omega-3 fatty acid production has led to studies on other valuable lipids from these fungi.** Thraustochytrids are now targets for **biofuel research** (Chang et al. 2012, 2014). Many thraustochytrids produce substantial amounts of saturated and mono-unsaturated fatty acids. These possess better oxidative and thermal stability and are therefore preferable in biodiesels. Many species contain a total of 30–40% of fatty acids in the form of SFAs and MUFAs. **Thraustochytrids also produce the powerful antioxidant squalene**, which is a hydrocarbon and triterpene. Squalene is presently obtained from the liver oils of deep-sea sharks. Several strains of *Aurantiochytrium* capable of producing nearly 1 g of squalene per liter culture have been found (Nakazawa et al. 2014).

Carotenoid pigments are important antioxidants that scavenge free radicals and are crucial in human health. **Many thraustochytrids produce the carotenoid β-carotene and oxygenated carotenoids such as the xanthophylls astaxanthin**

**and canthaxanthin, zeaxanthin, echinenone, and phoenicoxanthin**, probably to prevent oxidation of PUFAs present in their triglycerides (Aki et al. 2003; Armenta et al. 2006; Burja et al. 2006; Fan and Chen 2007). Astaxanthin is also used in aquaculture to enhance flesh coloration of salmonids and other animals. Up to 8 mg of β-carotene, 7 mg of astaxanthin, and 10 mg of canthaxanthin per liter medium have been reported under particular culture conditions, even in the absence of light. Expression of hemoglobin gene of the bacterium *Vitreoscilla stercoraria* has been shown to improve astaxanthin production in *Aurantiochytrium* sp. Thraustochytrids are also good expression vectors for useful products. Recombinant hemagglutinin proteins of influenza virus have been expressed in *Schizochytrium* sp. (Bayne et al. 2013).

**Thraustochytrids could be the source of novel applications and products in future.**

## 13.7 Other Applications

**A variety of other applications have been explored with marine fungi.**

- One of these is production of **biosurfactants**. Biosurfactants lower the surface tension of liquids and have a number of applications. These include dispersal of oil, detergents, bioremediation, food industries, etc. Interesting biosurfactants have been reported from yeasts such as *Yarrowia lipolytica* from coastal habitats and a yeast isolated from the deep-sea cold seep clam *Calyptogena soyoae* (see Damare et al. 2012). Biosurfactants have also been isolated from marine-derived fungi such as *Aspergillus ustus* from the marine sponge *Fasciospongia cavernosa,* collected from the peninsular coast of India and a species of *Penicillium.*
- Marine-derived fungi belonging to species of *Fusarium* and *Trichosporon* isolated from mangrove soils in China have been shown to **degrade phthalate esters**. These toxic compounds are used as plasticizers and are of environmental concern (Luo et al. 2012).
- **Marine, fungal exopolysaccharides** are another interesting group of compounds to study. Microbial polysaccharides have numerous potential benefits in food and beverage industries, drugs, oil industries, and many others (Rucocco et al. 2016). Pullulan is an important polysaccharide produced by the salt-tolerant black yeast *Aureobasidium pullulan.* Pullulan has a broad spectrum of use in the food and pharmaceutical industry (Leathers 2003; Singh and Saini 2008). Thraustochytids are another source of extracellular polysaccharides (EPS) (Jain et al. 2005; Chang et al. 2014). Since these straminipilan fungi are well known for their microbial PUFA-containing oils (see above), another interesting compound in the form of EPS is attractive. However, no potential applications from these are known.

- Marine fungi are adapted to in situ marine conditions of salinity, pH, temperature, and chemical nature of organic material available for their growth.
- Marine fungi respond in different ways to salinity for growth.
- Obligate marine lignicolous fungi grow and reproduce in media with seawater. Terrestrial species grow in such conditions, but their reproduction and sproe germination are affected.
- The "*Phoma* pattern" is a phenomenon in which several obligate as well as facultative mycetaen marine fungi grow better at higher salinities when growth temperature is higher.
- Straminipilan fungi belonging to Labyrinthulomycetes and Oomycetes are obligately marine and require seawater for growth.
- Fungi from hypersaline environments tolerate much higher levels of salinity compared to other marine fungi.
- The marine yeast *Debaryomyces hansenii* is a model organism for studying salinity tolerance.
- Fungi synthesize a mixture of solutes including polyols and amino compounds or accumulate ions to enable a suitable water potential by which water can flow into the cell. An efficient sodium extrusion or sequestration mechanism helps in preventing toxic levels of sodium within the cell, while maintaining sufficient levels of potassium.
- Marine fungi generally have a near-neutral pH optimum for growth.
- Fungi may be mesophilic, psychrotolerant, or psychrophilic depending on the habitats where they grow.
- Marine mycetaen fungi utilize a wide variety of carbon and nitrogen sources for growth.
- Many members of Labyrinthulomycetes have an obligate requirement for various vitamins.
- Unique enzymes and proteins from marine fungi that serve as survival tools in the environment are important in biotechnology.
- Many fungal enzymes such as proteases, lipases, chitinases, endopolygalacturonases, and lignocelluloytic enzymes act at extremes of temperature, salinity, or pH.
- It may be possible to discover novel genes by studying gene expression under specific conditions. Thus, deep-sea conditions may result in expression of novel genes.
- A number of obligate as well as facultatively marine fungi produce cellulases, xylanases, and lignin-degrading enzymes (LDEs).
- Some fungi are capable of mineralizing lignin to carbon dioxide by virtue of LDEs.
- Many facultative and obligate marine fungi produce high amounts of the LDE, laccase.

(continued)

- Many marine fungi produce high amounts of laccase and are capable of breaking down industrial effluents from paper and pulp industries, molasses based-alcohol distilleries, textile and dye industries, and polycyclic aromatic hydrocarbons (PAHs).
- A few fungi are known to degrade marine algal polymers such as alginates, laminarin, fucoidan, and carrageenan.
- Fungi from hydrothermal vents are ideal targets to study heavy metal tolerance.
- An isolate of the thraustochytrid *Ulkenia* sp., isolated from shallow water hydrothermal vents, grew better when Fe, Mn, or Pb were included in the culture medium than without.
- Some yeasts produce the organic chelators, siderophores, which increase the bioavailability of Fe(III).
- Many obligate as well as marine-derived fungi possess various levels of tolerance to different heavy metals.
- Oil or hydrocarbon pollution is a serious environmental problem in the marine ecosystem. Microbial populations act as a consortium in degradation of hydrocarbons in the environment.
- Yeasts are common in hydrocarbon polluted waters and are capable of degrading hydrocarbons.
- Thraustochytrids could be interesting fungi that are capable of degrading tar balls.
- Marine natural products are promising sources of novel drugs. Ascomycetes and basidiomycetes produce a number of secondary metabolites.
- Over 2100 marine fungal metabolites have been discovered so far.
- Marine fungal metabolites have been studied for antibacterial and anticancer properties, cell cycle inhibition, antagonism of platelet activating factor, antiviral activity, neuritogenic activity, phosphatise inhibition and kinase inhibition, and radical scavenging activities.
- Fungi that live endobiontically within animals and plants produce a number of novel and interesting metabolites.
- Most fungal metabolites are polyketides. Other important structures belong to alkaloids, terpenoids, and peptides.
- Most studies on marine fungal bioactive substances have in the last two decades focused on marine-derived fungi, including facultative marine fungi.
- A number of marine fungal metabolites, such as phenylahistin, gliotoxin, bostrycin, inophilins A and B, and ophiobolin O are promising anticancer agents.
- Novel approaches to produce marine fungal metabolites include cocultivation with another organism and culturing under different growth conditions.

(continued)

- The ω3 PUFAs, DHA and EPA are essential fatty acids that provide cardiovascular health, brain, and retinal development and prevent inflammation.
- *Schizochytrium* sp. belonging to Labyrinthulomycetes is the source of a commercial technology to produce DHA-rich biomass or oil. Commercial strains produce high biomass, lipid yield, and DHA.
- *Aurantiochytrium limacinum*, a typical mangrove inhabitant, has been studied by numerous researchers for DHA production.
- *Schizochytrium* sp. possesses an unusual, anaerobic PKS pathway for PUFA synthesis.
- Aquaculture nutrition products and human nutraceuticals comprising *Schizochytrium* oil are important products in the market.
- Thraustochytrids are now targets for biofuels, squalene, and carotenoids.
- A variety of other applications that have been explored with marine fungi include biosurfactants and fungal polysaccharides.

**Future Directions**

1. Salinity tolerance mechanisms of marine fungi are still not properly understood. Further studies will be useful in developing salt resistant plants.
2. Few studies have focused on psychrophilic fungi, of which there are sufficient indications. Physiological and molecular mechanisms of low temperature tolerance in fungi require attention.
3. Few studies have addressed degradative enzymes from fungi in marine sediments and extreme marine environments.
4. Most enzyme studies on fungi have focused on lignocellulases. Enzymes that degrade typical marine polymers have been inadequately addressed.
5. There is a large scope to understand production and properties of lignocellulases from obligate marine lignicolous fungi as well as facultative marine fungi and marine-derived fungi from mangroves.
6. Thraustochytrids appear to be interesting organisms from the point of view of hydrocarbon degradation, particularly the asphaltene fraction.
7. Despite a large volume of research on marine fungal metabolites, fungi from coastal and deep sea sediments have not been paid attention to.
8. Attempts should be made to explore the presence of novel, obligate marine fungi from coastal and deep sea sediments and study them for secondary metabolites.
9. There is an enormous scope to further develop innovative technologies for omega-3 PUFAs from thraustochytrids and also many other lipids and other metabolites.

# Chapter 14
# Origin and Evolution of Marine Fungi

*It is not the strongest of the species that survives, nor the most intelligent that survives. It is the one that is most adaptable to change.*

Charles Darwin

**What is the evolutionary origin of marine fungi?**

**Did fungi evolve in the sea or on land?**

These two fundamental questions have engaged the attention of mycologists for a long time and even more so since obligate marine fungi were discovered. These questions are also intimately related to the evolution of other opisthokont groups such as metazoans and choanozoans, since mycetaen fungi occupy a key position in the evolutionary tree of opisthokonts.

Most life forms evolved in the sea. In this context, it is interesting to address the evolutionary origin of the two groups of fungi, namely, the mycetaen and straminipilan fungi. **This question assumes importance particularly in the case of the mycetaen fungi, since the vast majority of these fungi are terrestrial. On the contrary, straminipilan fungi belonging to Labyrinthulomycetes are entirely marine. The oomycetan straminipilan fungi are highly diverse both in the marine and terrestrial environments.** The evolutionary history of these groups, therefore, needs to be considered separately.

## 14.1 Evolution of Marine Mycetaen Fungi

**The Kingdom Mycetae or Fungi and the Kingdom Metazoa or Animalia belong to the monophyletic supergroup Opisthokonta and share a common ancestor** (Fig. 1.1; Müller 2003; Cavalier-Smith 2004; James et al. 2006; Steenkamp et al. 2006). The choanoflagellates are the closest sister group of animals. The earliest animals, comprising the multicellular Porifera or sponges, evolved in the oceans probably around 600 Mya from choanoflagellate-like ancestors.



S. Raghukumar, *Fungi in Coastal and Oceanic Marine Ecosystems*,
DOI 10.1007/978-3-319-54304-8_14

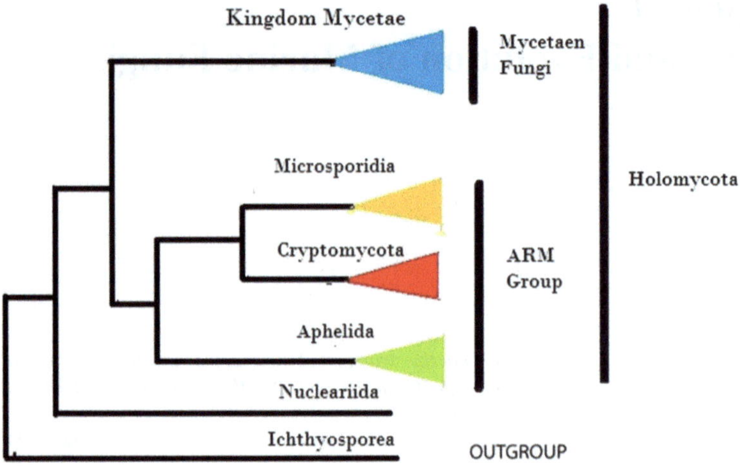


**Fig. 14.1** Holomycota comprising the Kingdom Mycetae and the ARM clade comprising Microsporidia, Cryptomycota, and Aphelida. The Nucleariida form the sister group (Adapted from: Karpov 2014)

Studies to understand the origin of fungi, whether marine, freshwater, or terrestrial, are based on various approaches. These are molecular techniques based on analysis of DNA and protein sequence data with molecular clocks (Berbee and Taylor 2010; Heckman et al. 2001) and fossil records.

**Certain points appear to be clear about the origin and phylogeny of mycetaen fungi.**

**Ancestors of fungi were unusual and predominantly phagotrophic organisms belonging to the ARM clade, comprising aphelids, the cryptomycota, and the microsporidia.** These constitute the basal, earliest diverging branches of Mycetae (Figs. 1.4, 14.1; Jones et al. 2011; Gleason et al. 2012b; James et al. 2013; Karpov et al. 2014). The aphelids, the cryptomycota, the microsporidia, and the nucleariids are predominantly freshwater organisms. These possess cysts that have a penetrative capability. Mitochondria possess tubular or flat cristae.

- Aphelids (Class Aphelidea) include two freshwater genera, *Aphelidium* and *Amoeboaphelidium*, and a marine genus, *Pseudaphelidium*. Aphelids are amoeboid endobiotic algal parasites. A plasmodium develops within the host, which subsequently divides into uninuclear amoeboid cells or uniflagellated zoospores.
- Cryptomycota or Rozellida were first retrieved as unique DNA sequences of unknown affiliation from a fresh water laboratory enclosure. These were provisionally termed the LKM11 clade. A single genus, *Rozella*, has been formally described from this clade, growing as a parasite of *Chytridium polysiphoniae*, which itself was growing on the red alga *Polysiphonia* (Jones et al. 2011; Gleason et al. 2012b). It was originally thought to be a chytrid. Many related environmental sequences of this clade have been retrieved from

different habitats including soils, marine, and freshwater sediments; freshwater planktonic samples; and oxygen-depleted environments in various geographical locations (Table 12.2). However, they seem to be absent in the upper marine water column (Jones et al. 2011). Additional members of the group were isolated in 2011 from samples dredged from a pond. Members of Cryptomycota lack the chitinous cell walls of mycetaen fungi and microsporidia. *Rozella* possesses amoeboid phases in its life cycle and seems to phagocytose the organelles of its host. They have an average size of about 3–5 μm and possess a typical, opisthokont posterior flagellum. *Rozella allomycis* is endobiontic and does not produce its own sporangium wall (Jones et al. 2011; James et al. 2013).

- Microsporidia are unicellular protistan parasites of animals, predominantly of insects and crustaceans (Stentiford et al. 2013). They produce nonmotile, walled spores and have highly reduced mitochondria and are poorly understood.
- **The mycetaen fungi and the ARM clade have together been called the Holomycota** (Lara et al. 2010). **The closest relatives of the Holomycota are the nucleariids**, a group of single-celled opisthokont amoeboid protists with filose pseudopodia that grow phagotrophically on algae and bacteria (Steenkamp et al. 2006; Fig. 14.2).
- **The osmotrophic nutrition of mycetaen fungi probably evolved during an early radiative transition from a phagotrophic life style, characteristic of the ARD clade** (Jones et al. 2011).

**Divergence of extant fungi probably took place from ancestral chytrids.**

- **Chytrids form the basal clade of the Kingdom Mycetae and are hence called the "lower fungi."** They are aquatic and are characterized by uniflagellate zoospores that are dispersed in water. The flagellated spores of chytrids, the

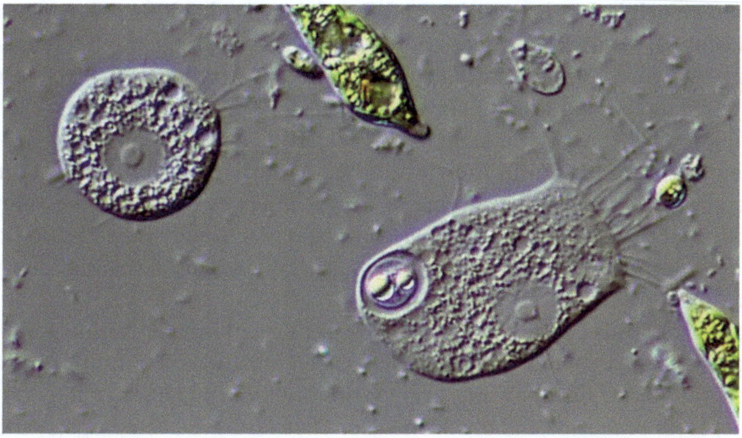

**Fig. 14.2** The nucleariid *Nuclearia thermophila*. Nucleariids are the closest relatives of Holomycota (Credit: Ferry Siemensma. Microworld—World of amoeboid organisms. http://www.arcella.nl/nuclearia-thermophila)

ancestral, and predominantly unicellular group of mycetaen fungi were lost leading to novel life styles of dispersal particularly microscopic, wind-dispersed spores in terrestrial fungi. This might have happened once or several times. Loss of flagella resulted in aflagellate groups of fungi, including ascomycetes and basidiomycetes. These are all predominantly filamentous and terrestrial (James et al. 2013).

**Establishment of eukaryotes on land was possible most probably as a consequence of the association of a fungus and a phototroph** (Le Calvez et al. 2009).

- Fossil evidence indicates that a major event in the colonization of land was the establishment of arbuscular mycorrhiza-like symbioses between fungi belonging to Glomeromycota and plant roots. Fossil hyphae in association with wood decay, fossil chytrids, and Glomales have been reported from the Devonian Period (408–360 Mya) (Vijaykrishna et al. 2006). This association is considered crucial in the colonization of land by plants. Extant members of the Glomeromycota live exclusively as obligate symbionts of photoautotrophs, including not only vascular plants and bryophytes but also cyanobacteria.
- What happened before the time of fossil evidence for fungi at about 400 Mya? Molecular clock estimates based on 18S rDNA have shown that the ascomycetes and basidiomycetes diverged about 600 Mya (Berbee and Taylor 1993; Vijaykrishna et al. 2006). **One estimate puts the origin of mycetaen fungi to be approximately between 700 Mya and 1.06 Bya** (Lücking et al. 2009).

**Did mycetaen fungi originate and undergo divergence in the sea, much before colonizing the land** or did it take place in freshwater environments and on land? This still remains a major, unresolved issue in evolution of fungi. Various views have been put forward.

- **A marine origin has been suggested** (Le Calvez et al. 2009). A large number of novel fungal sequences of Cryptomycota have been frequently recovered from deep-sea hydrothermal vent and anoxic regions hinting vaguely at a marine origin (Manohar and Raghukumar 2013).
- Members of the ARM clade and the nucleariids are predominantly freshwater organisms. **Therefore, did these events take place in freshwater and on land?** It has been suggested that the major fungal lineages evolved from limnic and marine macroalgae and early nonvascular land plants and diversified in line with the evolution and diversification of vascular plants and terrestrial ecosystems (Lücking et al. 2009).
- Origin of chytrids could provide clues regarding a marine or freshwater origin of fungi. Molecular studies of chytrids have been mostly of taxa from freshwater or terrestrial origin, since few marine chytrids are known (James et al. 2006). Only four species of marine and brackish water chytrids have been properly identified and partially characterized. These are *Rhizophydium littoreum* Amon, *Thalassochytrium gracilariopsis* Nyvall, Pedersen et Longcore, and *Chytridium polysiphoniae* Cohn. These species are either facultative or obligate parasites of

marine macroalgae and invertebrates. Also, some species of *Olpidium* and *Rhizophydium* are parasites of small marine green algae and diatoms (Gleason et al. 2012a). **There is no convincing evidence yet to show that ancestral chytrids were marine. The most likely evolutionary scenario is that chytrids evolved in freshwater habitats. Present-day marine chytrids may be secondary invaders from terrestrial, freshwater environments.** Recently, Bass et al. (2007) have recovered novel lineages of chytrids from environmental DNA from marine ecosystems. Their phylogenetic relationships with extant chytrids deserve further study.

Thus, the marine or freshwater origin and divergence of fungi is still a matter of debate.

### *14.1.1 Evolution of Obligate Marine Mycetaen Fungi*

**The Ascomycota are the largest group among Kingdom Mycetae with more than 32,000 species, of which more than 500 are obligately marine.**

**Several lineages of ascomycetes and basidiomycetes made independent transitions from terrestrial and freshwater to the marine ecosystem** (Jones et al. 2009; Pang 2012).

- **Members of Ascomycota were earlier believed to have evolved from marine red algae.** The insect parasites, Laboulbeniomycetes, were thought to be related to red algae. The marine algicolous ascomycete *Spathulospora* was believed to be a primitive, ancient fungus related to Laboulbeniomycetes and to represent the hypothetical ancestor of Ascomycetes. **This "Floridean hypothesis" has now been discarded** (Kohlmeyer and Kohlmeyer 1979; Kohlmeyer 1986; Vijaykrishna et al. 2006; Jones et al. 2009).
- **A pioneering study by** Spatafora et al. (1998) **showed that marine ascomycetes arose through several, independent migrations of terrestrial and freshwater ascomycetes to the sea** (Fig. 14.3). This migration was gradual and probably took place first from terrestrial to freshwater ecosystems. It appears likely that subsequently several independent transitions took place into the sea. Bitunicate and unitunicate ascomycetes may have followed different evolutionary pathways, the former preferably adapting to mangrove environments and the unitunicate forms to oceanic conditions (Sakayaroj et al. 2011).

  The transition may have brought about morphological diversity and changes in response to environmental conditions.
- **Migration to the marine environment is evidenced by the fact that marine ascomycetes occur as sister clades to terrestrial or freshwater species and also by the fact that several genera of ascomycetes contain both terrestrial and freshwater species together with marine species.** There are overlapping genera between terrestrial and freshwater, freshwater and mangrove, and

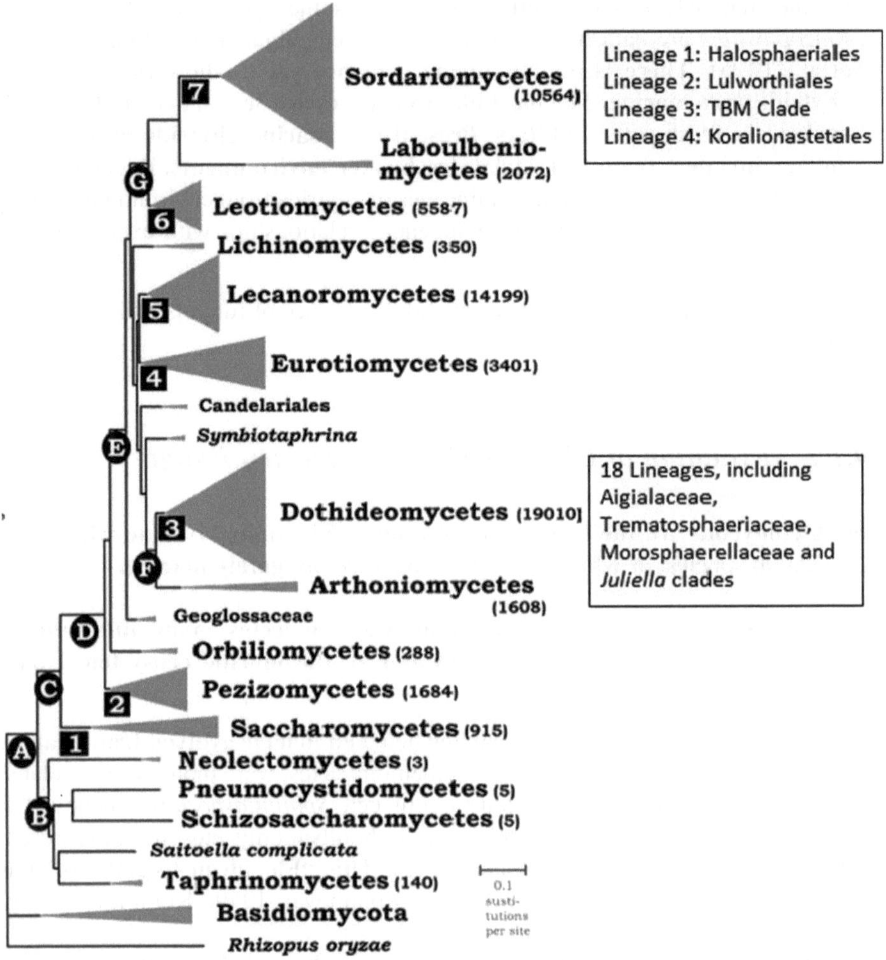


**Fig. 14.3** Marine lineages of Ascomycota (Source of the phylogenetic tree: Schoch, C.M. et al. The Ascomycota Tree of Life: A Phylum-wide Phylogeny Clarifies the Origin and Evolution of Fundamental Reproductive and Ecological Traits. Syst. Biol. 58: 224–239, 2009. By permission of Oxford University Press.). The two boxes on right show the marine lineages belonging to Sordariomycetes and Dothideomycetes

freshwater and oceanic fungi. Many terrestrial genera have marine members, e.g., *Mycosphaerella, Passeriniella, Lophiostoma, Massarina, Trematosphaeria, Phaeosphaeria, Leptosphaeria*, and *Savoryella* species (Pinruan et al. 2002; Jones et al. 2009; Suetrong et al. 2009; Sakayaroj et al. 2011).

- **Ascomycete taxa with unitunicate and bitunicate asci may have followed different evolutionary pathways.**
- Most unitunicate ascomycetes occur in temperate oceanic and coastal waters and possess deliquescing asci. They are characterized by a passive spore release and often possess ascospores with elaborate appendages. Conidia of their

anamorphic forms are frequently multibranched. The Sordariomycetes are unitunicate ascomycetes. Some of the major lineages are as follows (Fig. 14.3).

- The family Halosphaeriaceae of the Order Halosphaeriales is the largest group of marine ascomycetes, comprising more than 50 genera in nearly 130 species. A phylogenetic study of 12 taxa from 11 genera belonging to this family, based on the nuclear small subunit (SSU) and the nuclear large subunit (LSU) ribosomal RNAs (rDNA), confirmed their close phylogenetic relationship with terrestrial fungi of the Microascales (Spatafora et al. 1998). **The Halosphaeriaceae forms a distinct lineage of marine ascomycetes derived from terrestrial ancestors** (Fig. 14.3).
- Ascomycetes belonging to **Lulworthiales are another important group of marine ascomycetes derived from terrestrial counterparts** (Spatafora et al. 1998; Kohlmeyer et al. 2000). These were earlier placed in Halosphaeriales. This is a distinct lineage that is related to perithecial, terrestrial ascomycetes.
- **The coral inhabiting fungi Koralionastetales**, related to the *Lulworthia* clade, form another distinct lineage of marine ascomycetes derived from terrestrial species (Jones et al. 2009).
- **The TBM clade (*Torpedospora/Bertia/Melanospora*)** is a distinct lineage of ascomycetes that colonized the sea (Sakayaroj et al. 2005; Schoch et al. 2006, 2007). These fungi belong to the Hypocreales. This clade was detected through phylogenetic analyses of DNA sequences from protein coding and ribosomal nuclear loci. The TBM clade is associated with a well-supported terrestrial clade containing fungicolous species of *Melanospora* and wood inhabiting Coronophorales.
- **Several other minor lineages** within the Ascomycota comprising genera such as *Kallichroma* Kohlm. & Volkm.-Kohlm., *Heleococcum* Jørg. (Hypocreales), and others have been documented as secondary invaders of the sea (Rossman et al. 1999; Sakayaroj et al. 2005).

• Most of the bitunicate ascomycetes occur in tropical coastal waters. They possess an active spore discharge. Their ascospores lack appendages, but may possess a mucilaginous sheath that swells in water (Suetrong et al. 2009). **Bitunicate, marine ascomycetes belonging to Dothideomycetes have evolved several times from terrestrial counterparts and form distinct lineages.** Phylogenetic analyses of four nuclear genes, namely, the large and small subunits of the nuclear ribosomal RNA, transcription elongation factor 1-alpha, and the second largest RNA polymerase II subunit, established that the ecological group of marine bitunicate ascomycetes has representatives in the orders Capnodiales, Hysteriales, Jahnulales, Mytilinidiales, Patellariales, and Pleosporales (Jones et al. 2009; Suetrong et al. 2009). Eighteen out of 28 clades of Dothideomycetes have marine representatives, indicating that different lineages of these fungi colonized the sea independently. The most common among these were the Aigialaceae, Morosphaeriaceae, Trematosphaeriaceae, and *Julella* clades.

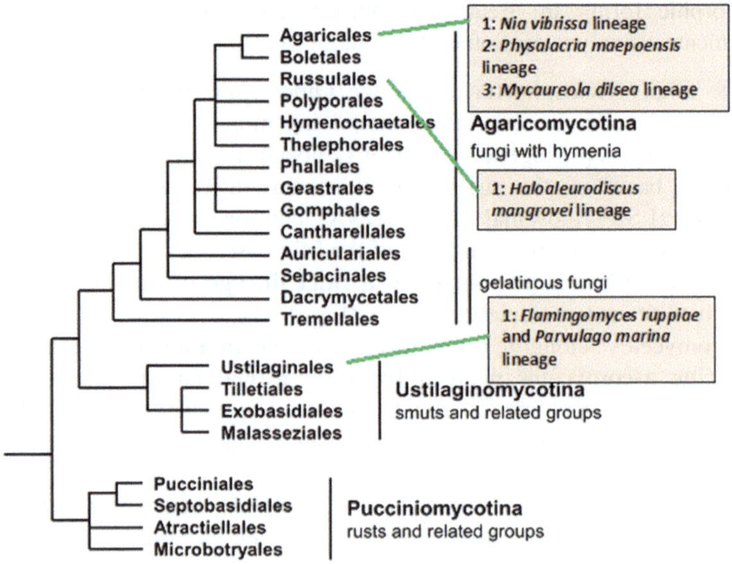


**Fig. 14.4** Marine lineages of Basidiomycetes (Source of the phylogenetic tree: Phylogenetic tree from M. Piepenbring. https://commons.wikimedia.org/wiki/File:03_01_01_phylogeny_Basidiomycota_(M._Piepenbring).png). The three boxes on right show the marine lineages belonging to Agaricales, Russulales, and Ustilaginales

**In addition to filamentous marine, ascomycete fungi, different lineages of ascomycete yeasts also appear to have migrated from land to sea.** One such lineage is that of the Saccharomycetes, Saccharomycetales lineage containing *Saccharomyces* spp. and *Metschnikowia* spp. (Jones et al. 2009).

**Members of the Basidiomycota also made a transition from land to sea, accompanied by several morphological changes, as with the Ascomycota** (Fig. 14.4). The most conspicuous group within the basidiomycetes is the "homobasidiomycetes." It comprises mushrooms and related forms belonging to Agaricomycetes and includes about 13,000 described species. Marine representatives of this group include eleven species of Agaricomycetes in eight genera (Jones et al. 2009). Yet another genus and species belong to the Ustilaginomycetes.

Marine basidiomycetes bear scant resemblance to macroscopic basidiomycetes such as the gilled mushrooms of Agaricomycetes. One of the changes that resulted in the migration from terrestrial to marine, aquatic realms is the reduction in the size of the basidiocarp as in *Halocyphina villosa* and *Calathella mangrovei*. Such "cyphelloid" basidiocarps are minute, cup-shaped structures, ranging from 0.3 to 1.0 mm in diameter. The other is the production of appendaged basidiospores as in *Nia vibrissa* and *Digitatispora* species (Binder et al. 2006; Jones and Choeyklin 2008). Reduced fruiting bodies with hairy surfaces and appendaged or tetradiate spores may function as floating devices to aid successful dispersal and adhesion to various substrates such as driftwood. This transition too took place several times, resulting in different lineages of marine basidiomycetes that colonized the sea (Binder et al. 2006).

**At least three lineages of the Basidiomycota migrated from freshwater and terrestrial to the marine ecosystem and possess relatives in the terrestrial ecosystem** (Binder et al. 2006; Jones et al. 2009). The Order Agaricales among Agaricomycetes contains three major marine clades of basidiomycetes. These are:

1. The monophyletic "Nia clade" comprises *Nia vibrissa, Halocyphina villosa* Kohlm., and *Calathella mangrovei* E.B.G. Jones and Agerer. This clade is nested within several terrestrial, cyphelloid fungi. The genus *Calathella* includes terrestrial species. *Halocyphina villosa* is closely related to the terrestrial cyphelloid genera *Henningsomyces* and *Cyphellopsis* that are closely related to marine homobasidiomycetes. *Nia vibrissa* is likewise closely related to the freshwater basidiomycete *Limnoperdon incarnatum*, both of which also bear a superficial resemblance to terrestrial puffballs (Hibbett and Binder 2001).
2. The Physalacriaceae clade comprising *Physalacria maipoensis* Inderbitzin and Desjardin produces minute, stipitate–capitate (inflated, headlike caps) basidiocarps and constitutes a second independent lineage of marine homobasidiomycetes in the euagarics.
3. Another Physalacriaceae clade comprising *Mycaureola dilseae*, which parasitizes the red alga *Dilsea carnosa*, is a second independent lineage of marine fungi in the Physalacriaceae clade that is not related to cyphelloid forms. Phylogenetic reconstructions based on rDNA data sets have suggested that *Mycaureola dilseae* represents an independent transition to marine habitat in homobasidiomycetes (Binder et al. 2006).

The mangrove ecosystem is probably an important transit point for evolution of marine basidiomycetes from terrestrial habitats. Two of the above clades, except *Mycaureola dilseae*, are found in mangroves.

In addition to the above, two more marine lineages appear to have taken place. One is that of the Ustilaginomycetes clade, comprising the smuts *Flamingomyces ruppiae* (Feldmann) R. Bauer, M. Lutz., Piatek, Vánky & Oberw. and *Parvulago marina* (Durieu) R. Bauer, M. Lutz., Piatek, Vánky & Oberw. The other is *Haloaleurodiscus mangrovei* N. Maek., Suhara & K. Kinjo, at the root of the "Peniophoraceae" clade of the Russulales belonging to Agaricomycetes.

**In addition to filamentous marine basidiomycetes, many marine basidiomycetous yeasts are also secondary invaders of the sea** (Jones and Choeyklin 2008; Jones et al. 2009). A number of phylogenetic lineages of such yeasts have been detected. These include (1) Tremellomycetes, Cystofilobasidiales lineage containing *Cystofilobasidium bisporidii*, *C. capitatum, Rhodosporidium diobovatum, R. paludigenum*, and *R. sphaerocarpum*; (2) Agaricostilbomycetes, Agaricostilbales lineage containing *Sterigmatomyces halophilus*; (3) Microbotryomycetes, Sporidiobolales lineage containing *Sakaguchia dacryoidea*, and (4) Leucosporidiales lineage containing *Leucosporidium* spp.

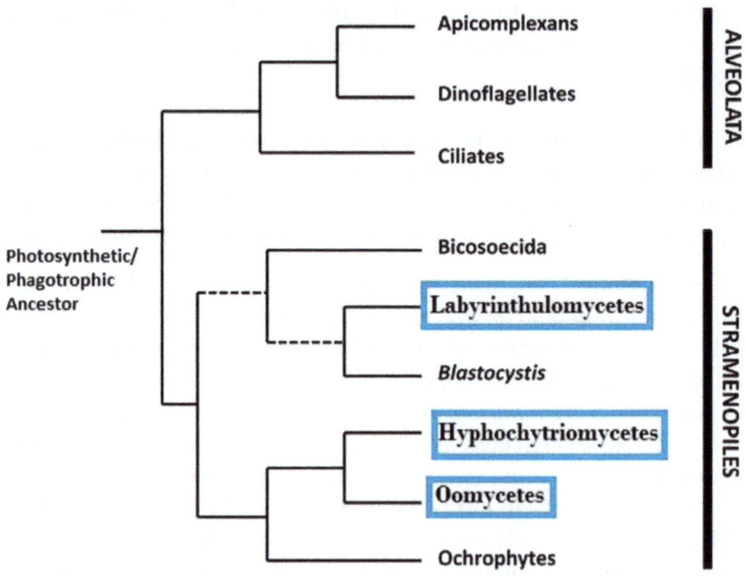


**Fig. 14.5** The common ancestry of the Alveolata and Straminipila (the Chromalveolates). Straminipilan fungi are enclosed in blue rectangles (Adapted from: Clement K. M. Tsui and Lilian L. P. Vrijmoed (2012). A Re-Visit to the Evolution and Ecophysiology of the Labyrinthulomycetes, Marine Ecosystems, Dr. Antonio Cruzado (Ed.), InTech, doi: 10.5772/35979. Available from: http://www.intechopen.com/books/marine-ecosystems/a-re-visit-to-the-evolution-and-ecophysiology-of-the-labyrinthulomycetes)

## 14.2 Evolution of Marine Straminipilan Fungi

**The monophyletic Kingdoms Straminipila and Alveolata are sister groups with a common ancestor. Together, there are called the "chromalveolates"** (Fig. 14.5; Tsui and Vrijmoed 2012).

**The straminipila are a highly diverse group and include microscopic and macroscopic organisms, whose nutritional mode may be photosynthesis or heterotrophy.** The photosynthetic organisms include brown algae, golden brown algae, and diatoms and the non-photosynthetic organisms consist of free-living protists and the osmoheterotrophic straminipilan fungi.

**The alveolates comprise the apicomplexans, dinoflagellates, and ciliates** (Keeling 2009).

Present evidence suggests that the ancestors **of the chromalveolates were probably photosynthetic/phagotrophic algae (mixotrophs). Photosynthetic and phagotrophic abilities of ancestral straminipiles appear to have been lost a few times during the evolution of the group** (Riisberg et al. 2009; Tsui et al. 2009).

**Fungi are found in three lineages of straminipiles** (Beakes et al. 2014). These are:

1. Class Labyrinthulomycetes
2. Class Hyphochytriomycetes
3. Class Oomycetes.

A number of studies based on 18S rRNA genes and nuclear protein coding genes have thrown light on the evolution of straminipiles and straminipilan fungi (Cavalier-Smith et al. 1994; Cavalier-Smith and Chao 2006; Keeling 2009; Tsui et al. 2009; Tsui and Vrijmoed 2012; Beakes et al. 2014).

**The three groups of straminipilan fungi belong to two distinct ancestral lineages within the Straminipila** (Fig. 14.5). A presumed photosynthetic and/or phagotrophic alga having zoospores with tubular, tripartite hairs and mitochondria with tubular cristae gave rise on the one hand to an ancestral line in which photosynthetic abilities were retained, while they were lost in the other.

- The lineage of ancestral, photosynthetic straminipiles might have given rise to the clade leading to ochrophytes, hyphochytriomycetes, and oomycetes.
  - **A secondary loss of photosynthesis in this lineage probably led to the clade of osmoheterotrophic hyphochytriomycetes and oomycetes** (Fig. 14.5). There is strong support to indicate that **hyphochytriomycetes and the oomycetes together form a sister group to the phototrophic stramenipiles, the ochrophytes.**
  - Traits of photosynthesis appear to be present in oomycetes and hyphochytriomycetes. The identification of an apparently plastid-derived 6-phosphogluconate dehydrogenase gene and genes of algal origin in the oomycete *Phytophthora infestans* suggest that it had a photosynthetic ancestor.
- The lineage in which phagotrophy was retained gave rise to the Labyrinthulomycetes and the bicosoecids. Sequences of the actin, beta-tubulin, and elongation factor 1-alpha gene fragments and ribosomal small subunit (SSU) genes have shown that the **labyrinthulomycetes and the bicosoecids are sister groups** in this lineage (Tsui and Vrijmoed 2012).
  - The labyrinthulomycete/bicosoecid lineage further evolved into two clades. Phagotrophy was lost in one of these, leading to the straminipilan fungal group of labyrinthulomycetes, while the lineage in which it was retained gave rise to the bicosoecids. Phagotrophy is the main mode of nutrition in bicosoecids, which feed on bacteria by the invagination of cell membrane. Thus, **the ancestors of labyrinthulomycetes were probably photosynthetic and phagotrophic**, traits of which may still be gleamed in these organisms (Tsui and Vrijmoed 2012).
  - Phagotrophy may have preceded the development of an ectoplasm and cell wall. In addition to their dominant walled, osmotrophic vegetative stage, labyrinthulomycetes including *Thraustochytrium striatum*, *Aurantiochytrium mangrovei*, *Ulkenia*, and *Labyrinthula* sp. can produce a transient phagotrophic amoeboid stage. Bacterivory has been noticed in one strain of thraustochytrid (Raghukumar 1992).

– The Labyrinthulomycetes also possess characteristics that indicate their early relation to photosynthetic ancestors. Many thraustochytrids produce omega-3 PUFA using desaturase and elongase which are usually located in chloroplasts. A few members can be phototactic (e.g., *Labyrinthula* sp. and *Ulkenia* sp.). The eyespot of *Labyrinthula* zoospores also resembles eyespots of other stramenipiles, and it may mark the remains of an ancestral chloroplast. Eyespots are absent in the basal thraustochytrids and aplanochytrids, and the phylogeny suggests that if these were the last remnants of chloroplasts/plastids, they must have undergone multiple convergent losses in the labyrinthulomycetes.

**Osmoheterotrophy of straminipilan fungi is a polyphyletic character.** This fungal mode of nutrition appears to have arisen from independent loss of photosynthesis and/or phagotrophy in the individual ancestors of labyrinthulomycetes on the one hand and the hyphochytriomycetes and oomycetes on the other (Tsui and Vrijmoed 2012). Oomycetes secrete enzymes and absorb dissolved nutrients across a continuous cell wall, while labyrinthulomycetes are believed to secrete enzymes and absorb dissolved nutrients across their wall-less ectoplasm, possibly reflecting the convergent origins of osmotrophy in these two groups.

**Straminipilan fungi perhaps originated about 600 to 800 Mya.** Origin of the Ochrophyta has been estimated to be around 571 Mya (a mean of estimates ranging from 735 to 434 Mya). Hyphochytrids and oomycetes presumably evolved after the ochrophyte line diverged, that is later than 570 Mya. The Labyrinthulomycetes are part of a parallel clade that presumably evolved around the same time or even earlier than the other osmotrophic stramenipiles.

**All three groups of straminipilan fungi appear to have originated in the sea. A marine origin of each of the three groups of straminipilan fungi is suggested by the following evidence** (Beakes et al. 2014).

- **Unlike mycetaen fungi which appear to have migrated from land to sea, many oomycetes could have migrated from sea into freshwater ecosystems and onto land.** Oomycetes are most closely related to the marine, straminipilan flagellate *Developayella* and marine environmental sequences termed the "MAST-1 clade." Most of the early diverging oomycete genera are marine. Many simple holocarpic oomycete genera, such as *Ectrogella* and *Lagenisma*, are parasites of marine diatoms. It has been surmised that the earliest oomycetes, which were holocarpic and morphologically simple, were necrotrophs of marine nematodes, crustaceans, and possibly algae, present in the Cambrian period around 550–500 Mya. The closest known relatives of hyphochytrids is the marine, phagotrophic, straminipilan protist *Pirsonia*, which is marine.
- **The Labyrinthulomycetes are of marine origin.** These fungi are obligately marine, and there has been little doubt about their marine origin. As with oomycetes, many thraustochytrids are also parasites of molluscs and gastropods.

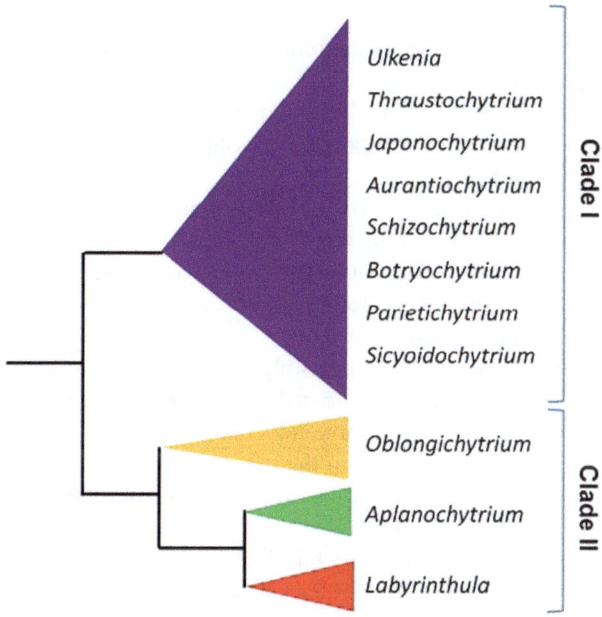


**Fig. 14.6** Phylogeny of Labyrinthulomycetes (Source: Clement K. M. Tsui and Lilian L. P. Vrijmoed (2012). A Re-Visit to the Evolution and Ecophysiology of the Labyrinthulomycetes, Marine Ecosystems, Dr. Antonio Cruzado (Ed.), InTech, DOI: 10.5772/35979. Available from: http://www.intechopen.com/books/marine-ecosystems/a-re-visit-to-the-evolution-and-ecophysiology-of-the-labyrinthulomycetes)

**Labyrinthulomycetes evolved into two distinct clades** (Tsui and Vrijmoed 2012). Daisuke Honda and others showed that the evolutionary history of Labyrinthulomycetes is not reflected in the morphological and life cycle features of individual species (Honda et al. 1999). For example, repeated binary division of the vegetative cell to form a cluster of zoosporangia, which was considered to be the distinguishing feature of the genus *Schizochytrium* Goldstein, was distributed across several distinct lineages (Yokoyama and Honda 2007). Similar was the case with the naked amoeboid stage of the vegetative cell prior to zoospore formation, considered the feature of the genus *Ulkenia* Gaertner (Yokoyama et al. 2007). Several studies based on DNA sequences later showed that straminipilan fungi evolved as two major clades (Fig. 14.6). Clade I includes thraustochytrids, which possess nonmotile vegetative cells that do not form colonies and non-gliding spores, which are flagellate and motile. Clade II includes the labyrinthulids, which are colonial with gliding vegetative cells, aplanochytrids with gliding reproductive spores, as well as the thraustochytrid *Oblongichytrium* R. Yokoyama & D. Honda. Aplanochytrids and labyrinthulids are sister clades.

- Fungi of the Kingdom Mycetae and the Kingdom Straminipila are polyphyletic groups.
- The Kingdom Mycetae or Fungi and the Kingdom Metazoa or Animalia belong to the monophyletic supergroup Opisthokonta and share a common ancestor.
- The closest relatives of the Kingdom Mycetae are the aphelids, the cryptomycota, and the microsporidia (ARM group).
- Holomycota, comprising mycetaen fungi and the ARM clade, are most closely related to nucleariids, a group of single-celled opisthokont amoeboid protists.
- Mycetaen fungi probably evolved from a phagotrophic life style.
- Mycetaen fungi are estimated to have evolved 760 Mya to 1.06 Bya.
- There is still a large uncertainty whether the fungi originated and underwent divergence in the sea, or on land.
- An association of a fungus with a phototroph helped establishment of eukaryotes on land.
- Chytrids form the basal clade of the Kingdom Mycetae and may have evolved in freshwater habitats, subsequently invading marine habitats.
- Marine ascomycetes arose through several independent migrations of terrestrial and freshwater ascomycetes to the sea.
- Different lineages of ascomycete yeasts also appear to have migrated from land to sea.
- At least three lineages of the Basidiomycota migrated from freshwater and terrestrial to the marine ecosystem.
- The monophyletic Kingdom Straminipila and the sister clade alveolates shared a common ancestor, which was a mixotroph with photosynthetic and phagotrophic nutrition.
- Straminipila includes microscopic and macrocospic organisms, whose nutritional mode may be photosynthesis or heterotrophy.
- Osmoheterotrophy evolved independently in the three groups of fungi among Straminipila, namely Oomycetes, Hyphochytriomycetes, and Labyrinthulomycetes.
- All three groups of straminipilan fungi appear to have originated in the sea.
- Ancestors of straminipiles were probably photosynthetic/phagotrophic algae (mixotrophs). These properties were lost several times in different lineages of straminipiles.
- Labyrinthulomycetes and the phagotrophic bicosoecids are sister groups.
- Hyphochytriomycetes and the oomycetes together are sister groups of photosynthetic ochrophytes.
- Freshwater and terrestrial oomycetes probably originated from marine ancestors.

**Future Directions**

- The relationship between the presumed ancestors of mycetaen fungi, the nucleariids, is still vague.
- It is not clear whether diverse groups of fungi evolved in the sea or in freshwater and land. Refinements of techniques as well as search for novel sequences and organisms from marine ecosystems will throw light on this.
- Several novel sequences of Kingdom Mycetae as well as Cryptomycota have been frequently recovered from deep-sea hydrothermal vent and anoxic regions. Their relevance to a marine origin of mycetaen fungi deserves to be studied.
- Many novel lineages of chytrids have been recovered from environmental DNA of marine ecosystems. Their phylogenetic relationships with extant chytrids deserve further study.
- Although a number of lineages of marine ascomycetes have been studied, most species of obligate marine ascomycetes has not yet been addressed. This will throw more light on the evolution of marine ascomycetes.
- Events leading to the loss of photosynthesis and/or phagotrophy in straminipilan fungi remains a mystery.
- Dating of straminipilan fungi has not been resolved.
- Known diversity of Labyrinthulomycetes is low, restricting our knowledge on their evolution. Recovery of novel species and sequences will help to understand this.

# Chapter 15
# Methods to Study Marine Fungi

*Man is a tool-using animal. Without tools he is nothing, with tools he is all.*

Thomas Carlyle

Appropriate methodological tools are required to study various aspects of marine fungi.

(1) An understanding of their diversity in any given habitat.
(2) Elucidation of their ecological role.
(3) Exploring their physiology, biochemistry, and biotechnological applications.

**The major sets of techniques required to accomplish this are culturing, microscopic identification, molecular identification, biomass estimations, and productivity studies** (Table 15.1). The study of marine fungi broadly uses standard mycological techniques (Srinivasan 2004). However, some of these have limitations when applied for marine fungi. Hence, in addition to such standard methods, marine mycology also requires the use of special ones. For example, dilution and plating methods are commonly used in studying fungi in terrestrial ecosystems. Since fungi are abundant in terrestrial ecosystems and their diversity is well documented, these methods are useful. Such methods, as well as metagenomic methods using DNA sequences when used to study marine habitats, almost often yield terrestrial species, which might have easily arisen from spores washed into the marine ecosystem or deposited by air. Hence, results obtained using such methods for marine fungi should be supported by other techniques to demonstrate their activity. Besides, mycetaen and straminipilan fungi require different methods to study them. A number of detailed accounts on culturing of marine fungi are available.

 
S. Raghukumar, *Fungi in Coastal and Oceanic Marine Ecosystems*,
DOI 10.1007/978-3-319-54304-8_15

**Table 15.1** Techniques to study various aspects of marine fungi

| Techniques | Diversity | Ecology | Physiology, Biochemistry, Biotechnology |
|---|---|---|---|
| Culturing | + | + | + |
| Microscopic identification | + | + | + |
| Metagenomics | + | + | + |
| Biomass estimations | – | + | – |
| Productivity | – | + | – |

## 15.1 Culturing of Marine Fungi

A variety of techniques are available to culture marine fungi. The choice depends upon the habitat of study and the kind of fungi that one wishes to study, namely, filamentous mycetaen fungi, yeasts, or straminipilan fungi (Table 15.2).

### *15.1.1 Direct Detection and Culturing*

**Direct detection method guarantees that the fungus which has been cultured was growing actively in the marine habitat.**

**Obligate marine, mycetaen fungi produce sporulating structures such as ascocarps (ascomycetes), basidiocarps (basidiomycetes), and conidia-bearing structures (anamorphic fungi) on the surface of large, persistent, particulate, organic substrates in which they grow actively in the form of hyphae.** Such substrates include allochthonous wood, mangrove wood, lignocellulosic material such as decaying leaves, macroalgae, coral, cuttlebone of squids, and exoskeleton of crustaceans. Most of the obligate marine, mycetaen fungi have been detected and cultured through direct detection and isolation (Kohlmeyer and Kohlmeyer 1979; Jones et al. 2009).

Samples from field collections should be immediately placed in plastic bags in order to avoid loss of moisture. The following steps are recommended.

1. Very often, samples such as wood may be covered by sediments or surface growth of barnacles and algae, making it difficult to observe fungal sporulating structures. Samples covered by sediment or debris are thoroughly washed in running tap water. Surface growth may be gently scraped off without extensive damage of the surface where fungal structures are present and then rinsed with tap water.
2. Samples are subjected to an initial examination under a stereomicroscope with magnifications between 10–40×. Fine stainless steel needles may be used to prod and pick up the fruiting bodies or conidia which are then mounted in seawater on a slide for examining under a compound microscope, prior to culturing. Fine glass needles may be drawn over a flame from thin glass rods.

**Table 15.2** A rough guide to culturing marine fungi from various habitats. Methods may be adopted and modified depending upon the need

| Habitat | Direct detection and culturing | Surface sterilization and plating | Particle plating | Baiting followed by culturing | Dilution plating/Direct plating |
|---|---|---|---|---|---|
| Saprotrophs in woody substrates | Obligate marine, mycetaen fungi | Facultative, marine, mycetaen fungi | Facultative, marine, mycetaen fungi | Labyrinthulomycetes and Oomycetes | Marine-derived fungi |
| Saprotrophs in mangrove, salt marsh grass, seagrass leaves, macroalgae | Obligate marine, mycetaen fungi | Facultative, marine, mycetaen fungi, Labyrinthulomycetes | Facultative, marine, mycetaen fungi, Labyrinthulomycetes | Labyrinthulomycetes and Oomycetes | Marine-derived fungi |
| Parasites in algae | Obligate marine, mycetaen fungi; | – | – | Labyrinthulomycetes and Oomycetes | – |
| Parasites in animal eggs | Oomycetes | Oomycetes | – | – | – |
| Corals; animal shells; exo-skeleton; marine invertebrates | Obligate marine, mycetaen fungi; | Facultative marine fungi; Labyrinthulomycetes | Facultative marine fungi; Labyrinthulomycetes | Labyrinthulomycetes; Oomycetes | Marine-derived fungi; Yeasts; Labyrinthulomycetes |
| Particles in water column | – | – | Facultative marine fungi; Labyrinthulomycetes | Labyrinthulomycetes | Marine-derived fungi; Yeasts; Labyrinthulomycetes |
| Water samples | – | – | – | Labyrinthulomycetes | Marine-derived fungi; Yeasts; Labyrinthulomycetes |
| Sediment samples | – | – | Facultative marine fungi; Labyrinthulomycetes | Labyrinthulomycetes | Marine-derived fungi; Yeasts; Labyrinthulomycetes |

A quick identification is often possible. Alternatively, microscope slides of sporulating structures and spores may be prepared for future reference and a quick note of the spore characteristics may be noted prior to culturing.
3. Fruit bodies of many species may not appear initially, but may come up only with incubation for a certain length of time. A bread box, sterilized internally by wiping with rectified spirit and lined with autoclaved, sterile blotting paper, serves well as an incubation chamber. The interior is kept moist by spraying the blotting paper with sterile seawater. There should be no standing water. Samples are then placed inside these boxes, which are then closed. The samples are periodically examined, roughly at intervals of every 15 days.
4. Single spore isolation of the fungi is carried out by picking fruit bodies with needles, transferring them to a small volume of sterile seawater on a sterile glass slide, and crushing them to release the spores. The spores are then transferred with a fine Pasteur pipette to a culture medium and the suspension allowed to flow by tilting the plates such that the spores are distributed on the agar surface. After overnight incubation, the plates are examined under a stereomicroscope or a compound microscope to locate isolated, germinating spores which are then picked up and transferred to fresh plates or tubes with agar media (Jones and Hyde 1988; Vrijmoed 2000). A low nutrient medium, such as corn meal agar made up with 50% seawater, may be used, fortified with an antibiotic mixture (e.g., a mixture of Penicillin G and Streptomycin in 1 g/l).
5. A variety of media are used for culturing these lignicolous fungi (Sect. 15.5).

**A number of oomycetes occur as parasites in algae and invertebrate eggs and larvae** (Raghukumar 2006; Li et al. 2010; Hatai 2012). These fungi produce vegetative cells, sporangia, or hyphae within filamentous algae or invertebrate eggs, which can be microscopically detected in freshly collected specimens. This enables their isolation and culturing. Very often, incubation of the samples for a period of days to weeks, depending on the nature of samples, promotes the production of sporulating structures which are then used for culturing.

- Freshly collected samples are examined under a stereomicroscope or compound microscope to detect the presence of fungal parasites. Aliquots of samples may be stained with the mycological stain lactophenol-cotton blue to detect fungi. Epifluorescence microscopy, following staining with the optical brightener Calcofluor or FITC-tagged lectin WGA (Wheat germ agglutinin used for detecting chitin), is also helpful. The incidence of infection may often be very low and not easily detectable (Raghukumar 2006).
- Samples may be incubated in sterile seawater containing antibiotics under low light conditions for 2 days to a week and examined regularly for the presence of parasites. The seawater is replaced daily to avoid bacterial overload.
- Certain parasites are culturable, and attempts should be made to grow them in vitro. For example, the chytrid *Rhizophydium littoreum* in algae and crustaceans as well as *Haliphthoros milfordensis* and *Lagenidium* in crustaceans are amenable to culturing. The chytrid *R. littoreum* has been cultured by streaking infected samples onto Modified Vishniac's medium (MV), containing antibiotics (Shields 1990). Members of Lagenidiales have been

cultured on PYGS medium containing antibiotics. The agar plates are incubated at ~25 °C for 5–7 d. Fungal colonies on agar plates are transferred onto fresh media and pure cultures are established.

- Many oomycete parasites in algae are biotrophs and are obligate parasites. They can only be maintained as dual cultures within their hosts. It is then important to maintain cultures of the host algae. Infected cultures are periodically inoculated onto healthy hosts. For example, the diatom *Licmophora* was maintained in F/2 culture medium independently and the parasite *Ectrogella perforans* maintained in dual cultures with it (Raghukumar 2006; Tsirigoti et al. 2014).
- In order to confirm pathogenicity of an organism, Koch's postulates must be fulfilled (Raghukumar 2006). These are: (1) Isolation of the causal organism from the host; (2) proof of constant association of the parasite with the host; (3) inoculation of the suspected pathogen and elicitation of disease symptoms; (4) reisolation of the suspected parasite from inoculated hosts and reidentification.

## 15.1.2 Plating

Many filamentous, mycetaen fungi often grow actively within a substrate in the form of vegetative hyphae, without producing sporulating structures on the surface. This endobiontic behavior is frequently seen in fungi growing in soft and delicate organic material, such as decaying mangrove leaves, endophytic fungi in living and dead seagrass leaves, and delicate macroalgae, as well as those in corals, animal shells, and chitinous exoskeleton of crustaceans. Yeasts, on the other hand, may grow epibiontically on the surfaces of organic particles, in water with high nutrient concentrations, or within guts of various animals.

**1. Surface sterilization and plating helps to isolate fungi actively growing within a substrate and which do not sporulate on their surfaces.**

Land run-offs into the sea and wind deposit spores into the marine environment, many of which may adhere to particulate material. Hence, caution has to be exercised in culturing filamentous, mycetaen fungi growing endobiontically within substrates to ensure that only actively growing fungi and not those arising out of such spore contamination grow out in culture. **A surface-sterilization method is mandatory in order to avoid such contaminants. Surface sterilization prior to plating is not necessary in the case of marine oomycetes such as *Halophytophthora* and labyrinthulomycetes because these are obligately marine.** Colonies arising from material plated without surface sterilization may develop from cells present on the surface of the material, as well as those from within. The following steps are recommended.

- Thoroughly wash the samples in sterile seawater supplemented with antibiotics to suppress bacterial growth. An antibiotic solution containing 1 mg $ml^{-1}$

each of penicillin G and streptomycin sulfate may be used. Various combinations of antibiotics are given later (Sect. 15.5).

- The samples may be cut to uniform sizes to enable quantification based on area or weight. For example, decaying mangrove or seagrass leaves or macroalgae may be cut to disks of 0.5 or 1 cm diameter with a cork borer.
- The material is then subjected to surface sterilization. As mentioned above, this step is optional for labyrinthulomycetes and yeasts but essential for filamentous mycetaen fungi. The following surface disinfectants are frequently used.
  - 0.001 g $L^{-1}$ mercury chloride in 5% ethanol and then washed twice with sterilized seawater (Menezes et al. 2010).
  - 0.5% sodium hypochlorite solution for 1 minute, followed by two to three thorough rinses with sterilized seawater (Raghukumar et al. 1994b).
  - A combination of rinsing with 70% ethanol for 5 s, followed by immersion in 4% sodium hypochlorite for 60 s and finally rinsing in sterile distilled water (Venkatachalam et al. 2015).
- Any of the following two methods may then be used.
  - The surface-sterilized material may be homogenized. The homogenates may be diluted in sterile seawater and then aliquots plated out in Petri plates with a suitable culture medium.
  - Small bits of 0.5–1 $cm^2$ bits of the material are plated out directly.
- Plates are regularly examined for cultures from about 2 days onwards.

**Surface sterilization is not recommended for yeasts because they are unicellular and may often grow on the surfaces of various particles.**

**2. The "particle plating method" aims to culture slow growing fungi.** Conventional mycological plating methods encourage rapid growth of generalist, ruderal fungi such as species of *Aspergillus*, *Penicillium*, *Cladosporium*, and many other such anamorphic fungi. This can happen when the particles plated are several millimeters in size. Many of the slow growing forms that follow a k-strategy will be suppressed and their diversity will not be revealed. Very often, such forms may actually contribute more to fungal biomass in various substrates and be biologically active. The "particle plating method" enables a more complete picture of the diversity of fungi in organic substrates (Bills and Polishook 1994). This method resembles the "extinction to dilution" method followed by marine microbiologists in recent years (Munn 2011). In the particle plating method, organic particles are homogenized to extremely small sizes of 100–200 μm prior to plating them. The method can be used for organic particles or soil samples containing organic particles. It follows the below given steps (Raghukumar et al. 2010).

- A known amount of surface-sterilized substrate (see above) is homogenized under sterile conditions to yield particles ranging in size from 100 to 200 μm. Sediment samples may be directly used following stirring and shaking.
- The homogenate or sediment samples are passed through a sterile filtration unit containing a 200 μm metallic mesh on top and a 100 μm mesh below. Particles

retained on the 100 μm are suspended in a small volume of water. Aliquots are spread plated in dilutions such that no more than 5 colonies grow in a plate. The presumption is that each of these colonies has grown from a particle that is 100–200 μm in size.

- Any suitable culture medium may be used (see below).

**3. Direct dilution plating may be used to isolate yeasts and marine-derived fungi.** The standard dilution plating method is recommended for yeasts in habitats such as the water column, sediments, surfaces of organic particles, or animal parts. Aliquots of diluted or undiluted water or sediment may be plated directly on a suitable medium containing antibiotics to suppress bacterial growth. Organic particles may be suspended in sterile seawater and vigorously shaken. The water may then be plated. Animal parts may be homogenized and then plated following suitable dilutions. Terrestrial species of fungi from the above substrates and habitats are also recovered by using a dilution plating technique. Such fungi may be active in the form of hyphae in such material or may have arisen from spores. These are marine-derived fungi.

Various media have been used by different authors. Polar ice samples that were melted have been plated on a general purpose DRBC medium or MYG agar media containing 10% glucose or 12% NaCl (Gunde-Cimerman et al. 2003).

### *15.1.3 Baiting and Culturing*

Baiting methods are commonly employed to isolate fungi from a variety of habitats. Any suitable organic material may be used as a bait to encourage fungal colonization and subsequent isolation in culture. For example, sterilized wooden panels may be placed in the sea and retrieved periodically to detect, identify, and isolate obligate marine, lignicolous mycetaen fungi.

**Baiting using pine pollen or brine shrimp (*Artemia*) larvae is an ideal method to isolate thraustochytrids and aplanochytrids, as well as chytrids from marine habitats** (Fig. 15.1; Bremer 2000). Pine pollen has the advantage of floating in water, which makes it easy to prepare microscope slides for examination. Pine pollen may be procured from forest departments and is also commercially available in recent times. *Artemia* eggs are available in aquarium pet stores. Pine pollen is a good, general bait for labyrinthulomycetes, but some of them may prefer *Artemia* larvae. It is advisable to use both baits simultaneously, whenever possible.

- Pine pollen is sterilized in an oven at 90–100 °C for 24–48 h. *Artemia* larvae are hatched from commercially available eggs, following instructions. The larvae are harvested and autoclaved in a small quantity of seawater for future use.

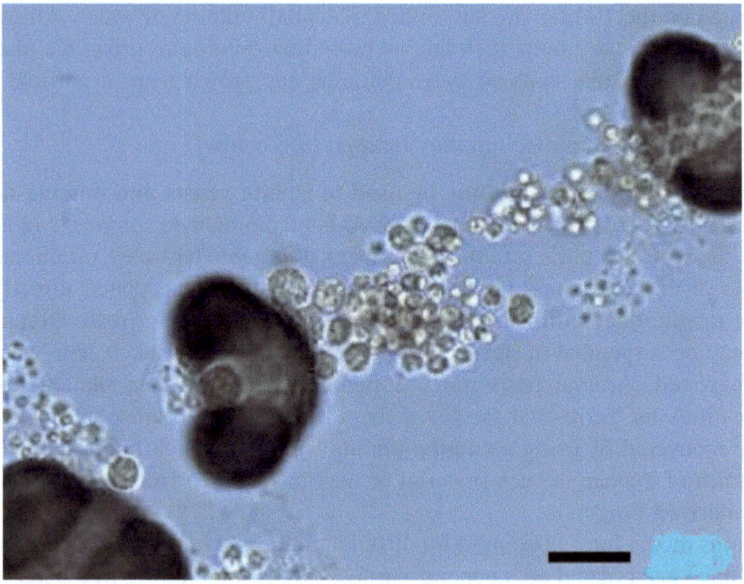

**Fig. 15.1** Cells of thraustochytrids growing on pine pollen. Several cells are also seen freely in seawater (Source: Raghukumar, C. (2008). Marine fungal biotechnology: an ecological perspective. Fungal Diversity 31: 19–35. Kind permission of Dr. Kevin Hyde, Fungal Diversity)

- Baiting is a preliminary isolation, prior to purification and establishment of axenic cultures. It is advisable not to use antibacterial antibiotics at this stage because of the risk of mycetaen fungal overgrowth in the absence of bacteria.
  - Seawater: For water samples, a small quantity of ca. 5 ml of seawater or a 1:10 dilution of the same is added to sterile, 5 cm Petri dishes or 15 ml screw-capped, flat-bottomed glass tubes. A small amount of sterile pine pollen is dusted on to the sample. In the case of *Artemia* larvae, a small aliquot from a stock containing autoclaved larvae is added to the seawater using a sterilized Pasteur pipette.
  - Sediments: A small amount, not exceeding what can be picked up with a small pair of forceps, is added to seawater and baited as above.
  - Organic particles: Small pieces of organic particles, such as decaying vegetation, or animal parts, approximately 0.5 cm in diameter, are picked up with sterile forceps and added to seawater and baited as above.
- Baited samples are incubated at approximately 25 °C for 4 days to 3 weeks. Samples are regularly examined for growth of labyrinthulomycetes on the baits. This preliminary isolation always results in heavy bacterial growth, ciliates, and flagellates, along with cells of labyrinthulomycetes.

- A loopful of this preliminary culture is transferred to fresh sterile seawater plates supplemented with antibiotics. Various combinations may be tried. One such is an end concentration of 0.5 mg of streptomycin and 1000 units of penicillin per ml. Alternatively, the preliminary culture may be streaked onto an agar medium containing antibiotics. The plates are examined after 2–3 days for absence of bacteria.
- Bacteria-free cultures are transferred onto seawater and pine pollen in tubes or plates or streaked once again on nutrient agar plates without antibiotics and then subsequently subcultured and preserved.
- The most commonly used medium for isolation and maintenance of thraustochytrids and aplanochytrids is the Modified Vishniac's medium (Sect. 15.5). *Labyrinthula* spp. (labyrinthulids) can generally be maintained only in the presence of bacteria or yeasts.
- Long-term preservation of thraustochytrids is done at −80 °C or under liquid nitrogen, in the presence of an anti-freezing compound such as dimethyl sulfoxide (DMS) or glycerol (Cox et al. 2009).
- Experimental work with thraustochytrids will require the production of large amounts of zoospores as inocula. Colonies growing on Petri plates are flooded with seawater. Large numbers of zoospores are released within a few hours using this method.
- Many other baits may be used to isolate zoosporic fungi, including chytrids, labyrinthulomycetes and oomycetes such as *Halophytophthora.* Such baits include sesame and other seeds, as well as insect wings.

## 15.2 Taxonomic Identification and Diversity Studies of Marine Fungi

A phylogenetically accurate taxonomic identification is crucial to studies on diversity, ecology, and biotechnology. **Fungal identifications are based on microscopic details of morphology and development as well as molecular sequences. Metagenomic tools are a great aid to unravel marine fungal diversity.**

### *15.2.1 Identification Based on Morphology*

**Morphological details of sporulating structures and spores are important criteria for identification of mycetaen fungi.** A vast amount of published information is available to identify mycetaen fungi based on a morphological and structural basis. The nearly 500 species of the obligate marine, mycetaen fungi are described in detail with illustrations in various monographs, beginning with the pioneering publication by Kohlmeyer and Kohlmeyer in 1979 (Kohlmeyer and Kohlmeyer 1979; Kohlmeyer 1984; Kohlmeyer and Volkmann-Kohlmeyer 1991;

Hyde and Sarma 2000; Hyde et al. 2000; Jones et al. 2009, 2015). The published literature covers various aspects of identification of marine ascomycetes, basidiomycetes, and anamorphic fungi.

Various characteristics of the ascocarps and asci as well as the structure and morphology of ascospores are the most important clues for the identification of ascomycetes. Anamorphic fungi are identified based on the morphology of conidia-bearing structures, the development of conidia, and the conidial morphology. Morphology of the basidiocarps and spore structures are essential to identify basidiomycetes. In addition, a number of ultrastructural characters based on transmission and scanning electron microscopy have been used to clarify identifications (Vrijmoed 2000; Jones et al. 2009).

**Labyrinthulomycetes are identified based on the life cycle of individual species** such as the zoosporangium formation and zoospore release. In addition, fatty acid and carotenoid profiles have been used extensively as aids to delineate genera and species (Porter 1990; Raghukumar 2002; Yokoyama et al. 2007; Yokoyama and Honda 2007).

### *15.2.2 Molecular Methods for Identification and Metagenomics*

The tremendous progress made in molecular biology in the last 30 years have contributed enormously to our understanding of the phylogeny, systematics, community structure, and functioning of fungi both on land and in the sea. Their impact has been twofold.

(1) Analysis of gene sequences has helped in the development of phylogenetic systematics by enabling an accurate placement of individual species with relation to others. (2) These advances have contributed to the development of a "culture-independent approach" to assess diversity and key organisms in various ecological habitats through environmental, metagenomic sequences. Thus, even organisms which cannot be cultured easily have been recognized in various ecosystems. We are also proceeding to understand the important functional genes that play a role in ecosystem processes. A number of reviews on the use of these techniques have been published (Pang and Mitchell 2005; Cuadros-Orellana et al. 2013; Lindahl et al. 2013; Manohar and Raghukumar 2013).

A broad scheme of the various steps involved is presented in Fig. 15.2. Numerous text books, reviews, and articles provide detailed methods of each of the steps involved. A number of commercial kits are available with accurate instructions for use. Several companies now manufacture equipments required for carrying out the necessary work.

- **Extraction of DNA:** It is essential to culture an organism prior to its identification based on molecular sequences. Samples of the fungus taken directly from natural substrates are fraught with the danger of contamination, which even if it

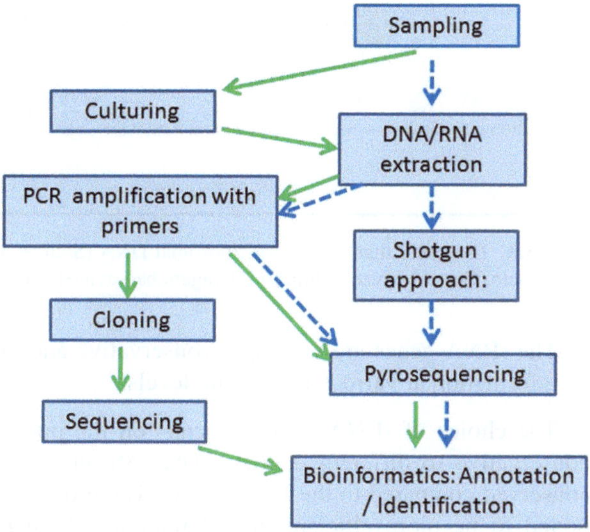


**Fig. 15.2** An overview of molecular methods for taxonomic and phylogenetic identification (*green arrows*) and community metagenomic diversity (*blue, broken arrows*)

occurs in minute quantities may yield false results. The organism may be cultured in any suitable nutrient medium.

An assessment of the total diversity of fungi in a given habitat using a culture-independent approach is made by extracting total DNA from the environmental sample. However, such DNA may also arise from dead or dormant organisms. Hence, the information generated from DNA is not totally reliable to understand which of the fungi are actually active in that environment. By contrast, RNA-based assessment is more reliable for the purpose because RNA has a shorter life span compared to DNA. RNA is produced by metabolically active organisms. Hence, it is more reliable to work on extracted RNA to identify organisms that are active in a community (Edgcomb et al. 2011).

- **Multiplexing or amplification of the gene of interest using PCR techniques**: A gene of interest is required to be amplified using appropriate primers in order to identify a given organism or a community of organisms. The use of appropriate primers to amplify fungal genes is extremely crucial. The choice of the gene depends upon various requirements.

  Ideally, the primers used should be common to all fungi, should be of sufficient base length to allow efficient amplification, and should have the required amount of variation to identify species. The ribosomal DNA (rDNA) operon contains several ribosomal RNA (rRNA) genes. In eukaryotes, including fungi, these are the 18S Small Subunit (18S SSU), 28S Large Subunit (28S LSU), the 5.8S Subunit (5.8 SSU), and the Internal Transcribed Spacer region (ITS) that are used as marker primers in fungi (Fig. 15.3). There are several advantages of using the rDNA as primers.
- The rDNA operon occurs in multiple copies, often in hundreds, in genomes and can be more easily amplified than genes that occur in single copies.

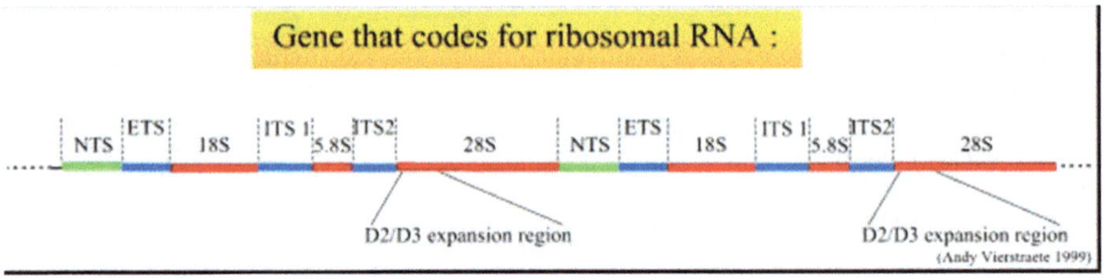


**Fig. 15.3** The genes that encode for ribosomal DNA (Source: With kind permission from Andy Vierstraete, Ghent Universit. http://users.ugent.be/~avierstr/principles/cell.html)

- The rRNA genes are varyingly conservative and provide reliable phylogenetic information at various taxonomic levels.

The choice of rRNA genes depends on the purpose. The different genes are conservative to different degrees. The 18S and 28S rDNA are relatively more conserved compared to the ITS regions. Therefore, these SSU and LSU genes are bound to be more efficient in differentiating higher taxonomic levels, such as families and genera. On the contrary, the ITS1 and ITS2 regions are characterized by a high degree of interspecific variability and are more useful to differentiate between species. They also have well conserved primer sites making them applicable over most fungi. The ITS regions have been highly recommended as universal DNA barcode markers for fungi. Other markers such as EF1-α (tef1), β-tubulin (tub1, tub2), actin (act1), or RNA polymerase II subunits (rpb1, rpb2) have also found use in phylogenetic determination of fungi (Cuadros-Orellana et al. 2013; Lindahl et al. 2013). In addition, ITS sequence data are well represented for fungi in databases. The ITS regions, however, have a limitation for higher taxonomic groups. Thus, the SSU and LSU units, which are more conserved than the ITS, are more useful at supra-species levels.

In order to identify metabolically active organisms, RNA is first isolated from the environment. It is then subjected to Reverse Transcription PCR (RT-PCR) to convert it to DNA. This complimentary DNA (cDNA) is amplified using the primers of choice (Edgcomb et al. 2011).

A variety of primers have been used to amplify the rRNA genes by various authors. Some of these are general, eukaryote-specific, or fungi-specific. The various genes used for identification and phylogeny of marine fungi are shown in Table 15.3.

Once amplified, The PCR products are analyzed by agarose gel electrophoresis and purified using any of the commercially available kits.

- **Sequencing the amplified gene**: The PCR products are cloned in a wide choice of vectors that facilitate sequencing. In recent years, the high-throughput sequencing method using 454 pyrosequencing technology has enabled a more rapid analysis of fungal community diversity in large environmental samples. This does not require cloning.

**Table 15.3** Ribosomal rDNA genes used for taxonomic identification and diversity studies of marine fungi

| Primers used | Identification: Culture-dependent | Identification: Culture-independent, environmental communities |
|---|---|---|
| 18S Small Subunit | Obligate marine ascomycetes and anamorphs (Jones et al. 2009) | Fungi in corals (Thurber et al. 2009) |
| | Marine Dothideomycetes (Suetrong et al. 2009) | Deep-sea methane cold seep sediments (Nagahama et al. 2011) |
| | Marine Halosphaeriales (Spatafora et al. 1998) | Deep-sea mussel (Van Dover et al. 2007) |
| | Marine Ascomycetes (Sakayaroj et al. 2005) | Deep-sea yeasts (Bass et al. 2007) |
| | Marine Basidiomycetes (Jones and Choeyklin 2007) | Deep-sea subsurface sediments (Edgcomb et al. 2011) |
| | Fungi cultured from the deep-sea (Rédou et al. 2015) | |
| | Thraustochytrids (Honda et al. 1999; Yokoyama and Honda 2007; Yokoyama et al. 2007) | |
| | Aplanochytrids (Leander et al. 2004; Damare and Raghukumar 2010) | |
| 28S Large Subunit | Obligate marine ascomycetes and anamorphs (Jones et al. 2009) | Fungi from deep-sea methane seep (Thaler et al. 2012) |
| | Marine Dothideomycetes (Suetrong et al. 2009) | Fungal community in corals (Amend et al. 2012) (D1 and D2 variable regions) |
| | Marine Halosphaeriales (Spatafora et al. 1998) | |
| | Marine Hypocreales (Schoch et al. 2007) | |
| | Deep-sea fungi (Rédou et al. 2015) | |
| ITS primers | Deep-sea fungi (Rédou et al. 2015) | Wood from Arctic waters (Rämä et al. 2016) |
| | Marine Hypocreales (Schoch et al. 2007) | Deep-sea sediments (Lai et al. 2007; Nagano et al. 2010; Zhang et al. 2014) |
| | | Active deep-sea volcano (Connell et al. 2009) |
| | | Deep-sea methane seep (Thaler et al. 2012) |

- **Phylogenetic analysis**: Bioinformatics tools are available to analyze information present in molecular databases to determine the phylogenetic position. Sequences may be programme-aligned using various algorithms.

Culture-independent methods are able to detect a larger diversity of fungi in different habitats than before. However, these methods have their own inaccuracies owing to bias in PCR primer selection and DNA extraction methods. Hence, it is important to combine various methods, including culturing, culture-independent methods, direct microscopy, and FISH, and biochemical methods such as ergosterol. Development of more powerful methodological tools is expected in the future (Nagahama and Nagano 2012).

## 15.3 Detection of Fungi and Biomass Estimations

Unlike unicellular bacteria, **filamentous fungi, including mycetaen, as well as oomycetan forms are cryptic within their substrates. Therefore, special techniques have to be employed to detect them and to estimate their biomass.** Brightfield as well as epifluorescence microscopy are employed to locate them, depending on the substrate in which they grow. Cellulose in vascular plants as well as chlorophyll in such plants and algae are autofluorescent. Brightfield microscopy is employed to detect fungi in these habitats. Epifluorescence microscopy enables detection of filamentous fungi living within substrates not belonging to plants or algae. The single-celled thraustochytrids and aplanochytrids are often planktonic, or grow upon the surfaces of various substrates, in addition to being endobiontic. These can be more easily detected than fungal hyphae using epifluorescence microscopy.

### *15.3.1 Detection and Estimation of Filamentous Fungi*

**Biomass of filamentous fungi may be estimated by microscopy or by biochemical means** (Raghukumar et al. 2010; Newell 1996b, 2000). Microscopic estimation is carried out as follows after clearing the tissues if required, as with hyphae residing within substrates (see below).

- The total lengths of hyphae present in a number of replicate areas of microscope field are first estimated. This may be carried out by capturing a number of photomicrographic images and manually estimating the lengths of the hyphae, with reference to a photographic image of a slide micrometer scale at the same magnification. Another method is to use an image analysis software which is programmed to estimate the length of a given object. The microscope object is highlighted using a threshold value to enhance the image. This is easier with epifluorescence microscopy where the contrasts are sharper.
- Extrapolate the total length of hyphae to a unit area or weight.

- Estimate the total biovolume for the unit area or weight based on the total length and the average width of hyphae, considering them to be cylindrical, using the equation: Biovolume = $(\pi/4) \times W^2 \times L$, where W is the width of the hyphae and L their length.
- Convert the biovolume to biomass based on a conversion factor.

Conversion values from biovolume to biomass vary among researchers. A commonly used value is a dry mass of 0.2 g $cm^{-3}$ biovolume (Newell et al. 1986).

**Clearing tissues and staining:** In order to visualize fungal filaments in plant tissues, fresh disks of known areas and wet weight are prepared. The dry weight may be calculated for independent samples. The fresh disks are then cleared with KOH and hydrogen peroxide according to the method described by Koske and Gemma (1989). The material is then stained with lactophenol-cotton blue. Fungal filaments are easily detected within tissues by this method. Examination of cleared material without homogenizing it helps to avoid destruction of delicate fungal hyphae and other fungal structures. Biomass may be calculated as outlined above.

**Calcofluor staining and epifluorescence microscopy:** Calcofluor (Sigma, St Louis, MO, USA) is an optical brightener that was first used to detect aquatic fungi (Müller and Sengbusch 1983).

- Fungal hyphae are extremely difficult to detect directly in sediments by bright field microscopy. However, they can be easily visualized by epifluorescence microscopy after staining sediments with Calcofluor White (Raghukumar 2008).
- In order to detect fungi living within calcareous substrates, such as molluscan shells and corals, the calcium carbonate is dissolved prior to staining using EDTA.
- Fungi within corals can be stained and biomass estimated as follows. A small sample of the material is added to a decalcification solution containing equal amounts of 20% citric acid and 50% formic acid. The material may be centrifuged, washed, and stained with 0.01% Calcofluor for 5–10 s. Fungal filaments can be detected using an epifluorescence microscope under UV or blue emission.
- In order to estimate the biomass, a pre-weighed sample of the material is decalcified and the material is gently homogenized at 5000 rpm. The homogenate is vacuum-filtered over a 0.22 μm pore size black Nuclepore polycarbonate filter paper such that the area of the filter paper can be related to the weight of the sample filtered over it. The filter is then stained with 0.01% Calcofluor for 5–10 s and the excess stain rinsed off with distilled water. The filter is mounted in nonfluorescent immersion oil and overlaid with a cover slip. The filter with stained fungal filaments may be examined immediately under an epifluorescence microscope as above or stored in the dark at 5 °C until further examination. Length, width and biomass of the hyphae per several microscope fields are determined (see above) and extrapolared to biomass in a given weight of sample, based on the total weight of the sample on the filter.

**Immunofluorescence detection:** Polyclonal antibodies raised against antigens of specific microorganisms are used in various assays to detect their presence in natural habitats. The basic steps are given in a wide range of literature.

Antibodies are generally raised by injecting extracted proteins from the target organism (antigen) subcutaneously or intramuscularly in New Zealand male white rabbits in the first and subsequent booster injections. Antibody values are regularly checked by estimating titer values against the antigen. The antiserum is subsequently recovered and the immunoglobulins (antibodies) recovered. There are two basic variations, namely, the Direct and Indirect method. In the direct method, antibodies against the antigen are directly tagged with the fluorescent stain Fluorescein Isothhiocyanate (FITC). The FITC antibodies are used to stain the sample, and the presence of the required microorganism is detected using an epifluorescence microscope. In the indirect method, the antibodies against the required antigen (e.g., rabbit immunoglobulin against the target organism) are applied to a natural substrate. This is then conjugated with antibodies raised against these (e.g., goat antibodies raised against rabbit immunoglobulin, namely, the goat-anti-rabbit-antiserum), which are tagged with FITC. This method gives enhanced fluorescence and better detection. It is important to check the specificity of the antiserum against the particular antigen by checking with several closely related species or strains.

In both methods, the antibodies may be tagged with enzymes and the microorganisms quantified using the Enzyme Linked Sorbent Assay (ELISA). Several studies on marine fungi have employed these techniques (Raghukumar 1988; Sathe-Pathak et al. 1993; Damare et al. 2006a).

It is easier to estimate the biomass of single-celled thraustochytrids using immunofluorescence method. For example, the biomass of a thraustochytrid on detritus of the brown alga *Sargassum cinereum* was estimated by the immunofluorescence method. In this, the number of cells in a microscope area were calculated and multiplied by two to cover both surfaces. The number was extrapolated to the area of the disk of material that was used and in turn estimated based on the dry weight of the disk (Sathe-Pathak et al. 1993).

There are some problems with immunological methods to detect microorganisms. The method is cumbersome; the process is lengthy and involves experimental animals. Cross-reactivity to other organisms may often occur, restricting its specificity. This problem may be tackled by using monoclonal antibodies, which is a more sophisticated method. In situ hybridization methods using molecular probes, particularly fluorescent in situ hybridization (FISH) methods, are now a quick and reliable method of detecting microbial cells in nature.

**DNA-based in-situ hybridization methods:** Marine microbiology made significant progress since methodologies based on molecular sequences were developed in the late 1980s (Munn 2011). **One of the powerful tools is the Fluorescent in situ Hybridization technique or FISH** (Zwirglmaier 2005; Amann and Fuchs 2008; Jensen 2014). This is an apt acronym because it literally helps to fish out specific organisms from the environment. FISH helps us to visually locate specific microorganisms in an ecosystem and estimate their numbers. Needless to say that this has led to significant findings regarding microbial diversity and ecosystem functioning. For example, FISH has been used to quantify the alphaproteobacterial group SAR11, which is the most abundant group of bacteria in the ocean water column. It has also helped to quantify the ANNAMOX or ammonium-oxidizing anaerobic bacteria and show their importance in anoxic waters. Marine

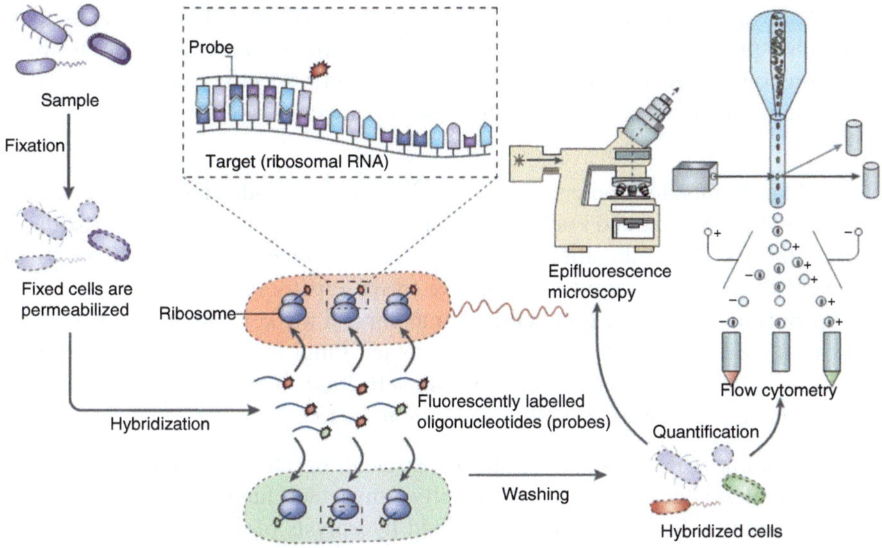


**Fig. 15.4** Basic steps of fluorescence in situ hybridization (Reprinted by permission from Macmillan Publishers Ltd.: [Nature Reviews Microbiology 6: 339–348] (Amann, R. and B.M. Fuchs. Single-cell identification in microbial communities by improved fluorescence in situ hybridization techniques), © 2008 Nature Publishing Group)

bacteriologists have made enormous use of FISH (Amann and Fuchs 2008). The technique has been used less often by marine mycologists. With our increasing knowledge of the diversity and potential role of marine fungi, FISH offers an excellent tool to further our understanding of these. As with phylogenetic systematics and metagenomic diversity studies, FISH makes use of rRNA and rDNA sequences. As explained above, these sequences are ideally suited because of their abundance within the cell. The basic steps involved in FISH are as follows (Fig. 15.4).

- Design rRNA or rDNA oligonucleotide probes that are specific to the target species or the target taxa.
- Tag the probes with a suitable fluorochrome for FISH for epifluorescence microscopy. Alternatively, tag them with an enzyme such as streptokinase or horseradish peroxidase for in situ hybridization using bright field detection.
- Collect samples and permeabilize cells therein such that the probes are able to penetrate the cells.
- Treat with probes.
- Visualize using an epifluorescence (FISH) or bright field microscopy (enzyme-based in situ hybridization). FISH can also be used in conjunction with a flow cytometer to count cells of the target organism in a sample.
- Chromogenic in situ hybridization (CISH) is used for bright field microscopy (Damare and Raghukumar 2010; Jensen 2014). Cells are permeabilized and

treated with biotinylated probes which are then conjugated with streptavidin-alkaline phosphatase (ALP) or streptavidinperoxidase under appropriate incubation conditions. The cells are then incubated in respective substrate solutions for the respective enzymes until they take up the colour of the reaction.

FISH has tremendous potentials in microbial ecology and is constantly evolving and improving. Amann and Fuchs (2008) discuss the use of horseradish peroxidase-labeled probes in combination with catalyzed reported deposition (CARD) of fluorescently labeled tyramides to improve FISH sensitivity.

**Estimations based on biochemical markers:** Two key biochemical markers have been used to estimate biomass of mycetaen fungi in environmental samples. These are glucosamine and ergosterol (Newell 2000; Newell and Porter 2001b; Joergensen and Wichern 2008; Wallander et al. 2013).

- Glucosamine (hexosamine) is the constituent of chitin. Chitin or N-acetyl glucosamine is characteristic of the cell of mycetaen fungi. Methods to extract glucosamine following deacetylation and hydrolysis of chitin are well known. Chitin persists even after death of the fungus. Hence, hexosamine methods yield the total biomass of living as well as dead fungi. Chitin-based biomass estimation of fungi poses a problem for the marine environment because of the ubiquitous presence of microscopic to large crustacean animals that contain chitin. This could lead to erroneous results. However, this method could still be useful for substrates such as decaying salt marsh grass.

  Ergosterol is a very useful biochemical marker for estimating living biomass of mycetaen fungi. It is a component of fungal membranes, and the amount of ergosterol is fairly constant in ascomycetes. Ergosterol can be extracted and estimated using High Performance Liquid Chromatography (HPLC). The method has been well standardized and used extensively to examine biomass of fungi in salt marsh grass (Newell 2001b). Ergosterol is fairly specific to mycetaen fungi and is found in a few green algae and protozoa.
- The method has further been adapted to estimate fungal productivity. The method involves incubation of the substrate containing mycetaen fungi in radio-labeled acetate, which is a precursor of ergosterol synthesis. Detailed methodologies are given by Gessner and Newell (2001) and Newell (2001b).

## *15.3.2 Detection and Estimation of Labyrinthulomycetes*

Labyrinthulomycetes can be detected using the Acriflavine Direct Detection (AfDD) method (Raghukumar and Schaumann 1993). This technique is based on the principle that acriflavine stains acid sulfated polysaccharides red and the nucleus yellow green. Since cell walls of thraustochytrids contain acidic polysaccharides, their cells can be stained with this fluorescent stain and visualized with an epifluorescence microscope, using a blue excitation filter. Both fresh samples as well as formalin-preserved samples may be used. It is advisable to standardize the method with a few thraustochytrid cultures before applying it to field samples. The basic steps are as follows.

- Samples are collected over a black Nuclepore filter with a maximum of 0.8 μm pore size in a filtration unit using vacuum.
- The sample on the filter is rinsed with filter-sterilized seawater and vacuum-drained.
- About 3–4 ml of a 0.05% solution of acriflavine in 0.1 M citrate buffer at pH 3.0 is added to the sample and left for 4 min.
- The stain is vacuum-drained, followed by an addition of 75% of isopropyl alcohol for differentiation.
- The alcohol is drained after 30–60 s by vacuum and the samples rinsed with distilled water.
- The sample may be post-stained with 0.025% Calcofluor for 1 min and vacuum-drained. This is an optional step.
- Samples are vacuum-filtered to dryness and mounted on a slide.
- A drop of water is placed over the filter and a coverslip placed over it.
- Slides are examined under an epifluorescence microscope with an excitation wavelength of 420–490 nm and a barrier filter of 515 nm.
- In order to enable both epifluorescence and transmitted microscopy, the material on the Nuclepore filter may be gently scraped and taken in a drop of water on a slide. The slide preparation can then be examined by both modes of microscopy.

The AfDD technique may not reveal all thraustochytrids. Zoospores with no cell wall and young cells with extremely thin walls may not be observable with this method.

## 15.4 Culturing Deep-Sea Fungi

Methods to study deep-sea fungi require the use of sophisticated equipments and ship time (Raghukumar et al. 2010). Deep-sea water samples can be collected using standard water samplers such as the Niskin samplers. These may be used individually or attached to a CTD equipment used to measure conductivity, temperature, and depth. Deep-sea sediment samples may be collected using box corers or other sediment samplers. However, such samplers cannot retain the in situ high hydrostatic pressure conditions of the deep sea, and the samples are exposed to atmospheric pressures of 1 bar once they are brought up. More sophisticated samplers that bring samples under in situ conditions are extremely expensive and available in special laboratories such as the Japan Agency for Marine-Earth Science and Technology (JAMSTEC ) in Japan. Deep-sea submersibles with robotic arms are often used.

Pressure tolerant microorganisms survive when exposed to atmospheric pressure, and such fungi can be cultured on board the ship. Stringent measures to prevent aerial contamination and controls to check for such contamination should be followed.

Sediment or water samples may be cultured using standard mycological procedures as outlined above, such as dilution plating, particle plating, and baiting. Cultures may be incubated at room temperature and pressure. However, it is preferable to incubate cultures under high hydrostatic pressure. In the pressure incubation method, a known quantity of sediment is placed in sterile plastic pouches (4 × 4 cm) containing 2 ml of a nutrient medium made up with seawater. The open ends of the pouches are sealed with an electrical sealing machine and the bags are placed in a deep-sea culture vessel (Tsurumi and Seiki Co., Yokohama, Japan), filled with sterile water and pressurized. The pressure vessel is immediately placed at 5 °C and incubated for 30 days. At the end of this incubation period, 100 μl of the sediment suspension is spread-plated on nutrient media prepared in seawater, and the plates are incubated at 0.1 MPa (1 bar or atmospheric pressure) and 30 °C of 5 °C until fungal colonies appear.

## 15.5 Culture Media

Conventional mycological culture media are frequently used to isolate and culture marine fungi. These include potato dextrose agar (PDA), corn meal agar (CMA), Emerson's YpSS agar, and Malt Extract Agar (MEA). A few special media compositions have also been often used. These media may be used as solid media with agar or as liquid media. **All media are made up with 50% seawater or with full strength seawater.**

Some of the media are as follows.

- Lignicolous marine fungi:

    1. Glucose-yeast extract agar: Glucose: 0.1%; Yeast extract: 0.01%; Aged seawater 100 ml (Kohlmeyer and Kohlmeyer 1979).
    2. Modified Sguros et al.'s medium: Glucose: 0.5%; Magnesium sulfate: 0.24%; Ammonium nitrate: 0.24%; Yeast extract: 0.1%; Tris buffer: 0.121 g (pH 7.5); Natural seawater: 100 ml (Kohlmeyer and Kohlmeyer 1979).
    3. GYP sea water broth: Glucose: 0.4%; Yeast extract: 0. 4%; Peptone: 0.2% (Jones et al. 2009).

Kohlmeyer and Kohlmeyer (1979) describe a number of other methods which are very useful in culturing and maintenance of marine, lignicolous fungi. For example, many of them may be maintained well on sterile wooden chips or tooth picks and filter paper immersed in seawater containing 0.1% yeast extract.

- **Filamentous, mycetaen fungi from other sources.** In addition to the conventional Corn meal agar, Potato dextrose agar, and malt extract agar, the following two have also been used.

  1. Modified malt extract agar (MMEA): Malt extract: 1.0%; Yeast extract: 0.1%; Bacteriological Peptone: 1.5%; Artificial seawater: 100 ml (Morrison-Gardiner 2002).
  2. Yeast peptone agar (YPA; Yeast extract: 0.19%; Bacteriological Peptone: 0.19%; Artificial seawater: 100 ml (Morrison-Gardiner 2002).

- **Oomycetes and Labyrinthulomycetes**

  1. Thraustochytrid medium: Glucose—0.01%; Peptone: 0.01%; Yeast extract: 0.05%; Vitamin mix; Seawater: 100 ml.
  2. Modified Vishniac's medium—Glucose: 0.1%; Gelatin hydrolysate: 0.1%; Bacto-Peptone: 0.01%; Yeast extract: 0.01%; Seawater: 100 ml.
  3. Yeast-extract-peptone (YEP medium)—Yeast extract: 0.1 g; Mycological peptone: 0.1 g; 15% natural seawater: 100 ml (Fan et al. 2002).
  4. PYGS medium for parasitic oomycetes—Glucose: 0.3%; Bacto-peptone: 0.125%; Yeast extract: 0.125%; Seawater: 100 ml (Nakamura et al. 1995; Hatai et al. 2000)

**Antibacterial antibiotics need to be incorporated in the culture media during the initial isolation from natural samples**. A variety of combinations have been used by different researchers. These are:

- Chloramphenicol (50–150 mg $L^{-1}$)
- Streptomycin sulfate and ampicillin, each at a concentration of 500 μg $ml^{-1}$
- Streptomycin sulfate (10 mg $L^{-1}$)
- Penicillin G (100,000 units) 0.5 to 1.0 g of and 0.5 g to 1.0 g streptomycin sulfate $L^{-1}$ medium.

# Abbreviations

| | |
|---|---|
| AfDD | Acriflavine direct detection |
| AODC | Acridine orange direct count |
| ARA | Arachidonic acid |
| BPE | Bleach plant effluent |
| Bya | Billion years ago |
| CFU | Colony forming unit |
| DGGE | Denaturing gradient gel electrophoresis |
| DHA | Docosahexaenoic acid |
| DOC | Dissolved organic carbon |
| DOM | Dissolved organic matter |
| DSE | Dark septate endophyte |
| EN | Ectoplasmic net |
| EPA | Eicosapentaenoic acid |
| FITC | Fluorescein isothiocyanate |
| GPP | Gross primary production |
| ITS | Internal transcribed spacer |
| LDE | Lignin-degrading enzyme |
| LSU | Large subunit |
| MAA | Mycosporine-like amino acid |
| MBSF | Metres below surface |
| Mpa | Megapascal |
| MPN | Most probable number |
| MSW | Molasses spent wash |
| MUFA | Monounsaturated fatty acid |
| Mya | Million years ago |
| NPP | Net primary production |
| OMZ | Oxygen minimum zone |
| OUT | Operational taxonomic unit |
| PAH | Polyaromatic hydrocarbon |
| PCR | Polymerase chain reaction |



S. Raghukumar, *Fungi in Coastal and Oceanic Marine Ecosystems*,
DOI 10.1007/978-3-319-54304-8

| | |
|---|---|
| POC | Particulate organic carbon |
| POM | Particulate organic matter |
| PP | Primary production |
| Ppt | Parts per thousand |
| PUFA | Polyunsaturated fatty acid |
| QPX | Quahog parasite unknown |
| SAR | Stramenopile Alveolate Rhizaria |
| SFA | Saturated fatty acid |
| SSU | Small subunit |
| TEPS | Transparent Extracellular Polysaccharides |

# References


Abdel-Waheb MA, El-Sharouny HM (2002) Ecology of subtropical mangrove fungi with emphasis on Kandelia candel mycota. In: Kevin D (ed) Fungi in marine environments. Fungal Diversity Press, Hong Kong, pp 247–265

Abe F, Miura T, Nagahama T (2001) Isolation of highly copper-tolerant yeast, *Cryptococcus* sp., from the Japan Trench and the induction of superoxide dismutase activity by $Cu^{2+}$. Biotechnol Lett 23:2027–2034

Abe F, Minegishi H, Miura T, Nagahama T, Usami R, Horikoshi K (2006) Characterization of cold- and high-pressure-active polygalacturonases from a deep-sea yeast, *Cryptococcus liquefaciens* strain N6. Biosci Biotechnol Biochem 70:296–299

Addepalli MK, Fumita Y (2002) Regulatory role of external calcium on *Pythium porphyrae* (Oomycota) zoospore release, development and infection in causing red rot disease of *Porphyra yezoensis* (Rhodophyta). FEMS Microbiol Lett 21:253–257

Adl SM, Simpson AGB, Lane CE et al (2012) The revised classification of eukaryotes. J Eukaryot Microbiol 59:429–493

Agarwal PK, Shukla PS, Gupta K, Jha B (2013) Bioengineering for salinity tolerance in plants: State of the Art. Mol Biotechnol 54:102–123

Aki T, Hachida K, Yoshinaga M, Katai Y, Yamasaki T et al (2003) Thraustochytrid as a potential source of carotenoids. J Am Oil Chem Soc 80:789–794

Alexander E, Stock A, Breiner H-W, Behnke A, Bunge J, Yakimov MM, Stoeck T (2009) Microbial eukaryotes in the hypersaline anoxic L'Atalante deep-sea basin. Environ Microbiol 11:360–381

Alias SA, Jones EBG (2000) Vertical distribution of marine fungi in *Rhizophora apiculata* at Morib mangrove, Selangor, Malaysia. Mycoscience 41:431–436

Alker AP, Smith GW, Kim K (2001) Characterization of *Aspergillus sydowii* (Thom et Church), a fungal pathogen of Caribbean sea fan corals. Hydrobiologia 460:105–111

Alldredge K, Youngbluth M (1995) The significance of macroscopic aggregates (marine snow) as sites for heterotrophic bacterial production in the mesopelagic zone of the subtropical Atlantic. Deep Sea Res Part A 32:1445–1456

Al-Nasrawi HG, Hughes AR (2012) Fungal Diversity associated with salt marsh plants *Spartina alterniflora* and *Juncus roemerianus* in Florida. Jordan J Biol Sci 5:247–254

Alva P, McKenzie EHC, Pointing SB, Pena-Muralla R, Hyde KD (2002) Do seagrasses harbour endophytes? In: Hyde KD (ed) Fungi in marine environment, Fungal Diversity Research series, vol 7. Fungal Diversity Press, Hong Kong, pp 167–178

Amann R, Fuchs BM (2008) Single-cell identification in microbial communities by improved fluorescence *in situ* hybridization techniques. Nat Rev Microbiol 6:339–348




S. Raghukumar, *Fungi in Coastal and Oceanic Marine Ecosystems*,
DOI 10.1007/978-3-319-54304-8

Amend AS, Barshis DJ, Oliver TA (2012) Coral-associated marine fungi form novel lineages and heterogeneous assemblages. ISME J 6:1291–1301

Amon JP (1978) Thraustochytrids and labyrinthulids of terrestrial, aquatic and hypersaline environments of the Great Salt Lake, USA. Mycologia 70:1299–1301

Amon JP (1984) *Rhizophydium littoreum*: a chytrid from siphonaceous marine algae—an ultrastructural examination. Mycologia 76:132–139

Anahid S, Yaghmaei S, Ghobadinejad Z (2011) Heavy metal tolerance of fungi. Sci Iran C 18:502–508

Ananda K, Sridhar KR (2004) Diversity of filamentous fungi on decomposing leaf and woody litter of mangrove forests of southwest coast of India. Curr Sci 87:1431–1437

Ananda K, Prasannarai K, Sridhar KR (1998) Occurrence of higher marine fungi on marine animal substrates of some beaches along the west coast of India. Indian J Mar Sci 27:233–236

Ananda K, Sridhar KR, Raviraja NS, Bärlocher F (2008) Breakdown of fresh and dried *Rhizophora mucronata* leaves in a mangrove of Southwest India. Wetl Ecol Manag 16:1–9

Apt KE (1988) Galls and tumor-like growths on marine macroalgae. Dis Aquat Org 4:211–217

Arfi Y, Marchand C, Wartel M, Record E (2012) Fungal diversity in anoxic-sulfidic sediments in a mangrove soil. Fungal Ecol 5:282–285

Armenta RE, Burja A, Radianingtyas H, Barrow CJ (2006) Critical assessment of various techniques for the extraction of carotenoids and co-enzyme Q10 from the thraustochytrid strain ONC-T18. J Agric Food Chem 54:9752–9758

Arotsker L, Kramarsky-Winter E, Kushmaro A (2011) Coral-associated heterotrophic protists. In: Rosenberg E, Gophna U (eds) Beneficial organisms in multicellular life forms. Springer, Berlin, Heidelberg, pp 151–161

Asmaniadou A, Lipiatou E (eds) (2000) Extreme marine environments. European Commission Community Research Report, Luxembourg

Babich H, Stotzky G (1983) Nickel toxicity to estuarine marine fungi and its amelioration by magnesium in sea water. Water Air Soil Pollut 19:193–202

Bahnweg G (1979a) Studies on the physiology of thraustochytriales I. Growth requirements and nitrogen nutrition of *Thraustochytrium* spp., *Schizochytrium* sp., *Japonochytrium* sp., *Ulkenia* spp., and *Labyrinthuloides* spp. Veröff Inst Meeresforsch Bremerh 17:245–268

Bahnweg G (1979b) Studies on the physiology of thraustochytriales 2. Carbon nutrition of *Thraustochytrium* spp., *Schizochytrium* sp., *Japonochytrium* sp., *Ulkenia* spp., and *Labyrinthuloides* spp. Veröff Inst Meeresforsch Bremerh 17:269–273

Bahnweg G, Bland CE (1980) Comparative physiology and nutrition of *Lagenidium callinectes* and *Haliphthoros milfordensis*, fungal parasites of marine crustaceans. Bot Mar 23:689–698

Bahnweg G, Gotelli D (1980) Physiology and nutrition of *Lagenidium callinectes*, fungal parasites of the blue crab (*Callinectes sapidus*). Bot Mar 23:219–225

Bahnweg G, Sparrow FK (1974a) Four new species of *Thraustochytrium* from Antaractic regions, with notes on the distribution of zoosporic fungi in the Antarctic marine ecosystems. Am J Bot 61:754–766

Bahnweg G, Sparrow FK (1974b) Occurrence, distribution and kinds of zoosporic fungi in Subantarctic and Antarctic waters. Veröff Inst Meeresforsch Bremerh Suppl 5:149–157

Bak RPM, Laane RWPM (1987) Annual black bands in skeletons of reef corals (Scleractinia). Mar Ecol Prog Ser 38:169–175

Balabanova LK, Pivkin MV, Rasskazov VA (2012) The distribution and substrate specificity of extracellular nuclease activity in marine fungi. Open J Mar Sci 2:188–195

Barata M (2002) Fungi on the halophyte *Spartina maritima* in salt marshes. In: Hyde KD (ed) Fungi in marine environments. Fungal Diversity Press, Hong Kong, pp 179–193

Barclay W, Weaver C, Metz J, Hansen J (2010) Development of a docosahexaenoic acid production technology using *Schizochytrium*: historical perspective and update. In: Cohen Z, Ratledge C (eds) Single cell oils: microbial and algal oils. AOCS Press, Urbana, pp 75–96

Barghoorn ES, Linder DH (1944) Marine fungi: their taxonomy and biology. Farlowia 1:395–467

Bärlocher F, Newell SY (1994) Growth of the saltmarsh periwinkle *Littoraria irrorata* on fungal and cordgrass diets. Mar Biol 118:109–114

Bärlocher F, Newell SY, Arsuffi TL (1989) Digestion of *Spartina alterniflora* Loisel material with and without fungal constitutes by the periwinkle *Littorina irrorata* Say (Mollusca: Gastropoda). J Exp Mar Biol Ecol 130:45–53
Barr DJS (1992) Evolution and kingdoms of organisms from the perspective of a mycologist. Mycologia 84:1–8
Bass D, Howe A, Brown N, Barton H, Demidova M, Michelle H, Li L, Sanders H, Watkinson SCC, Willcock S, Richards TAA (2007) Yeast forms dominate fungal diversity in the deep oceans. Proc R Soc B 274:3069–3077
Bayne A-CV, Boltz D, Owen C, Betz Y, Maia G (2013) Vaccination against influenza with recombinant hemagglutinin expressed by *Schizochytrium* sp. confers protective immunity. PLoS One 8:e61790
Beakes GW, Honda D, Thines M (2014) Systematics of the straminipila: labyrinthulomycota, hyphochytriomycota, and oomycota. In: McLaughlin DJ, Spatafora JW (eds) The Mycota VII. Part A. Systematics and evolution, 2nd edn. Springer, Berlin, Heidelberg, pp 39–96
Benner R, Hodgson RE (1985) Microbial degradation of the leachable and lignocellulosic components of leaves and wood *Rhizophora mangle* in *a* tropical mangrove swamp. Mar Ecol Prog Ser 23:221–230
Bentis CJ, Kaufman L, Golubic S (2000) Endolithic fungi in reef-building corals (Order: Scleractinia) are common, cosmopolitan and potentially pathogenic. Biol Bull 198:254–260
Berbee ML, Taylor JW (1993) Dating the evolutionary radiations of the true fungi. Can J Bot 71:1114–1127
Berbee ML, Taylor JW (2010) Dating the molecular clock in fungi – how close are we? Fungal Biol Rev 24:1–16
Bergbauer M, Newell SY (1992) Contribution to lignocellulose degradation and DOC formation from a salt marsh by the ascomycete *Phaeospheria spartinicola*. FEMS Microbiol Ecol 86:341–348
Besitulo A, Moslem MA, Hyde KD (2010) Occurrence and distribution of fungi in a mangrove forest on Siargao Island, Philippines. Bot Mar 54:535–544
Bhadury P, Wright PC (2004) Exploitation of marine algae: biogenic compounds for potential antifouling applications. Planta 219:561–578
Bhadury P, Mohammad BT, Wright PC (2006) The current status of natural products from marine fungi and their potential as anti-infective agents. J Ind Microbiol Biotechnol 33:325–337
Bhatnagar I, Kim S-K (2010) Immense essence of excellence: marine microbial bioactive compounds. Mar Drugs 8:2673–2701
Bigelow DM, Olsen MW, Gilbertson RL (2005) *Labyrinthula terrestris* sp. nov., a new pathogen of turf grass. Mycologia 97:185–190
Bills GF, Polishook JD (1994) Abundance and diversity of microfungi in leaf litter of a lowland rain forest in Costa Rica. Mycologia 86:187–198
Binder M, Hibbett DS, Wang Z, Farnham WF (2006) Evolutionary relationships of Mycaureola dilseae (Agaricales), a basidiomycetes pathogen of a subtidal Rhodophyte. Am J Bot 93:547–556
Blouin NA, Brodie JA, Grossman AC, Xu P, Brawley SH (2011) *Porphyra*: a marine crop shaped by stress. Trends Plant Sci 16:29–37
Blum LK, Mills AL, Zieman JC, Zieman RT (1988) Abundance of bacteria and fungi in seagrass and mangrove detritus. Mar Ecol Prog Ser 42:73–78
Blunt JW, Copp BR, Keyzers RA, Munroa MHG, Prinsep MR (2015) Marine natural products. Nat Prod Rep 2015(32):116
Blunt JW, Copp BR, Keyzers RA, Munro MH, Prinsep MR (2016) Marine natural products. Nat Prod Rep 32:16–211
Bodaker I, Sharon I, Suzuki MT, Feingersch R, Shmoish M, Béja R (2010) Comparative community genomics in the Dead Sea: an increasingly extreme environment. ISME J 4:399–407

Bongiorni L (2012) Thraustochytrids, a neglected component of organic matter decomposition and food webs in marine sediments. In: Raghukumar C (ed) Biology of marine fungi. Springer, Berlin, Heidelberg, pp 1–14

Bongiorni L, Dini F (2002) Distribution and abundance of thraustochytrids in different Mediterranean coastal habitats. Aquat Microb Ecol 30:49–56

Bongiorni L, Mirto S, Pusceddu A, Danovaro R (2005a) Response of benthic protozoa and thraustochytrid protists to fish-farm impact in seagrass (*Posidonia oceanica*) and soft bottom sediments. Microb Ecol 50:268–276

Bongiorni L, Pusceddu A, Danovaro R (2005b) Enzymatic activities of epiphytic and benthic thraustochytrids involved in organic matter degradation. Aquat Microb Ecol 41:299–305

Bonugli-Santos RC, dos Santos Vasconcelos MR, Passarini MRZ, Vieira GAL, Lopes VCP, Mainardi PH, dos Santos JA, de Azevedo Duarte L, Otero IVR, da Silva Yoshida AM, Feitosa VA, Pessoa A Jr, Sette LD (2015) Marine-derived fungi: diversity of enzymes and biotechnological applications. Front Microbiol 6:269. doi:10.3389/fmicb201500269

Bouillon S, Borges AV, Castañeda-Moya E, Diele K, Dittmar T, Duke NC, Kristensen E, Lee SY, Marchand C, Middelburg JJ, Rivera-Monroy VH, Smith TJ III, Twilley RR (2008) Mangrove production and carbon sinks: a revision of global budget estimates. Global Biogeochem Cycles 22:GB2013. doi:10.1029/2007GB003052

Bower SM (1987) *Labyrinthuloides haliotidis* n. sp. (Protozoa: Labyrinthomorpha), a pathogenic parasite of small juvenile abalone in a British Columbia mariculture facility. Can J Zool 65:1996–2007

Bower SM (2000) Infectious diseases of abalone (Haliotis spp.) and risks associated with transplantation. In: Campbell A (ed) Workshop on rebuilding abalone stocks in British Columbia. NRC Research Press, Ottawa, pp 111–122

Bower SM, Meyer GM (2005) Synopsis of infectious diseases and parasites of commercially exploited shellfish: *Labyrinthuloides haliotidis* of Abalone Fisheries and Oceans Canada Government of Canada. http://www.dfo-mpo.gc.ca/science/aah-saa/diseases-maladies/labhalab-eng.html

Brad T, Braster M, van Breukelen BM, van Straalen NM, Roling WFM (2008) Eukaryotic diversity in an anaerobic aquifer polluted with land fill leachate. Appl Environ Microbiol 74:3959–3968

Bremer GB (2000) Isolation and culture of thraustochytrids. In: Hyde KD, Pointing SB (eds) Marine mycology – a practical approach, Fungal Diversity Research Series, vol 1. Fungal Diversity Press, Hong Kong

Bremer GB, Talbot G (1995) Cellulolytic enzyme activity in the marine protist *Schizochytrium aggregatum*. Bot Mar 38:37–41

Brink KH (2004) The grass is greener in the coastal ocean. Oceanus. http://oceanusmag.whoi.edu/v42n3/brink.html

Buchalo AS, Nevo E, Wasser SP, Oren A, Molitoris HP (1998) Fungal life in the extremely hypersaline water of the Dead Sea: first records. Proc R Soc Lond 265:1461–1465

Buchan A, Newell SY, Moreta JIL, Moran MA (2002) Analysis of Internal Transcribed Spacer (ITS) regions of rRNA genes in fungal communities in a Southeastern US salt marsh. Microb Ecol 43:329–340

Buchan A, Newell SY, Butler M, Biers EJ, Hollibaugh JT, Moran MA (2003) Dynamics of bacterial and fungal communities on decaying salt marsh grass. Appl Environ Microbiol 69:6676–6687

Bucher VVC, Hyde KD, Pointing SB, Reddy CA (2004) Production of wood decay enzymes, mass loss and lignin solubilization in wood by marine ascomycetes and their anamorphs. Fungal Divers 15:1–14

Buchsbaum RN, Short FT, Cheney DP (1990) Phenolic-nitrogen interactions in eelgrass, *Zostera marina* L.: possible implications for disease resistance. Aquat Bot 37:291–297

Bugni TS, Ireland CM (2004) Marine-derived fungi: a chemically and biologically diverse group of microorganisms. Nat Prod Rep 21:143–163

Bull AT, Ward AC, Goodfellow M (2000) Search and discovery strategies for biotechnology: the paradigm shift. Microbiol Mol Biol Rev 64:573–606

Burdick DM, Short FT, Wolf J (1993) An index to assist and monitor the progression of wasting disease in eelgrass *Zostera marina*. Mar Ecol Prog Ser 94:83–90

Burgaud G, Calvez TL, Arzur D, Vandenkoornhuyse P, Barbier G (2009) Diversity of culturable marine filamentous fungi from deep-sea hydrothermal vents. Environ Microbiol 11:1588–1600

Burgaud G, Arzur D, Durand L, Cambon-Bonavita MA, Barbier G (2010) Marine culturable yeasts in deep-sea hydrothermal vents: species richness and association with fauna. FEMS Microbiol Ecol 73:121–133

Burja AM, Radianingtyas H, Windust A, Barrow C (2006) Isolation and characterization of polyunsaturated fatty acid producing *Thraustochytrium* species: screening of strains and optimization of omega-3 production. Appl Microbiol Biotechnol 72:1161–1169

Burkill PH, Mantoura RFC, Owens NJP (1993) Biogeochemical cycling in the northwestern Indian Ocean: a brief overview. Deep-Sea Res 40:643–649

Burtseva YV, Sova VV, Pivkin MV, Anastyuk SD, Gorbach VI, Zvyagintseva TN (2010) Distribution of O-glycosylhydrolases in marine fungi of the sea of Japan and the sea of Okhotsk: characterization of exocellular N-acetyl-beta-D-glucosaminidase of the marine gungus *Penicillium canescens*. Appl Biochem Microbiol 46:648–656

Butinar L, Santos S, Spencer-Martins I, Oren A, Gunde-Cimermane N (2005a) Yeast diversity in hypersaline habitats. FEMS Microbiol Lett 244:229–234

Butinar L, Sonjak S, Zalar P, Plemenitaš A, Gunde-Cimerman N (2005b) Melanized halophilic fungi are eukaryotic members of microbial communities in hypersaline waters of solar salterns. Bot Mar 48:73–79

Butinar L, Zalar P, Frisvad JC, Gunde-Cimerman N (2005c) The genus *Eurotium* – members of indigenous fungal community in hypersaline waters of salterns. FEMS Microbiol Ecol 51:155–166

Castro P, Freitas H (2000) Fungal biomass and decomposition in *Spartina maritima* leaves in the Mondego salt marsh (Portugal). Hydrobiologia 428:171–177

Cavalier-Smith T (2004) Only six kingdoms of life. Proc R Soc Lond B 271:1251–1262

Cavalier-Smith T, Chao EE (2006) Phylogeny and megasystematics of phagotrophic heterokonts (kingdom Chromista). J Mol Evol 62:388–420

Cavalier-Smith T, Allsopp MTEP, Chao EE (1994) Thraustochytrids are chromists, not fungi: 18S rDNA signatures of heterokonta. Philos Trans R Soc Lond Ser B Biol Sci 346:387–397

Cavicchioli R, Siddiqui KS, Andrews D, Sowers KR (2002) Low-temperature extremophiles and their applications. Curr Opin Biotechnol 13:253–261

Chang KJL, Dunstan GA, Abell GC, Clementson LA, Blackburn SI, Nichols PD, Koutoulis A (2012) Biodiscovery of new Australian thraustochytrids for production of biodiesel and long-chain omega-3 oils. Appl Microbiol Biotechnol 93:2215–2231

Chang KJL, Nichols CM, Blackburn SI, Dunstan GA, Koutoulis A, Nichols PD (2014) Comparison of thraustochytrids *Aurantiochytrium* sp., *Schizochytrium* sp., *Thraustochytrium* sp., and *Ulkenia* sp. for production of biodiesel, long-chain omega-3 oils, and exopolysaccharide. Mar Biotechnol 16:396–411

Cheng R, Ge Y, Yang B, Zhong X, Lin X, Huang Z (2013) Cloning and functional analysis of putative malonyl-CoA:acyl-carrier protein transacylase gene from the docosahexaenoic acid-producer Schizochytrium sp TIO1101. World J Microbiol Biotechnol 29:959–967

Chukanhom K, Borisuthpeth P, Khoa LV, Hatai K (2003) *Haliphthoros milfordensis* isolated from black tiger prawn larvae (*Penaeus monodon*) in Vietnam. Mycoscience 44:123–127

Colaço A, Raghukumar C, Mohandass C, Cardigos F, Santos RS (2006) Effect of shallow-water venting in Azores on a few marine biota. Cah Biol Mar 47:359–364

Connell L, Barrett A, Templeton A, Staudigel H (2009) Fungal diversity associated with an active deep sea volcano: Vailulu'u Seamount, Samoa. Geomicrobiol J 26:597–605

Cooke RC, Rayner ADM (1984) Ecology of saprotrophic fungi. Longman, London, New York

Cooney JJ, Doolittle MM, Grahl-Nielsen O, Haaland IM, Kirk PW (1993) Comparison of fatty acids of marine fungi using multivariate statistical analysis. J Ind Microbiol 12:373–378

Cox SL, Hulston D, Maaset EW (2009) Cryopreservation of marine thraustochytrids (Labyrinthulomycetes). Cryobiology 59:363–365

Craven KD, Peterson PD, Windham DE, Mitchell TK, Martin SB (2005) Molecular identification of the turf grass rapid blight pathogen. Mycologia 97:160–166

Cuadros-Orellana S, Leite LR, Smith A, Medeiros JD, Badotti F (2013) Assessment of fungal diversity in the environment using metagenomics: a decade in review. Fungal Genomics Biol 3:110

Cundell AM, Brown MS, Stanford R, Mitchell R (1979) Microbial degradation of *Rhizophora mangle* leaves immersed in the sea. Estuar Coast Mar Sci 9:281–286

Cuomo V, Vazanella F, Fresi E, Cinelli F, Mazella L (1985) Fungal flora of *Posidonia oceanica* and its ecological significance. Trans Br Mycol Soc 84:35–40

Curran PMT (1980) The effect of temperature, pH, light and dark on the growth of fungi from Irish coastal waters. Mycologia 72:350–358

Cury JC, Araujo FV, Coelho-Souza SA, Peixoto RS, Oliveira JAL (2011) Microbial diversity of a Brazilian coastal region influenced by an upwelling system and anthropogenic activity. PLoS One 6:e16553

D'Souza DT, Tiwari R, Sah AK, Raghukumar C (2006) Enhanced production of laccase by a marine fungus during treatment of colored effluents and synthetic dyes. Enzym Microb Technol 38:504–511

da Luz CM, Barata M (2012) Salt marsh fungi. In: Jones EBG, Pang K-L (eds) Marine mycology-marine fungi and fungal-like organisms. De Gruyter, Berlin, Germany, pp 345–381

Dalmaso GZL, Ferreira D, Vermelho AB (2015) Marine extremophiles: a source of hydrolases for biotechnological applications. Mar Drugs 13:1925–1965. doi:10.3390/md13041925

Damare S, Raghukumar C (2008a) Fungi and macroaggregation in deep-sea sediments. Microb Ecol 56:168–177

Damare V, Raghukumar S (2008b) Abundance of thraustochytrids and bacteria in the equatorial Indian Ocean, in relation to transparent exopolymeric particles (TEPs). FEMS Microbiol Ecol 25:40–49

Damare V, Raghukumar S (2010) Association of the stramenopilan protists, the aplanochytrids, with zooplankton of the equatorial Indian Ocean. Mar Ecol Prog Ser 399:53–68

Damare V, Raghukumar S (2012) Marine aggregates and transparent exopolymeric particles (TEPs) as substrates for the stramenopilan fungi, the thraustochytrids: roller table experimental approach. Kavaka 40:22–31

Damare S, Raghukumar C, Raghukumar S (2006a) Fungi in deep-sea sediments of the Central Indian Basin. Deep-Sea Res Pt I 53:14–27

Damare S, Raghukumar C, Muraleedharan UD, Raghukumar S (2006b) Deep-sea fungi as a source of alkaline and cold-tolerant proteases. Enzyme Microb Technol 39:172–181

Damare S, Singh P, Raghukumar S (2012) Biotechnology of marine fungi. In: Raghukumar C (ed) Biology of marine fungi. Springer, Berlin, Heidelberg, pp 277–298

Damare V, Damare S, Ramanujam P, Mina RM, Raghukumar S (2013) Preliminary studies on the association between zooplankton and the stramenopilan fungi, aplanochytrids. Microb Ecol 65:955–963

Dassarma P, Klebahn G, Klebahn H (2010) Translation of Henrich Klebahn's 'Damaging agents of the klippfish – a contribution to the knowledge of the salt-loving organisms'. Saline Syst 6:7

Davis SE III, Corronado-Molinab C, Childersa DL, Day JW Jr (2003) Temporally dependent C, N, and P dynamics associated with the decay of *Rhizophora mangle* L leaf litter in oligotrophic mangrove wetlands of the Southern Everglades. Aquat Bot 75:199–215

Dawson SC, Pace NR (2002) Novel kingdom-level eukaryotic diversity in anoxic environments. Proc Natl Acad Sci USA 99:8324–8329

de Araujo FV, Soares CA, Hagler AN, Mendonça-Hagler LC (1995) Ascomycetous yeast communities of marine invertebrates in a southeast Brazilian mangrove ecosystem. Antonie van Leeuw Microb 68:91–99

DeLong EF, Pace NR (2001) Environmental diversity of bacteria and archaea. Syst Biol 50:470–478
den Hartog C (1987) Wasting disease and other dynamic phenomena in *Zostera* beds. Aquat Bot 27:3–14
Dick GJ, Anantharaman K, Baker BJ, Li M, Reed DC, Sheik CS (2013) The microbiology of deep-sea hydrothermal vent plumes: ecological and biogeographic link ages to seafloor and water column habitats. Front Microbiol 4:124. doi:10.3389/fmicb201300124. eCollection 2013
Dix NJ, Webster J (1995) Fungal ecology. Springer, The Netherlands
Domart-Coulon LJ, Sinclair CS, Hill RT, Tambutte S, Puverel S, Ostrander GK (2004) A basidiomycete isolated from the skeleton of *Pocillopora damicornis* (Scleractinia) selectively stimulates short-term survival of coral skeletogenic cells. Mar Biol 144:583–592
Donachie SP, Zdanowski MK (1998) Potential digestive function of bacteria in krill *Euphausia superba* stomach. Aquat Microb Ecol 14:129–136
Douglas NL, Mullen KM, Talmage SC, Harvell CD (2007) Exploring the role of chitinolytic enzymes in the sea fan coral, *Gorgonia ventalina*. Mar Biol 150:1137–1144
Duarte CM, Cebrian J (1996) The fate of marine autotrophic production. Limnol Oceanogr 41:1758–1766
Duarte CM, Middelburg J, Caraco N (2005) Major role of marine vegetation on the oceanic carbon cycle. Biogeosciences 2:1–8
Duc PM, Wada S, Kurata O, Hatai K (2010) *In vitro* and *in vivo* efficacy of antifungal agents against *Acremonium* sp. Fish Pathol 45:109–114
Ducklow H (2000) Bacterial production and biomass in the oceans. In: Kirchman DL (ed) Microbial ecology of the oceans. Wiley-Liss, New York, pp 85–120
Duffy E (2006) Biodiversity and the functioning of seagrass ecosystems. Mar Ecol Prog Ser 311:233–250
Dunlap WC, Shick JM (1998) Ultraviolet radiation-absorbing mycosporine-like amino acids in coral reef organisms: a biochemical and environmental perspective. J Phycol 34:418–430
Dupont J, Magnin S, Rousseau F, Zbinden M, Frebourg G, Samadi S, Richer de Forges B, Jones EBG (2009) Molecular and ultrastructural characterization of two ascomycetes found on sunken wood off Vanuatu Islands in the deep Pacific Ocean. Mycol Res 113:1351–1364
Durako MJ, Kuss KM (1994) Effects of *Labyrinthula* infection on the photosynthetic capacity of *Thalassia testudinum*. Bull Mar Sci 54:727–732
Eaton RA (1985) Preservation of marine timbers. In: Findlay WPK (ed) Preservation of timber in the tropics. Martinus Nijhoff/DrWJunk Publishers, Dordrecht, pp 157–188
Ebel R (2012) Natural products from marine-derived fungi. In: EBG J, Pang KL (eds) Marine fungi and fungal-like organisms. De Gruyter, Berlin, pp 411–440
Ebersberger I, de Matos SR, Kupczok A, Gube M, Kothe E, Voigt K, von Haeseler A (2012) A consistent phylogenetic backbone for the fungi. Mol Biol Evol 29:1319–1334
Edgcomb VP, Kysela DT, Teske A, de Vera GA, Sogin ML (2002) Benthic eukaryotic diversity in the Guaymas Basin hydrothermal vent environment. Proc Natl Acad Sci USA 99:7658–7662
Edgcomb VP, Beaudoin D, Gast R, Biddle JF, Teske A (2011) Marine subsurface eukaryotes: the fungal majority. Environ Microbiol 13:172–183
Ein-Gil N, Ilan M, Carmeli S, Smith GW, Pawlik JR, Yarden O (2009) Presence of *Aspergillus sydowii*, a pathogen of gorgonian sea-fans in the marine sponge *Spongia obscura*. ISME J 3:752–755
Emmerson WD, McGynne G (1992) Feeding and assimilation of mangrove leaves by crab *Sesarma meinerti* in relation to leaf litter production in Mgazana, a warm temperate southern African mangrove swamp. J Exp Mar Biol Ecol 157:41–53
Enriquez S, Duarte CM, Sand-Jensen K (1993) Patterns in decomposition rates among photosynthetic organisms: the importance of detritus C:N:P content. Oecologia 94:457–471
Fan KW, Chen F (2007) Production of high-value products by marine microalgae thraustocytrids. In: Yang S-T (ed) Bioprocessing for value-added products from renewable resources New Technologies and Applications. Elsevier, Amsterdam, pp 293–324

Fan KW, Vrijmoed LLP, EBG J (2002) Physiological studies of subtropical mangrove thraustochytrids. Bot Mar 45:50–57

Fan KW, Jiang Y, Faan YW, Chen F (2007) Lipid characterization of mangrove thraustochytrid – *Schizochytrium mangrovei*. J Agric Food Chem 55:2906–2910

Fell JW (2012) Yeasts in marine environments. In: Jones EBG, Pang KL (eds) Marine fungi and fungal-like organisms. Walter de Gruyter GmbH & Co KG, Berlin/Boston, pp 91–102

Fell JW, Master IM (1980) The association and potential role of fungi in mangrove detrital systems. Bot Mar 23:257–263

Fell JW, Statzell AC, Hunter IL, Hunter IL, Phaff HJ (1969) *Leucosporidium* gen n., the heterobasidiomycetous stage of several yeasts of the genus *Candida*. Antonie Van Leeuwenhoek 35:433–462

Findlay RH, Fell JW, Coleman NK, Vestal JR (1986) Biochemical indicators of the role of fungi and thraustochytrids in mangrove detrital systems. In: Moss ST (ed) The Biology of Marine fungi. Cambridge University Press, Cambridge, pp 91–103

Fize A, Manier JF, Maurand J (1970) Sur un cas d'infestation du Copepode *Eurytemora velox* (Lillj) par une levure du genre *Metschnikowia* (Kamienski). Ann Parasitol Hum Comp 45:357–363

Fletcher K, Uljević A, Tsirigoti A, Antolić B, Katsaros C, Nikolić V, van West P, Küpper FC (2015) New record and phylogenetic affinities of the oomycete *Olpidiopsis feldmanni* infecting *Asparagopsis* sp. (Rhodophyta). Dis Aquat Organ 117:45–57

Ford T, Ryan D (1995) Toxic metals in aquatic ecosystems: a microbiological perspective. Environ Health Perspect 103:25–28

Gachon CMM, Küpper H, Küpper FC, Šetlik I (2006) Single-cell chlorophyll fluorescence kinetic microscopy of *Pylaiella littoralis* (Phaeophyceae) infected by *Chytridium polysiphoniae* (Chytridiomycota). Eur J Phycol 41:395–403

Gachon CMM, Sime-Ngando T, Strittmatter M, Chambouvet A, Kimet GH (2010) Algal diseases: spotlight on a black box trends. Plant Sci 15:633–640

Gadanho M, Sampaio JP (2005) Occurrence and diversity of yeasts in the mid-Atlantic ridge hydrothermal fields near the Azores Archipelago. Microb Ecol 50:408–417

Gadanho M, Almeida JMGCF, Sampaio JP (2003) Assessment of yeast diversity in a marine environment in the south of Portugal by microsatellite-primed PCR. Antonie Van Leeuwenhoek 84:217–227

Gaertner A (1968) Eine Methode des quantitativen Nachweises niederer mit Pollen köderbarer Pilze im Meerwasser und im Sediment. Veröff Inst Meeresforsch Bremerh Suppl 3:75–92

Gaertner A (1979) Some fungal parasites found in the diatom populations of the Rosfjord area (South Norway) during March 1979. Veröff Inst Meeresforsch Bremerh 18:29–33

Gaertner A (1982) Lower marine fungi from the Northwest African upwelling areas and from the Atlantic off Portugal. Meteor Forsch Ergebn 34:9–30

Gaertner A, Raghukumar S (1980) Ecology of the thraustochytrids (lower marine fungi) in the Fladen Ground and other parts of the North Sea. I "Meteor" Forschungsergebnisse, Reihe A, No 22:165–185

Galstoff PS (1942) Wasting disease causing mortality of sponges in the West Indies and Gulf of Mexico. Proceedings of the Eighth American Science Congress 3:411–412

Ganuza E, Izquierdo MS (2007) Lipid accumulation in *Schizochytrium* G13/2S produced in continuous culture. Appl Microbiol Biotechnol 76:985–990

Gao Z, Li B, Zheng C, Wang G (2008) Molecular detection of fungal communities in the Hawaiian marine sponges *Suberites zeteki* and *Mycale armata*. Appl Environ Microbiol 74:6091–6101

Gao Z, Johnson ZI, Wang G (2010) Molecular characterization of the spatial diversity and novel lineages of mycoplankton in Hawaiian coastal waters. ISME J 4:111–120

Garcia-Vedrenne AE, Groner M, Page-Karjian A, Siegmund G-F, Singhal S, Sziklay J, Roberts S (2013) Development of genomic resources for a thraustochytrid pathogen and investigation of temperature infl uences on gene expression. PLoS One 8(9):e74196

Geiser DM, Taylor JW, Ritchie KB, Smith GW (1998) Cause of sea fan death in the West Indies. Nature 394:137–138

Genilloud O, Pelaez F, Gonzalez I, Diez MT (1994) Diversity of actinomycetes and fungi on seaweeds from the Iberian coasts. Microbiologia 10:413–422

GESAMP (2007) Estimates of oil entering the marine environment from sea-based activities. International Maritime Organization, London
Gessner RV (1977) Seasonal occurrence and distribution of fungi associated with *Spartina alterniflora* from a Rhode Island estuary. Mycologia 69:477–491
Gessner RV (1980) Degradative enzyme production by salt-marsh fungi. Bot Mar 23:133–139
Gessner RV, Goos R (1973) Fungi from decomposing *Spartina alterniflora*. Can J Bot 51:51–55
Gessner MO, Newell SY (2001) Biomass, growth rate, and production of filamentous fungi in plant litter. In: Hurst CJ, Mc Ineme M, Stetzenbach L, Knudsen G, Walter M (eds) Manual of environmental microbiology, 2nd edn. ASM Press, Washington, DC, pp 390–408
Gisbert C, Rus AM, Bolarín MC, López-Coronado JM, Arrillaga I, Montesinos C, Caro M, Serrano R, Moreno V (2000) The yeast HAL1 gene improves salt tolerance of transgenic tomato. Plant Physiol 123:393–402
Gleason FH, Kagami M, Lefevre E, Sime-Ngando T (2008) The ecology of chytrids in aquatic ecosystems: roles in food web dynamics. Fungal Biol Rev 22:17–25
Gleason FH, Frithjof CK, Glöckling SL (2012a) Zoosporic true fungi. In: Jones EBG, Pang KL (eds) Marine fungi and fungal-like organisms. Walter de Gruyter, Berlin/Boston, pp 101–114
Gleason FH, Carney LT, Lilje O, Glockling SL (2012b) Ecological potentials of species of Rozella (Cryptomycota). Fungal Ecol 5:651–656
Glockling SL, Marshall WL, Gleason FH (2013) Phylogenetic interpretations and ecological potentials of the Mesomycetozoea (Ichthyosporea). Fungal Ecol 6:237–247
Goldstein S (1973) Zoosporic marine fungi (Thraustochytriaceae and Dermocystidiaceae). Annu Rev Microbiol 27:13–25
Golubic S, Perkins RD, Lukas KJ (1975) Boring microorganisms and microborings in carbonate substrates. In: Frey RW (ed) The study of trace fossils. Springer, Berlin, Heidelberg, pp 229–259
Golubic S, Radtke G, Le Campion-Alsumard TL (2005) Endolithic fungi in marine ecosystems. Trends Microbiol 13:229–235
Gomes NGM, Lefranc F, Kijjoa A, Kiss R (2015) Can some marine-derived fungal metabolites become actual anticancer agents? Mar Drugs 2015(13):3950–3991
Gostinčar C, Grube M, de Hoog S, Zalar P, Gunde-Cimerman N (2010) Extremotolerance in fungi: evolution on the edge. FEMS Microbiol Ecol 71(1):2–11
Gotelli D (1974) The morphology of *Lagenidium callinectes* I vegetative development. Mycologia 66:639–647
Guillot J, Petit T, Degorce-Rubiales F, Guého E, Chermette R (1998) Dermatitis caused by *Malassezia pachydermatis* in a California sea lion (*Zalophus californianus*). Vet Rec 142:311–312
Gunde-Cimerman N, Zalar P, de Hoog S, Plemenitaš A (2000) Hypersaline waters in salterns: natural ecological niches for halophilic black yeasts. FEMS Microbiol Ecol 32:235–340
Gunde-Cimerman N, Sonjak S, Zalar P, Frisvad JC, Diderichsen B, Plemenitas A (2003) Extremophilic fungi in arctic ice: a relationship between adaptation to low temperature and water activity. Phys Chem Earth 28:1273–1278
Gutiérrez M, Pantoja S, Quiñones R, González R (2010) First record of filamentous fungi in the coastal upwelling ecosystem off central Chile. Gayana 74:66–73
Gutiérrez MH, Pantoja S, Tejos E, Quiñones RÃ (2011) The role of fungi in processing marine organic matter in the upwelling ecosystem off Chile. Mar Biol 158:205–219
Gutiérrez MH, Jara AM, Pantoja S (2016) Fungal parasites infect marine diatoms in the upwelling ecosystem of the Humboldt current system off central Chile. Environ Microbiol 18:1646–1653
Hanic LA, Sekimoto S, Bates SS (2009) Oomycete and chytrid infections of the marine diatom *Pseudonitzschia pungens* (Bacillariophyceae) from Prince Edward Island, Canada. Botany 87:1096–1105
Harel M, Ben-dov E, Rasoulouniriana D, Siboni N, Kramarsky-Winter E, Loya Y, Barak Z, Weisman Z, Kushmaro A (2008) A new thraustochytrid, strain Fng1, isolated from the surface mucus of the hermatypic coral *Fungia granulose*. FEMS Microbiol Ecol 64:378–387

Harrison KE (1990) The role of nutrition in maturation, reproduction and embryonic development of decapod crustaceans: a review. J Shellfish Res 9:1–28
Harvell CD, Kim K, Burkholder JM, Colwell RR, Epstein PR, Grimes DJ, Hoffmann EE, Lipp EK, Osterhaus ADME, Overstreet RM, Porter JW, Smith GW, Vasta GR (1999) Emerging marine diseases—climate links and anthropogenic factors. Science 285:1505–1510
Hasle GR, Lange CB, Syvertsen EE (1996) A review of *Pseudonitzschia*, with special reference to the Skagerrak, North Atlantic, and adjacent waters. Helgoländer Meeresun 50:131–175
Hatai K (2012) Diseases of fish and shell fish caused by marine fungi. In: Raghukumar C (ed) Biology of marine fungi. Springer, Berlin, pp 15–52
Hatai K, Bian BZ, Baticados MCL, Egusa S (1980) Studies on the fungal diseases in Crustaceans II *Haliphthoros philippinensis* sp nov isolated from cultivated larvae of the jumbo tiger prawn (*Penaeus monodon*). Trans Mycol Soc Jpn 21:47–55
Hatai K, Roza D, Nakamura K (2000) Identification of lower fungi isolated from larvae of mangrove crab, *Scylla serrata*, in Indonesia. Mycoscience 41:565–572
Hauvermale A, Kuner J, Rosenzweig B, Guerra D, Diltz S, Metz JG (2006) Fatty acid production in *Schizochytrium* sp.: involvement of a polyunsaturated fatty acid synthase and a type I fatty acid synthase. Lipids 41:739–747
Hawksworth DL (1988) The variety of fungal-algal symbioses, their evolutionary significance, and the nature of lichens. Bot J Linn Soc 96:3–20
Hawksworth DL (2000) Freshwater and marine lichen-forming fungi. In: Hyde KD, Ho WH, Pointing SB (eds) Aquatic mycology across the millennium. Fungal Diversity Press, Hong Kong, pp 1–7
Heckman DS, Geiser DM, Eidell BR, Stauffer RL, Kardos NL, Hedges SB (2001) Molecular evidence for the early colonization of land by fungi and plants. Science 293:1129–1131
Hemminga MA, Duarte CM (2000) Seagrass ecology. Cambridge University Press, Cambridge
Hibbett DS, Binder M (2001) Evolution of marine mushrooms. Biol Bull 201:319–322
Hibbett DS, Binder M, Bischoff JF, Blackwell M, Cannon PF (2007) A higher-level phylogenetic classification of the Fungi. Mycol Res 111:509–547
Hibbits J (1978) Marine Eccrinales (Trichomycetes) found in crustaceans of the San Juan archipelago, Washington. Syesis 11:213–261
Hodgkiss IJ, Leung HC (1986) Cellulase associated with mangrove leaf decomposition. Bot Mar 29:467–469
Hodson RE, Christian RR, Maccubbin AE (1984) Lignocellulose and lignin in the salt marsh grass, *Spartina alterniflora*: initial concentrations and short-term post-depositional changes in detrital material. Mar Biol 81:1–7
Höhnk W (1969) Über den pilzlichen Befall Kalkiger Hartteile von Meerestieren. Ber Dtsch Wiss Komm Meeresforsch 20:129–140
Höller U, Wright AD, Matthee GF, Konig GM, Draeger S, Aust HJ (2000) Fungi from marine sponges: diversity, biological activity and secondary metabolites. Mycol Res 104:1354–1365
Hollibaugh JT, Lovejoy C, Murray AE (2007) Microbiology in Polar Oceans. Oceanography 20:14–144
Honda D, Yokochi T, Nakahara T, Raghukumar S, Nakagiri A, Schaumann K, Higashihara T (1999) Molecular phylogeny of labyrinthulids and thraustochytrids based on the sequencing of 18S ribosomal RNA gene. J Eukaryot Microbiol 46:637–647
Hong DD, Anh HTL, Thu NTH (2011) Study on biological characteristics of heterotrophic marine microalga—*Schizochytrium mangrovei* Pq6 isolated from Phu Quoc Island, Kien Giang Province, Vietnam. J Phycol 47:944–954
Horner R, Schrader GC (1962) Relative contributions of ice algae, phytoplankton, and benthic microalgae to primary production in nearshore regions of the Beaufort Sea. Arctic 35:485–503
Hou Y-H, Wang T-H, Long H, Zhu H-Y (2006) Novel cold-adaptive *Penicillium* strain FS010 secreting thermo-labile xylanase isolated from Yellow Sea. Acta Biochim Biophys Sin 38:142–149
https://www.britannica.com/science/fungus/Annotated-classification

http://courses.washington.edu/ocean101/Lex/Lecture26.pdf
http://ocean.otr.usm.edu/~w529014/index_files/Page319.htm
http://shigen.nig.ac.jp/algae_tree/FungiE.html
http://syst.bio.konan-u.ac.jp/labybase/index_en.html
http://www.nhm.ku.edu/~fungi/
Huckabone SE, Gulland FMD, Johnson SM, Colegrove KM, Dodd EM et al (2015) Coccidiomycosis and other systemic mycoses of marine mammals stranding along the Central California, USA Coast: 1998–2012. J Wildl Dis 51:295–308
Hulvey J, Telle S, Nigrelli L, Lamour K, Thines M (2010) Salisapiliaceae – a new family of oomycetes from marsh grass litter of southeastern North America. Persoonia 25:109–116
Hutchinson J, Spalding M, zu Ermgassen P (2014) The role of mangroves in fisheries enhancement. The Nature Conservancy and Wetlands International
Hyde KD (1988) Observations on the vertical distribution of marine fungi in *Rhizophora* spp at Kampong Danau mangrove, Brunei. Asian Mar Biol 5:77–81
Hyde KD (1990) A study of vertical zonation of intertidal fungi on *Rhizophora apiculata* at Kampong Kapok mangrove, Brunei. Aquat Bot 36:255–262
Hyde KD (1991) Fungal colonization of *Rhizophora apiculata* and *Xylocarpus granatum* poles in Kampong Kapok mangrove, Brunei. Sydowia 43:31–38
Hyde KD (ed) (2002) Fungi in marine environments, Fungal Diversity Research Series. Fungal Diversity Press, Hong Kong
Hyde KD, Jones EBG (1988) Marine mangrove fungi. PSZNI Mar Ecol 9:15–33
Hyde KD, Jones EBG (1989) Observations on ascospore morphology in marine fungi and their attachment to surfaces. Bot Mar 32:205–218
Hyde KD, Lee SY (1995) Ecology of mangrove fungi and their role in the nutrient cycling: what gaps occur in our knowledge? Hydrobiologia 295:107–118
Hyde KD, Pointing SP (eds) (2000) Marine mycology: a practical approach. Fungal Diversity Press, Hong Kong
Hyde KD, Sarma VV (2000) Pictorial key to higher marine fungi. In: Hyde KD, Pointing SB (eds) Marine mycology: a practical approach. Fungal Diversity Press, Hong Kong, pp 205–270
Hyde KD, Chalermongse A, Boonthavikoon T (1993) The distribution of intertidal fungi on *Rhizophora apiculata*. In: Morton JB (ed) The marine biology of South China Sea Hong Kong. Hong Kong University Press, Hong Kong, pp 643–652
Hyde KD, Jones EBG, Leano E, Pointing S, Poonyth ÃD, Vrijmoed L (1998) Role of fungi in marine ecosystem. Biodivers Conserv 7:1147–1161
Hyde KD, Sarma VV, Jones EBG (2000) Morphology and taxonomy of higher marine fungi. In: Hyde KD, Pointing SB (eds) Marine mycology a practical approach. Fungal Diversity Press, Hong Kong, pp 172–204
Imhoff JI (2016) Natural products from marine fungi—still an underrepresented resource. Mar Drugs 14:19
Inoue Y, Matsuda T, Sugiyama K, Izawa S, Kimura A (1999) Genetic analysis of glutathione peroxidase in oxidative stress response of Saccharomyces cerevisiae. J Biol Chem 274:27002–27009
Jain R, Raghukumar S, Tharanathan R, Bhosle NB (2005) Extracellular polysaccharide production by thraustochytrid protists. Mar Biotechnol 7:184–192
James TY, Kauff F, Schoch C, Matheny PB, Hoffstetter V, Cox C, Vilgalys R (2006) Reconstructing the early evolution of Fungi using a six-gene phylogeny. Nature 443:818–822
James TY, Pelin A, Bonen L, Ahrendt S, Sain D, Corradi N, Stajich JE (2013) Shared signatures of parasitism and phylogenomics unite Cryptomycota and Microsporidia. Curr Biol 23:1548–1553
Jebaraj CS, Raghukumar C (2009) Anaerobic denitrification in fungi from the coastal marine sediments off Goa, India. Mycol Res 113:100–109
Jebaraj CS, Raghukumar C, Behnke A, Stoeck T (2010) Fungal diversity in oxygen-depleted regions of the Arabian Sea revealed by targeted environmental sequencing combined with cultivation. FEMS Microbiol Ecol 71:399–412

Jebraj CS, Forster D, Kauff F, Stoeck T (2012) Molecular diversity of fungi from marine Oxygen-Deficient Environments (ODEs). In: Raghukumar C (ed) Biology of marine fungi. Springer, Germany, pp 189–208

Jennings DH (1983) Some aspects of the physiology and biochemistry of marine fungi. Biol Rev 58:423–459

Jennings DH (1986) Fungal growth in the sea. In: Moss ST (ed) The biology of marine fungi. Cambridge University Press, London, pp 1–10

Jensen E (2014) Technical review: in situ hybridization. Anat Rec 297:1349–1353

Jensen PR, Fenical W (2002) Secondary metabolites from marine fungi. In: Hyde KD (ed) Fungi in marine environments. Fungal Diversity Press, Thailand, pp 293–315

Jensen PR, Jenkins KM, Porter D, Fenical W (1998) Evidence that a new antibiotic flavone glycoside chemically defends the Sea Grass *Thalassia testudinum* against Zoosporic fungi. Appl Environ Microbiol 64:1490–1496

Jin Y, Weining S, Nevo E (2005) A MARK gene from Dead Sea fungus confers stress tolerance to lithium salt and freezing-thawing: prospects for saline agriculture. Proc Natl Acad Sci USA 102:18992–18997

Joergensen RG, Wichern F (2008) Quantitative assessment of the fungal contribution to microbial tissue in soil. Soil Biol Biochem 40:2977–2991

Johnson TW, Sparrow FK (1961) Fungi in oceans and estuaries. J Cramer, Germany

Jones EBG (ed) (1976) Recent advances in aquatic mycology. Elek, London

Jones EBG (2000) Marine fungi: some factors influencing biodiversity. Fungal Divers 4:53–73

Jones EBG (2006) Form and function of fungal spore appendages. Mycoscience 47:167–183

Jones EBG, Byrne P (1976) Physiology of the higher marine fungi. In: Jones EBG (ed) Recent advances in aquatic mycology. Wiley, New York, pp 1–51

Jones EBG, Choeyklin R (2008) Ecology of marine and freshwater basidiomycetes. In: Boddy L, Frankland JC, van West P (eds) Ecology of saprotrophic basidiomycetes. Elsevier, London, pp 301–324

Jones EBG, Fell JW (2012) Basidiomycota. In: Jones EBG, Pang KL (eds) Marine fungi and fungal-like organisms. Walter de Gruyter GmbH & Co KG, Berlin/Boston, pp 49–63

Jones EBG, Harrison JL (1976) Physiology of marine phycomycetes. In: Jones EBG (ed) Recent advances in aquatic mycology. ELEK Science, London, pp 261–278

Jones EBG, Hyde KD (1988) Methods for the study of marine fungi from the mangroves. In: Agate AD, Subramanian CV, Vannucci M (eds) Mangrove microbiology role of microorganisms in nutrient cycling of mangrove soils and waters. UNDP/UNESCO, New Dehli, India, pp 9–27

Jones EBG, Hyde KD (2002) Succession: where do we go from here? In: Hyde KD, Jones EBG (eds) Fungal succession, Fungal Diversity, vol 10, pp 241–253

Jones EBG, Jennings DH (1965) The effect of cations on the growth of fungi. New Phytol 70:511–518

Jones EBG, Le Campion-Alsumard T (1970) Marine fungi on polyurethane covered plates submerged in the sea. Nova Hedwigia 19:567–582

Jones EBG, Pang KL (eds) (2012) Marine fungi and fungal-like organisms. Walter de Gruyter GmbH & Co KG, Berlin

Jones EBG, Sakayaroj J, Suetrong S, Somrithipol S, Pang KL (2009) Classification of marine Ascomycota, anamorphic taxa and Basidiomycota. Fungal Divers 35:1–187

Jones MDM, Forn I, Gadelha C, Egan MJ, Bass D, Massana R (2011) Discovery of novel intermediate forms redefines the fungal tree of life. Nature 474:200–203

Jones EBG, Pang K-L, Stanley SJ (2012) Fungi from marine algae. In: Jones EBG, Pang K-L (eds) Marine fungi and fungal-like organisms. Walter de Gruyter, Berlin/Boston, pp 329–344

Jones EBG, Suetrong S, Sakayaroj J, Bahkali AH, Abdel-Wahab MA, Boekhout T, Pang K-L (2015) Classification of marine ascomycota, basidiomycota, blastocladiomycota and chytridiomycota. Fungal Divers 73:1–72

Jørgensen BB, Boetius A (2007) Feast and famine—microbial life in the deep-sea bed. Nat Rev Microbiol 5:770–781

Kachalkin AV (2014) Yeasts of the White Sea intertidal zone and description of *Glaciozyma litorale* sp. nov. Antonie Van Leeuwenhoek 105(6):1073–1083

Kagami M, de Bruin A, Ibelings BW, Van Donk E (2007) Parasitic chytrids: their effects on phytoplankton communities and food-web dynamics. Hydrobiologia 578:113–129

Kanchana R, Muraleedharan UD, Raghukumar S (2011) Alkaline lipase activity from the marine protists, thraustochytrids. World J Microbiol Biotechnol 27:2125–2131

Karl DM (1999) A sea of change: biogeochemical variability in the North Pacific Subtropical Gyre. Ecosystems 2:181–214

Karpov S, Mamkaeva MA, Aleoshin V, Nassonova E, Lilje O, Gleason FH (2014) Morphology, phylogeny, and ecology of the aphelids (Aphelidea, Opisthokonta) and proposal for the new superphylum Opisthosporidia. Front Microbiol 5:112

Kathiresan K, Saravanakumar K, Anburaj R, Gomathi V, Abirami G, Sahu SK, Anandhan S (2011) Microbial enzyme activity in decomposing leaves of mangroves. Int J Adv Biotechnol Res 2:382–389

Kawamura Y, Yokoo K, Tojo M, Hishiike M (2005) Distribution of *Pythium porphyrae*, the causal agent of red rot disease of *Porphyra* spp, in the Ariake Sea, Japan. Plant Dis 89:1041–1047

Keeling PJ (2009) Chromalveolates and the evolution of plastids by secondary endosymbiosis. J Eukaryot Microbiol 56:1–8

Kendrick B (2000) The Fifth Kingdom. Mycologue Publications British Columbia, Canada

Kendrick B, Risk MJ, Michaelides J, Bergman K (1982) Amphibious microborers: bioeroding fungi isolated from live corals. Bull Mar Sci 32:862–867

Khoa LV, Hatai K (2005) First case of *Fusarium oxysporum* infection in cultured kuruma prawn *Penaeus japonicus* in Japan. Fish Pathol 40:195–196

Kiiskinen LL, Rättö M, Kruus K (2004) Screening for novel laccase producing microbes. J Appl Microbiol 97:640–646

Kim K, Harvel CD, Kim PD, Smith GW, Merkel SM (2000) Fungal disease resistance of Caribbean sea fan corals (*Gorgonia* spp). Mar Biol 136:259–267

Kim GH, Moo K-H, Kim j-Y, Shim J, Klochkov TA (2014) A revaluation of algal diseases in Korean *Pyropia* (*Porphyra*) sea farms and their economic impact. Algae 29:249–265

Kimura H, Fukura T, Naganuma T (1999) Biomass of thraustochytrid protoctists in coastal water. Mar Ecol Prog Ser 189:27–33

Kimura H, Sato M, Sugiyama C, Naganuma T (2001) Coupling of thraustochytrids and POM, and of bacterio- and phytoplankton in a semi-enclosed coastal area: implication for different substrate preference by the planktonic decomposers. Aquat Microb Ecol 25:293–300

Kiørboe T (2001) Formation and fate of marine snow: small-scale processes with large-scale implications. Sci Mar 65:57–71

Kis-Papo T (2005) Marine fungal communities. In: Dighton J, White JF, Oudemans P (eds) The fungal community, its organization and role in the ecosystem, 3rd edn. CRC Press, Boca Raton, pp 61–92

Kis-Papo T, Oren A, Wasser SP, Nevo E (2003) Survival of filamentous fungi in hypersaline Dead Sea water. Microb Ecol 45:183–190

Koh LI, Tan TK, Chou LM, Goh NKC (2000) Fungi associated with gorgonians in Singapore. Proceedings of the ninth international Coral Reef Symposium 1:521–526

Kohlmeyer J (1977) New genera and species of higher fungi from the deep sea (1615-5315m). Rev Mycol 41:189–206

Kohlmeyer J (1984) Tropical marine fungi. PSZNI Mar Ecol 5:329–378

Kohlmeyer J (1986) Taxonomic studies of the marine Ascomycotina. In: Moss ST (ed) The biology of marine fungi. Cambridge University Press, Cambridge, pp 99–210

Kohlmeyer J, Kohlmeyer E (1979) Marine mycology: the higher fungi. Academic Press, New York

Kohlmeyer J, Volkmann-Kohlemyer B (1988) *Halographis* (Opegraphales) a new endolithic lichenoid from corals and snails. Can J Bot 66:1138–1141

Kohlmeyer J, Volkmann-Kohlmeyer B (1975) Biology and geographical distribution of *Spathulospora* species. Mycologia 67:629–637

Kohlmeyer J, Volkmann-Kohlmeyer B (1991) Illustrated key to the filamentous higher marine fungi. Bot Mar 34:1–61

Kohlmeyer J, Volkmann-Kohlmeyer B (1992) Two Ascomycotina from coral reefs in the Caribbean and Australia. Cryptogam Bot 2:367–374

Kohlmeyer J, Volkmann-Kohlmeyer B (1998) *Mycophycias*, a new genus for the mycobionts of *Apophlaea, Ascophyllum* and *Pelvetia*. Systema Ascomycetum 16:1–7

Kohlmeyer J, Volkmann-Kohlmeyer B (2001) The biodiversity of fungi on *Juncus roemerianus*. Mycol Res 105:1409–1412

Kohlmeyer J, Volkmann-Kohlmeyer B (2003) Marine ascomycetes from algae and animal hosts. Bot Mar 46:285–306

Kohlmeyer J, Spatafora JA, Volkmann-Kohlmeyer B (2000) Lulworthiales, a new order of marine ascomycota. Mycologia 92:453–458

Koske RE, Gemma JN (1989) A modified procedure for staining roots to detect VA mycorrihzae. Mycol Res 92:486–488

Kristiansen HB (2014) Characterization of marine fungal communities using next generation sequencing techniques. Master's Thesis, University of Svalbard. http://urnnbno/URN:NBN:no-44144

Kubanek J, Jensen PR, Keifer PA, Sullards MC, Collins DO, Fenical W (2003) Seaweed resistance to microbial attack: a targeted chemical defence against marine fungi. Proc Natl Acad Sci USA 100:6916–6921

Kühn S, Medlin L, Eller G (2004) Phylogenetic position of the parasitoid nanoflagellate *Pirsonia* inferred from nuclear-encoded small subunit ribosomal DNA and a description of *Pseudopirsonia* n. gen. and *Pseudopirsonia mucosa* (Drebes) comb. nov. Protist 155:143–156

Kumaresan V, Suryanarayanan TS (2002) Endophyte assemblages in young, mature and senescent leaves of *Rhizophora apiculata*: evidence for the role of endophytes in mangrove litter degradation. Fungal Divers 9:81–91

Kunčič MK, Kogej T, Drobne D, Gunde-Cimerman N (2010) Morphological response of the halophilic fungal genus *Wallemia* to high salinity. Appl Environ Microbiol 76:329–337

Kuo J, McComb AJ, Cambridge ML (1981) Ultrastructure of the seagrass rhizosphere. New Phytol 89:139–143

Küpper FC, Müller DG (1999) Massive occurrence of the heterokont and fungal parasites *Anisolpidium*, *Eurychasma* and *Chytridium* in *Pylaiella littoralis* (Ectocarpales, Phaeophyceae). Nova Hedwigia 69:381–389

Kutty SN, Philip R (2008) Marine yeasts-a review. Yeast 25:465–483

Laby Base: http://syst.bio.konan-u.ac.jp/labybase/index_en.html

Lai X, Cao L, Tan H, Fang S, Huang Y, Zhou S (2007) Fungal communities from methane hydrate bearing deep-sea marine sediments in South China Sea. ISME J 1:75–762

Lalli CM, Parsons TR (1997) Biological oceanography: an introduction, 2nd edn. Elsevier Butterworth-Heinemann, Oxford

Lara E, Moreira D, López-Garcia P (2010) The environmental clade LKM11 and *Rozella* form the deepest branchingclade of Fungi. Protist 161:116–121

Le Calvez T, Burgaud G, Mahe S, Barbier G, Vandenkoornhuyse P (2009) Fungal diversity in deep-sea hydrothermal ecosystems. Appl Environ Microbiol 75:6415–6421

Le Campion-Alsumard T, Golubic S, Priess K (1995) Fungi in corals – symbiosis or disease – interaction between polyps and fungi causes pearl-like skeleton biomineralization. Mar Ecol Prog Ser 117:137–147

Leaño EM (2001) Straminipilous organisms from fallen mangrove leaves from Panay Island, Philippines. Fungal Divers 6:75–81

Leaño EM (2002) *Haliphthoros* spp. from spawned eggs of captive mud crab, *Scylla serrata*, broodstocks. Fungal Divers 9:93–103

Leaño EM, Damare V (2012) Labyrinthulomycota. In: Jones EBG, Pang K-L (eds) Marine fungi and fungal-like organisms. de Gruyter, Berlin, Boston, pp 245–249

Leaño EM, Vrijmoed LLP, Jones EBG (1998) Zoospore chemotaxis of two mangrove strains of *Halophytophthora vesicula* from Mai Po, Hong Kong. Mycologia 90:1001–1008

Leaño EM, Jones EBG, Vrijmoed LLP (2000) Why are *Halophytophthora* species well adapted to mangrove habitats? In: Hyde KD, Ho WH, Pointing SB (eds) Aquatic mycology across the millennium, Fungal Diversity, vol 5, pp 131–151

Leathers TD (2003) Biotechnological production and applications of pullulan. Appl Microbiol Biotechnol 62:468–473

Lee SY (1995) Mangrove outwelling: a review. Hydrobiologia 295:203–212

Leightley LE (1980) Wood decay activities of marine fungi. Bot Mar 23:387–395

Leong WF, Tan TK, Jones EBG (1991) Fungal colonization of submerged *Bruguiera cylindrica* and *Rhizophora apiculata* wood. Bot Mar 34:69–76

Levin L (2005) Ecology of cold seep ecosystems: interactions of fauna with flow, chemistry and microbes. Oceanogr Mar Biol Annu Rev 43:1–46

Li Q, Wang G (2009) Diversity of fungal isolates from three Hawaiian marine sponges. Microbiol Res 164:233–241

Li L, Kato C, Horikoshi K (1999) Bacterial diversity in deep-sea sediments from different depths. Biodivers Conserv 8:659–677

Li X, Kondo R, Sakai K (2003) Studies on hypersaline-tolerant whiterot fungi IV: effects of $Mn^{2+}$ and $NH_4^+$ on manganese peroxidase production and poly R-478 decolorization by the marine isolate *Phlebia* sp. MG-60 under saline conditions. J Wood Sci 49:355–360

Li W, Zhang T, Tang X, Wang B (2010) Oomycetes and fungi: important parasites on marine algae. Acta Oceanol Sin 29:74–81

Li Q, Wang X, Liu X, Jiao N, Wang G (2013) Abundance and novel lineages of thraustochytrids in Hawaiian waters. Microb Ecol 66:823–830

Li J-L, Sun X, Chen L, Guo L-D (2016) Community structure of endophytic fungi of four mangrove species in Southern China. Mycology 7:180–190

Lindahl BD, Nilsson RH, Tedersoo L, Abarenkov K, Carlsen T, Kjøller R, Kõljalg U, Pennanen T, Rosendahl S, Stenlid J, Kauserud H (2013) Fungal community analysis by highthroughput sequencing of amplified markers—a user's guide. New Phytol 199:288–299

Littler MM, Littler DS (1998) An undescribed fungal pathogen of reef-forming crustose coralline algae discovered in American Samoa. Coral Reefs 17:144

Littman R, Willis BL, Bourne DG (2011) Metagenomic analysis of the coral holobiont during a natural bleaching event on the Great Barrier Reef. Environ Microbiol Rep 3:651–660

Loilong A, Sakayaroj J, Rungjindamai N, Choeyklin R, Jones EBG (2012) Biodiversity of fungi on the palm *Nypa fruticans*. In: Jones EBG, Pang KL (eds) Marine fungi and fungal-like organisms. Walter de Gruyter GmbH & Co KG, Berlin/Boston, pp 267–284

López-García P, Rodriguez-Valera F, Pedrós-Alió C, Moreira D (2001) Unexpected diversity of small eukaryotes in deep-sea Antarctic plankton. Nature 409:603–660

López-García P, Philippe H, Gail F, Moreira D (2003) Autochthonous eukaryotic diversity in hydrothermal sediment and experimental microcolonizers at the Mid-Atlantic Ridge. Proc Natl Acad Sci USA 100(2):697–702

López-García P, Vereshchaka A, Mozeira D (2007) Eukaryotic diversity associated with carbonates and fluid-seawater interface in Lost City hydrothermal field. Environ Microbiol 9:546–554

Loque CP, Medeiros AO, Pellizzari EM, Oliveira EC, Rosa CA, Rosa LH (2010) Fungal community associated with marine macroalgae from Antarctica. Polar Biol 33:641–648

Lorenz R, Molitoris HP (1992) Combined influence of salinity and temperature (Phoma-pattern) on growth of marine fungi. Can J Bot 70:2111–2115

Lorenz R, Molitoris HP (1997) Cultivation of fungi under simulated deep-sea conditions. Mycol Res 11:1355–1365

Lucas MI, Newell RC, Velimirov B (1981) Heterotrophic utilisation of mucilage released during fragmentation of kelp (*Ecklonia maxima* and *Laminaria pallida*) II Differential utilisation of dissolved organic components from kelp mucilage. Mar Ecol Prog Ser 4:43–55

Lücking R, Huhndorf S, Pfister DH, Plata ER, Lumbsch HT (2009) Fungi evolved right on track. Mycologia 101:810–822
Luo W, Vrijmoed LLP, Jones EBG (2005) Screening of marine fungi for lignocellulose-degrading enzyme activities. Bot Mar 48:379–386
Luo Z-H, Pang KL, Wu Y-R, Gu J-D, Chow RKK, Vrijmoed LLP (2012) Degradation of phthalate esters by *Fusarium* sp. DMT-5-3 and *Trichosporon* sp. DMI-5-1 isolated from mangrove sediments. In: Raghukumar C (ed) Biology of marine fungi. Springer, Berlin, Heidelberg, pp 299–328
Lyons JI, Newell SY, Buchan A, Moran MA (2003) Diversity of ascomycete laccase gene sequences in a Southeastern US salt marsh. Microb Ecol 45:270–281
Madhupratap M, Prasanna Kumar S, Bhattahiri PMA, Dileep Kumar M, Raghukumar S, Nair KKC, Ramaiah N (1996) Mechanisms of the biololgical response to winter cooling in the northeastern Arabian Sea. Nature 384:549–552
Maldonado M, Cortadellas N, Trillas I, Rützler K (2005) Endosymbiotic yeast maternally transmitted in a marine sponge. Biol Bull 209:94–106
Maldonado-Ramírez SL, Torres-Pratts H (2005) First report of *Clathrus* cf *crispus* (Basidiomycota: Clathraceae) occurring on decomposing leaves of *Rhizophora mangle* in Puerto Rico. Caribb J Sci 41:357–359
Manikan M, Kalil MS, Hamid AA (2015) Response surface optimization of culture medium for enhanced docosahexaenoic acid production by a Malaysian thraustochytrid. Sci Rep 5:8611
Mann KH (1988) Production and use of detritus in various freshwater, estuarine, and coastal marine ecosystems. Limnol Oceanogr 33:910–930
Manohar CS, Raghukumar C (2013) Fungal diversity from various marine habitats deduced through culture-independent studies. FEMS Microbiol Lett 341:69–78
Manohar CS, Boekhout T, Muller WH, Stoeck T (2014) *Tritirachium candoliense* sp. nov., a novel basidiomycetous fungus isolated from the anoxic zone of the Arabian Sea. Fungal Biol 118:139–149
Marano AV, Pires-Zottarelli CLA, de Souza JI, Glockling SL, Leano EM, Gachon CMM, Strittmatter M, Gleason FH (2012) Hyphochytriomycota, oomycota and perkinsozoa (Supergroup Chromalveolata). In: Jones EBG, Pang K-L (eds) Marine mycology-marine fungi and fungal-like organisms. De Gruyter, Berlin/Boston, pp 167–213
Marchand C, Disnar J-R, Lallier-Vergès R, Lottier N (2005) Early diagenesis of carbohydrates and lignin in mangrove sediments subject to variable redox conditions (French Guiana). Geochim Cosmochim Acta 69:131–142
Maria GL, Sridhar K (2003) Diversity of filamentous fungi on woody litter of five mangrove plant species from the southwest coast of India. Fungal Divers 14:109–126
Maria GL, Sridhar KR, Bärlocher F (2006) Decomposition of dead twigs of *Avicennia officinalis* and *Rhizophora mucronata* in a mangrove in southwestern India. Bot Mar 49:450–455
Massana R, Pedrós-Alió C (2008) Unveiling new microbial eukaryotes in the surface ocean. Curr Opin Microbiol 11:213–218
Massana R, Castresana J, Balague V, Guillou L, Romari K, Groisillier A, Valentin K, Pedrós-Alió C (2004a) Phylogenetic and ecological analysis of novel marine stramenopiles. Appl Environ Microbiol 70:3528–3534
Massana R, Balague V, Guillou L, Pedros-Alio C (2004b) Picoeukaryotic diversity in an oligotrophic coastal site studied by molecular and culturing approaches. FEMS Microbiol Ecol 50:231–243
McGenity TJ, Folwell BD, McKew BA, Sanni GO (2012) Marine crude-oil biodegradation: a central role for interspecies interactions. Aquat Biol 8:10
McLean IN, Porter D (1987) Lesions produced by a thraustochytrid in *Tritonia diomedea* (Mollusca: Gastropoda: Nudibranchia). J Invertebr Pathol 49:223–225
Menezes CBA, Bonugli-Santosa RC, Miquelettoa PB, Passarinia MRZ, Silvaa CHD, Justoa MR, Leala RR, Fantinatti-Garbogginia F, Oliveiraa VM, Berlinck RG, Sette LD (2010) Microbial diversity associated with algae, ascidians and sponges from the north coast of Saõ Paulo state, Brazil. Microbiol Res 165:466–482

Meyers SP, Reynolds ES (1960) Occurrence of lignicolous fungi in Northern Atlantic and Pacific marine localities. Can J Bot 38:217–226

Meyers SP, Ahearn DG, Grunkel W, Roth FJ Jr (1967) Yeasts from the North Sea. Mar Biol 1:118–123

Mfilinge PL, Meziane T, Bachok Z, Tsuchiya M (2003) Fatty acids in decomposing mangrove leaves: microbial activity, decay and nutritional quality. Mar Ecol Prog Ser 265:97–105

Miao L, Kwong TFN, Qian P-Y (2006) Effect of culture conditions on mycelial growth, antibacterial activity, and metabolite profiles of the marine-derived fungus *Arthrinium* cf *saccharicola*. Appl Microbiol Biotechnol 72:1063–1073

Miller DJ, Jones EBG (1983) Observations on the association of thraustochytrid marine fungi with decaying seaweed. Bot Mar 26:345–351

Misra JK, Lichtwardt RW (2000) Illustrated genera of trichomycetes: fungal symbionts of insects and other arthropods. Science Publishers, New Delhi

Molitoris HP, Buchalo AS, Kurchenko I, Nevo E, Rawal BS, Wasser SP, Oren A (2000) Physiological diversity of the first filamentous fungi isolated from the hypersaline Dead Sea. In: Hyde DK, Ho WH, Pointing SB (eds) Aquatic mycology across the millennium, vol 5. Fungal Diversity Press, Hong Kong, pp 55–70

Moore MM, Strom MS (2003) Infection and mortality by the yeast *Metschnikowia bicuspidata* var *bicuspidata* in chinook salmon fed live adult brine shrimp (*Artemia franciscana*). Aquaculture 220:43–57

Moreira D, López-García P (2003) Are hydrothermal vents oases for parasitic protists? Trends Parasitol 19:556–558

Morrison JM, Codispoti LA, Gaurin S, Jones B, Manghnani V, Zheng Z (1998) Seasonal variation of hydrographic and nutrient fields during the US JGOFS Arabian Sea process study. Deep-Sea Res II 45:2053–2101

Morrison-Gardiner S (2002) Dominant fungi from Australian coral reefs. Fungal Divers 9:105–121

Motta FL, Andrade CCP, Santana MHA (2011) A review of xylanase production by the fermentation of xylan: classification, characterization and applications. In: Anuj Chandel, Anuj and da Silva, SS (Eds) Sustainable degradation of lignocellulosic biomass – techniques, applications and commercialization. InTech, doi: 10.5772/53544

Mouzouras R (1986) Patterns of timber decay caused by marine fungi. In: Moss ST (ed) The biology of marine fungi. Cambridge University Press, Cambridge, UK, pp 341–353

Mouzouras R (1989) Soft rot decay of wood by marine fungi. J Inst Wood Sci 11:193–201

Mtui G, Nakamura Y (2004) Lignin-degrading enzymes from mycelial cultures of basidiomycetes fungi isolated in Tanzania. J Chem Eng Jpn 37:113–118

Muehlstein LK, Porter D, Short FT (1988) *Labyrinthula* sp, a marine slime mold producing the symptoms of wasting disease in eelgrass, *Zostera marina*. Mar Biol 99:465–472

Müller WEG (2003) The origin of metazoan complexity: porifera as integrated animals. Integr Comp Biol 43:3–10

Müller U, Sengbusch P (1983) Visualization of aquatic fungi (Chytridiales) parasitizing on algae by means of induced fluorescence. Arch Hydrobiol 97:471–485

Munn CB (2011) Marine microbiology: ecology and applications, 2nd edn. Garland Science, Taylor & Francis Group, New York & London

Muraosa Y, Lawhavinit O-R, Hatai K (2006) *Lagenidium thermophilum* isolated from eggs and larvae of black tiger shrimp *Penaeus monodon* in Thailand. Fish Pathol 41:35–40

Mydlarz LD, Harvell CD (2007) Peroxidase activity and inducibility in the sea fan coral exposed to a fungal pathogen. Comp Biochem Physiol A 146:54–62

Nagahama T, Nagano Y (2012) Cultured and uncultured fungal diversity in deep-sea environments. In: Raghukumar C (ed) Biology of marine fungi. Springer, Berlin, Heidelberg, pp 173–187

Nagahama T, Hamamoto M, Nakase T, Horikoshi K (2001) *Rhodotorula lamellibrachii* sp nov, a new yeast species from a tubeworm collected at the deep-sea floor in Sagami bay and its phylogenetic analysis A van Leeuw. J Microbiol 80:317–323

Nagahama T, Hamamoto M, Nakase T, Horikoshi K (2003) *Rhodotorula benthica* sp nov and *Rhodotorula calyptogenae* sp nov, novel yeast species from animals collected from the deep-sea floor, and *Rhodotorula lysiniphila* sp nov, which is related phylogenetically. Int J Syst Evol Microbiol 53:897–903

Nagahama T, Takahashi E, Nagano Y, Abdel-Wahab MA, Miyazaki M (2011) Molecular evidence that deep-branching fungi are major fungal components in deep-sea methane cold-seep sediments. Environ Microbiol 13:2359–2370

Nagano Y, Nagahama T, Hatada Y, Nunoura T, Takami H, Miyazaki J, Takai K, Horikoshi K (2010) Fungal diversity in deep-sea sediments-the presence of novel fungal groups. Fungal Ecol 3:316–325

Nagano N, Matsui S, Kuramura T, Taoka Y, Honda D, Hayashi M (2011) The distribution of extracellular cellulase activity in marine eukaryotes, thraustochytrids. Marine Biotechnol 13:133

Naganuma T, Takasugi H, Kimura H (1998) Abundance of thraustochytrids in coastal plankton. Mar Ecol Prog Ser 162:105–110

Naganuma T, Kimura H, Karimoto R, Pimenov NV (2006) Abundance of planktonic thraustochytrids and bacteria and the concentration of particulate ATP in the Greenland and Norwegian Seas. Polar Biosci 20:37–45

Nakagaki K, Hata K, Iwata E, Takeo K (2000) *Malassezia pachydermatis* isolated from a South American sea lion (*Otaria byronia*) with dermatitis. J Vet Med Sci 62:901–903

Nakagiri A (2000) Ecology and diversity of *Halophytophthora* species. In: Hyde KD, Ho WH, Pointing SB (eds) Aquatic mycology across the millennium, Fungal Diversity, vol 5, pp 153–164

Nakagiri A, Newell SY, Ito T (1994) Two new *Halophytophthora* species, *H. tartarea* and *H. masteri*, from intertidal decomposing leaves in saltmarsh and mangrove regions. Mycoscience 35:223–232

Nakagiri A, Newell SY, Ito T, Tan TK, Pek CL (1996) Biodiversity and ecology of the oomycetous fungus, *Halophytophthora*. In: Turner IM, Diong CH, Urn SSL, Ng PKL (eds) Biodiversity and the Dynmics o Ecosystems. DIWPA Series, vol 1. pp 273–280

Nakai R, Naganuma T (2015) Diversity and ecology of thraustochytrid protists in the marine environment. In: Ohtsuka S, Suzaki T, Horiguchi T, Suzuki N, Not F (eds) Marine Protists. Springer, Japan, pp 331–346

Nakai R, Nakamura K, Jadoon WA, Kashihara K, Naganuma T (2013) Genus-specific quantitative PCR of thraustochytrid protists. Mar Ecol Prog Ser 486:1–12

Nakamura K, Hatai K (1995) *Atkinsiella dubia* and its related species. Mycoscience 36:431–438

Nakamura K, Wada S, Hatai K, Sugimoto T (1994) *Lagenidium myophilum* infection in the coonstripe shrimp, *Pandalus hypsinotus*. Mycoscience 35:99–104

Nakamura K, Nakamura M, Hatai K, Zafran (1995) *Lagenidium* infection in eggs and larvae of mangrove crab (*Scylla serrata*) produced in Indonesia. Mycoscience 36:399–404

Nakazawa A, Kokubun Y, Matsuura H, Yonezawa N, Kose R et al (2014) TLC screening of thraustochytrid strains for squalene production. J Appl Phycol 26:29

Namikoshi M, Akano K, Kobayashi H, Koike Y, Kitazawa A, Rondonuwu AB, Pratasik S (2002) Distribution of marine filamentous fungi associated with marine sponges in coral reefs of Palau and Bunaken Island, Indonesia. J Tokyo Univ Fish 88:15–20

Naqvi SWA (1994) Denitrification processes in the Arabian Sea. Proc Indian Acad Sci Earth Planet Sci 103:279–300

Naqvi SWA, Naik H, Pratihary A, D'Souza W, Narvekar PV, Jayakumar DA (2006) Coastal versus open-ocean denitrification in the Arabian Sea. Biogeosciences 3:621–633

Nayak SS, Gonsalves V, Nazareth S (2011) Isolation and salt tolerance of halophilic fungi from mangroves and solar salterns – India. Indian J Geo-Mar Sci 41:164–172

Nazareth S (2014) The world of halophilic fungi. Kavaka 42:131–144

Neuhauser S, Gleason FH, Kirchmair M (2012) Phytomyxea (Super-group Rhizaria). In: Jones EBG, Pang KL (eds) Marine mycology-marine fungi and fungal-like organisms. De Gruyter, Berlin, Germany, pp 245–249

Newell SY (1976) Mangrove fungi: the succession in the mycoflora of red mangroves (*Rhizophora mangle* L) seedlings. In: Jones EBG (ed) Recent advances in aquatic mycology. Elek Science, London, pp 93–124

Newell SY (1984) Bacterial and fungal productivity in the marine environment: a contrastive overview. Colloque Int Cent Natn Rech Scient (Marseille) 331:133–139

Newell SY (1992) Autumn distribution of marine Pythiaceae across a mangrove-salt marsh boundary. Can J Bot 70:1912–1916

Newell SY (1993) Decomposition of shoots of a saltmarsh grass; methodology and dynamics of microbial assemblages. Adv Microb Ecol 13:301–326

Newell SY (1996a) Established and potential impacts of eukaryotic mycelial decomposers in marine/terrestrial ecotones. J Exp Mar Biol Ecol 200:187–206

Newell SY (1996b) The [$^{H}$C]acetate-to-ergosterol method: factors for conversion from acetate incorporated to organic fungal mass synthesized. Soil Biol Biochem 28:681–683

Newell SY (2000) Methods for determining biomass and productivity of mycelial marine fungi. In: Hyde KD, Pointing SB (eds) Marine mycology – a practical approach. Fungal Diversity Press, Hong Kong, pp 69–91

Newell SY (2001a) Multiyear patterns of fungal biomass dynamics and productivity within naturally decaying smooth cordgrass shoots. Limnol Oceanogr 46:573–583

Newell SY (2001b) Fungal biomass and productivity. In: Methods in microbiology, Vol 3. Academic Press, pp 357–372

Newell SY, Bärlocher F (1993) Removal of fungal and total organic matter from decaying cordgrass leaves by shredder snails. J Exp Mar Biol Ecol 171:39–49

Newell SY, Fell JW (1992) Ergosterol content of living and submerged, decaying leaves and twigs of red mangrove. Can J Microbiol 38:979–982

Newell SY, Porter D (2000) Microbial secondary production from saltmarsh-grass shoots, and its known and potential fates. In: Weinstein MP, Kreeger DA (eds) Concepts and controversies in tidal marsh ecology. Kluwer, Amsterdam, The Netherlands, pp 159–185

Newell RC, Lucas MI, Velimirov B, Setderer LJ (1980) Quantitative significance of dissolved organic losses following fragmentation of kelp (*Ecklonia maxima* and *Laminaria pallida*). Mar Ecol Prog Ser 2:45–59

Newell RC, Field JG, Griffiths CL (1982) Energy balance and significance of micro-organisms in a kelp bed community. Mar Ecol Prog Ser 8:103–113

Newell SY, Fallon RD, Miller JD (1986) Measuring fungal biomass dynamics in standing-dead leaves of a salt-marsh vascular plant. In: Moss ST (ed) The biology of marine fungi. Cambridge University Press, New York, pp 19–25

Newell SY, Miller JD, Fell JW (1987) Rapid and pervasive occupation of fallen mangrove leaves by a marine zoosporic fungus. Appl Environ Microbiol 53:2464–2469

Newell SY, Fallon RD, Miller JD (1989) Decomposition and microbial dynamics for standing, naturally positioned leaves of the salt-marsh grass *Spartina alterniflora*. Mar Biol 101:471–481

Newell RIE, Marshall N, Sasekumar A, Chong VC (1995) Relative importance of benthic microalgae, phytoplankton, and mangroves as sources of nutrition for penaeid prawns and other coastal invertebrates from Malaysia. Mar Biol 123:595–606

Newell SY, Porter D, Lingle WL (1996) Lignocellulolysis by ascomycetes (Fungi) of a saltmarsh grass (smooth cordgrass). Microsc Res Tech 33:32–46

Newsham KK, Hopkins DW, Carvalhais LC, Fretwell PT, Rushton SP, Donnell AGO, Dennis PG (2015) Relationship between soil fungal diversity and temperature in the maritime Antarctic. Nat Clim Chang 6:182–186

Norton JH, Thomas AD, Barker JR (1994) Fungal infection in the cultured juvenile boring clam *Tridacna crocea*. J Invertebr Pathol 64:273–275

O'Brien HE, Parrent JL, Jackson JA, Moncalvo J-M, Vilgalys R (2005) Fungal community analysis by large-scale sequencing of environmental samples. Appl Environ Microbiol 71:5544–5550

Oren A (1999) Microbiological studies in the Dead Sea: future challenges toward the understanding of life at the limit of salt concentrations. Hydrobiologia 405:1–9

Oren A (2002) Diversity of halophilic microorganisms: environments, phylogeny, physiology, and applications. J Ind Microbiol Biotechnol 28:56–63

Oren A, Gunde-Cimerman N (2012) Fungal life in the Dead Sea. In: Raghukumar C (ed) Biology of marine fungi. Springer, Germany, pp 89–114

Overy DP, Bayman P, Kerr RG, Bills GF (2014) An assessment of natural product discovery from marine (*sensu strictu*) and marine-derived fungi. Mycology 5:145–167

Pang KL (2012) Phylogeny of the marine Sordariomycetes. In: Jones EBG, Pang K-L (eds) Marine fungi and fungal-like organisms. Walter de Gruyter GmbH & Co KG, Berlin/Boston, pp 35–47

Pang K-L, Mitchell JI (2005) Molecular approaches for assessing fungal diversity in marine substrata. Bot Mar 48:332–347

Pang K-L, Chow RKK, Chan C-W, Vrijmoed LLP (2010) Diversity and physiology of marine lignicolous fungi in Arctic waters: a preliminary account. Polar Res 30:5859–5863

Panno L, Voyron S, Anastasi A, Sartor RM, Varese GC (2011) Biodiversity of marine fungi associated with the seagrass *Posidonia oceanica*: an ecological and biotechnological perspective. Biol Mar Mediterr 18:85–88

Passarini MZR, Rodrigues MV, da Silva M, Sette LD (2011) Marine-derived filamentous fungi and their potential application for polycyclic aromatic hydrocarbon bioremediation. Mar Pollut Bull 62:364–370

Perhar G, Arhonditsis GB, Brett MT (2013) Modelling the role of highly unsaturated fatty acids in planktonic food web processes: sensitivity analysis and examination of contemporary hypotheses. Eco Inform 13:77–98

Perkins FO (1973) A new species of marine labyrinthulid *Labyrinthuloides yorkensis* gennov spec nov – cytology and fine structure. Arch Mikrobiol 90:1–17

Perovic-Ottstadt S, Adell T, Proksch P, Wiens M, Korzhev M, Gamulin V (2004) A (1-3)-beta-D-glucan recognition protein from the sponge *Suberites domuncula*-mediated activation of fibrinogen-like protein and epidermal growth factor gene expression. Eur J Biochem 271:1924–1937

Petes L, Harvell CD, Peters EC, Webb MAH, Mullen KM (2003) Pathogens compromise reproduction and induce melanization in Caribbean sea fans. Mar Ecol Prog Ser 264:167–171

Phillips NW (1984) Role of different microbes and substrates as potential supplies of specific essential nutrients to marine detritivores. Bull Mar Sci 35:283–298

Phongpaichit S, Preedanan S, Rungiindama N, Sakayroj J, Benzies C, Chuaypat J, Plathong S (2006) Aspergillosis of the gorgonian sea fan *Annella* sp after the tsunami at Mu Ko Similan National Park, Andaman Sea, Thailand. Coral Reefs 25:296

Pinruan U, Jones EBG, Hyde KD (2002) Aquatic fungi from peat swamp palms: *Jahnula appendiculata* sp. Nova Sydowia 54:242–247

Pitt JI, Hocking AD (1997) Fungi and food spoilage, 2nd edn. Blackie Academic & Professional, London

Pöggeler S, Nowrousian M, Kück U (2006) Fruiting-body development in ascomycetes. In: Kües U, Fischer R (eds) The mycota, Growth, differentiation and sexuality, vol I. Springer, Berlin, Heidelberg, pp 325–355

Pointing SB, Hyde KD (2000) Lignocellulose-degrading marine fungi. Biofouling 15:221–229

Pointing SB, Vrijmoed LLP, Jones EBG (1998) A qualitative assessment of lignocelluloses degrading activity in marine fungi. Bot Mar 41:290–298

Pointing SB, Buswell JA, Jones EBG, Vrijmoed LLP (1999) Extracellular cellulolytic enzyme profiles of five lignicolous mangrove fungi. Mycol Res 103:696–700

Polglase JL, Alderman DJ, Richards RH (1986) Aspects of the process of mycotic infections in marine animals. In: Moss ST (ed) The biology of marine fungi. Cambridge University Press, Cambridge, pp 155–164

Pollegioni F, Tonin F, Rosini E (2015) Lignin-degrading enzymes. FEBS J 282:1190–2013

Pollock CG, Rohrbach B, Ramsay EC (2000) Fungal dermatitis in captive pinnipeds. J Zoo Wildl Med 31:374–378

Poon MOK, Hyde KD (1998) Biodiversity of intertidal estuarine fungi on *Phragmites* at Mai Po marshes, Hong Kong. Bot Mar 41:141–155

Porter D (1986) Mycoses of marine organisms: an overview of pathogenic fungi. In: Moss ST (ed) The biology of marine fungi. Cambridge University Press, Cambridge, pp 141–153

Porter D (1990) Phylum Labyrinthulomycota. In: Margulis L, Corliss JO, Melkonian M, Chapman DJ (eds) Handbook of Protoctista. Jones and Bartlett Publishers, Boston

Porter D, Lingle WL (1992) Endolithic thraustochytrid marine fungi from planted shell fragments. Mycologia 84:289–299

Poulicek M, Machiroux R, Toussaint C (1986) Chitin diagenesis in deep-water sediments. In: Muzzarelli R et al (eds) Chitin in nature and technology. Plenum Press, New York, pp 523–530

Powell, H (2005) A mysterious disease is infecting northeast clam beds but a new technique is revealing the secrets of QPX ('Quahog Parasite Unknown'). Oceanus Magazine, 44(3), Woods Hole Oceanographic Institution web page: http://wwwwhoiedu/oceanus/viewArticledo?id=7566

Prasannarai K, Sridhar KR (2001) Diversity and abundance of higher marine fungi on woody substrates along the west coast of India. Curr Sci 81:304–311

Priess K, Le Campion-Alsumard T, Golubic S, Gadel F, Thomassin BA (2000) Fungi in corals: black bands and density-banding of *Porites lutea* and *P. lobata* skeleton. Mar Biol 136:19–27

Prista C, Michán C, Miranda IM, Ramos J (2016) The halotolerant *Debaryomyces hansenii*, the Cinderella of non-conventional yeasts. Yeast 33:523–533

Pugh GJF, Allsopp D (1982) Microfungi in Signy Island, South Orkney Islands. Brit Antarct Surv Bull 57:55–67

Pugh GJF, Jones EBG (1986) Antarctic marine fungi: a preliminary account. In: Moss ST (ed) The biology of marine fungi. Cambridge University Press, Cambridge, pp 323–330

Quick JA Jr (1974) *Labyrinthuloides schizochytrops* nsp, a new marine *Labyrinthula* with spheroid "spindle" cells. Trans Am Microsc Soc 93:344–365

Rabinowitz C, Douek J, Weisz R, Shabtay A, Rinkevich B (2006) Isolation and characterization of four novel thraustochytrid strains from a colonial tunicate. Indian J Mar Sci 35:341–350

Raghukumar C (1980a) An ultrastructural study of the marine diatom *Licmophora hyalina* and its parasite *Ectrogella perforans*. I. Infection of host cells. Can J Bot 58:1280–1290

Raghukumar C (1980b) An ultrastructural study of the marine diatom *Licmophora hyalina* and its parasite *Ectrogella perforans*. II. Development of the fungus in the host. Can J Bot 58:2557–2574

Raghukumar S (1985) Enumeration of the thraustochytrids (heterotrophic microorganisms) from the Arabian Sea. Mahasagar, Bull Natl Inst Oceanogr 18:457–465

Raghukumar C (1987) Fungal parasites of marine algae from Mandapam (South India). Dis Aquat Org 3:137–145

Raghukumar S (1988) Detection of the thraustochytrid protist *Ulkenia visurgensis* in a hydroid, using immunofluorescence. Mar Biol 97:253–258

Raghukumar S (1990) Speculations on niches occupied by fungi in the sea with relation to bacteria. Proc Indian Acad Sci Earth Planet Sci 100:129–138

Raghukumar S (1992) Bacterivory: a novel dual role for thrausochytrids in the sea. Mar Biol 113:165–169

Raghukumar S (2002) Ecology of the marine protists, the Labyrinthulomycetes (Thraustochytrids and Labyrinthulids). Eur J Protistol 38:127–136

Raghukumar S (2005) The role of fungi in marine detrital processes. In: Ramaiah N (ed) Marine microbiology: facets and opportunities. National Institute of Oceanography, Goa, India, pp 91–101

Raghukumar C (2006) Algal-fungal interactions in the marine ecosystem: symbiosis to parasitism. In: Tewari A (ed) Recent advances on applied aspects of Indian marine algae with reference to global scenario, vol 1. Gujarat, Central Salt & Marine Chemicals Research Institute, pp 366–385

Raghukumar S (2007) Marine eukaryote diversity, with particular reference to fungi: lessons learnt from prokaryotes. Indian J Mar Sci 35:388–398

Raghukumar C (2008) Marine fungal biotechnology: an ecological perspective. Fungal Divers 31:19–35
Raghukumar S (2009) Thraustochytrid marine protists: production of PUFAs and other emerging technologies. Mar Biotechnol 10:631–640
Raghukumar C (ed) (2012) Biology of marine fungi. Springer, Berlin
Raghukumar S, Balasubramanian R (1991) Occurrence of thraustochytrid fungi in corals and coral mucus Indian. J Mar Sci 20:176–181
Raghukumar S, Damare V (2011) Increasing evidence for the important role of Labyrinthulomycetes in marine ecosystems. Bot Mar 54:3–11
Raghukumar S, Gaertner A (1980) Ecology of the thraustochytrids (lower marine fungi) in the Fladen Ground and other parts of the North Sea II. Veröff Inst Meeresforsch Bremerh 18:289–308
Raghukumar C, Lande V (1988) Shell disease of rock oyster *Crassostrea cucullata*. Dis Aquat Org 4:77–81
Raghukumar C, Raghukumar S (1991) Fungal invasion of massive corals. Mar Ecol PSZNI 12:251–260
Raghukumar C, Raghukumar S (1998) Barotolerance of fungi isolated from deep-sea sediments of the Indian Ocean. Aquat Microb Ecol 15:153–163
Raghukumar S, Raghukumar C (1999) Thraustochytrid fungoid protists in faecal pellets of the tunicate *Pegea confoederata*, their tolerance to deep-sea conditions and implication in degradation processes. Mar Ecol Prog Ser 190:133–140
Raghukumar C, Ravindran J (2012) Fungi and their role in corals and coral reef ecosystems. In: Raghukumar C (ed) Biology of marine fungi. Springer, Berlin, pp 89–114
Raghukumar S, Schaumann K (1993) An epifluorescence microscopy method for direct detection and enumeration of the fungi-like marine protists, the thraustochytrids. Limnol Oceanogr 38:182–187
Raghukumar C, Raghukumar S, Sharma S, Chandramohan D (1992) Endolithic fungi from deep-sea calcareous substrata Isolation and laboratory studies. In: Desai BN (ed) Oceanography of the Indian Ocean. Oxford & IBH, New Delhi, pp 3–9
Raghukumar C, Raghukumar S, Chinnaraj A, Chandramohan D, D'Souza TM, Reddy CA (1994a) Laccase and other lignocellulose modifying enzymes of marine fungi isolated off the coast of India. Bot Mar 37:515–523
Raghukumar S, Sharma S, Raghukumar C, Sathe-Pathak V (1994b) Thraustochytrid and fungal component of marine detritus. IV. Laboratory studies on decomposition of leaves of the mangrove *Rhizophora apiculata* Blume. J Exp Mar Biol Ecol 183:113–131
Raghukumar S, Sathe-Pathak V, Sharma S, Raghukumar C (1995) Thraustochytrid and fungal component of marine detritus. III. Field studies on decomposition of leaves of the mangrove *Rhizophora apiculata* Blume. Aquat Microb Ecol 9:117–125
Raghukumar C, D'Souza TM, Thorn RG, Reddy CA (1999) Lignin modifying enzymes of *Flavodon flavus*, a basidiomycete isolated from a coastal marine environment. Appl Environ Microbiol 65:2100–2111
Raghukumar S, Ramaiah N, Raghukumar C (2001) Dynamics of thraustochytrid protists in the water column of the Arabian Sea. Aquat Microb Ecol 24:175–186
Raghukumar C, Muraleedharan UD, Gaud VR, Mishra R (2004a) Xylanases of marine fungi of potential use for biobleaching of paper pulp. J Ind Microbiol Biotechnol 31:433–441
Raghukumar C, Raghukumar S, Sheelu G, Gupta SM, Nath BN, Rao BR (2004b) Buried in time: culturable fungi in a deep-sea sediment core from the Chagos Trench, Indian Ocean. Deep Sea Res, Part I 51:1759–1768
Raghukumar C, Mohandass C, Cardigo F, Santos RS, D'Costa PM, Colaço A (2008a) Assemblage of benthic diatoms and culturable heterotrophs in shallow-water hydrothermal vent of the D João de Castro Seamount, Azores in the Atlantic Ocean. Curr Sci 95:1715–1723
Raghukumar C, D'Souza-Ticlo D, Verma AK (2008b) Treatment of colored effluents with lignin-degrading enzymes: an emerging role of marine-derived fungi. Crit Rev Microbiol 34:189–206
Raghukumar C, Damare S, Singh P (2010) A review on deep-sea fungi: occurrence, diversity and adaptions. Bot Mar 53:479–492

Rai JN, Tewari JP, Mukerji KG (1969) Mycoflora of mangrove mud. Mycopathol Mycol Appl 38:17–31

Raikar MT, Raghukumar S, Vani V, David JJ, Chandramohan D (2001) Thraustochytrid protists degrade hydrocarbons. Indian J Mar Sci 30:139–145

Rajendran N, Kathiresan K (2007) Microbial flora associated with submerged mangrove leaf litter in India. Rev Biol Trop 55:393–400

Ralph PJ, Short FT (2002) Impact of the wasting disease pathogen, *Labyrinthula zosterae*, on the photobiology of eelgrass *Zostera marina*. Mar Ecol Prog Ser 226:265–271

Rämä T, Nordén J, Davey ML, Mathiassen GH, Spatafora JW, Kauserud H (2014) Fungi ahoy! Diversity on marine wooden substrata in the high North. Fungal Ecol 8:46–58

Rämä T, Davey ML, Nordén J, Halvorsen R, Blaalid R, Mathiassen GH, Alsos IG, Kauseru H (2016) Fungi sailing the Arctic Ocean: speciose communities in North Atlantic Driftwood as Revealed by high-throughput amplicon sequencing. Microb Ecol 72(2):295–304

Ramaiah N (2006) A review on fungal diseases of algae, marine fishes, shrimps and corals. Indian J Mar Sci 35:380–387

Ramos-Flores T (1983) Lower marine fungus associated with black line disease in star corals (*Montasrea annularis* E & S). Biol Bull 165:429–435

Rangel M, Falkenberg M (2015) An overview of the marine natural products in clinical trials and on the market. J Coastal Life Med 3:421–428

Rateb ME, Ebel R (2011) Secondary metabolites of fungi from marine habitats. Nat Prod Rep 28:290–344

Ravichandran S, Kannupandi T, Kathiresan K (2006) Mangrove leaf litter processing by sesarmid crabs. Cey J Sci (Bio Sci) 35:107–114

Ravindran J, Raghukumar C, Raghukumar S (2001) Fungi in *Porites lutea*: association with healthy and diseased corals. Dis Aquat Org 47:219–228

Ray S, Straškraba M (2001) The impact of detritivorous fishes on a mangrove estuarine system. Ecol Model 140:207–218

Rédou V, Navarri M, Meslet-Cladière L, Barbier G, Burgaud G (2015) Species richness and adaptation of marine fungi from deep-subseafloor sediments. Appl Environ Microbiol 81:3571–3583

Rice DL (1982) The detritus nitrogen problem: new observations and perspectives from organic geochemistry. Mar Ecol Prog Ser 9:153–162

Richards TA, Jones MDM, Leonard G, Bass D (2012) Marine fungi: their ecology and molecular diversity. Annu Rev Mar Sci 4:495–522

Richter W (1985) Marine sponges as substrate for Thraustochytriaceae marine lower fungi. Veröff Inst Meeresforsch Bremerh 20:141–150

Riemann F, Schaumann K (1993) Thraustochytrid protists in Antarctic fast ice? Antarct Sci 5:279–280

Rieper-Kirchner M (1989) Microbial degradation of North Sea macroalgae: field and laboratory studies. Bot Mar 32:241–252

Rieper-Kirchner M (1990) Macroalgal decomposition: laboratory studies with particular regard to microorganisms and meiofauna. Helgoländer Meeresun 44:397–410

Riisberg I, Orr RJ, Kluge R, Shalchian-Tabrizi K, Bowers HA, Patil V, Edvardsen B, Jakobsen KS (2009) Seven gene phylogeny of heterokonts. Protist 160:191–204

Ritchie D (1957) Salinity optima for marine fungi affected by temperature. Am J Bot 44:870–874

Robertson AI (1988) Decomposition of mangrove leaf litter in tropical Australia. J Exp Mar Biol Ecol 116:235–247

Rodríguez-Jasso R, Mussatto SI, Pastrana L, Aguilar CN, Teixeira JA (2010) Fucoidan-degrading fungal strains: screening, morphometric evaluation, and influence of medium composition. Appl Biochem Biotechnol 162:2177–2188

Rohrmann S, Molitoris HP (1992) Screening for wood decay enzymes in marine fungi. Can J Bot 70:2116–2123

Rossman AY, Samuels GJ, Rogerson CT, Lowen R (1999) Genera of bionectriaceae, hypocreaceae and nectriaceae (hypocreales, ascomycetes). Stud Mycol 42:3–238

Roth FJ, Orpurt PA, Ahearn DG (1964) Occurrence and distribution of fungi in a subtropical marine environment. Can J Bot 42:375–383

Rucocco N, Costantini S, Guariniello S, Costantini M (2016) Polysaccharides from the marine environment with pharmacological, cosmeceutical and nutraceutical potential. Molecules 21 (5): pii: E551

Ruisi S, Barreca D, Selbmann L, Zucconi L, Onofri S (2007) Fungi in Antarctica. Rev Environ Sci Biotechnol 6:127–141

Rypien KL (2008) African dust is an unlikely source of *Aspergillus sydowii*, the causative agent of sea fan disease. Mar Ecol Prog Ser 367:125–131

Rypien KL, Baker DM (2009) Isotopic labelling and antifungal resistance as tracers of gut passage of the sea fan pathogen *Aspergillus sydowii*. Dis Aquat Org 86:1–7

Sajith S, Priji P, Sreedevi S, Benjamin S (2016) An overview on fungal cellulases with an industrial perspective. J Nutr Food Sci 6:461

Sakaguchi K, Matsuda T, Kobayashi T, Ohara J, Hamaguchi R, Abe E, Nagano N, Hayashi M, Ueda M, Honda D, Okita Y, Taoka Y, Sugimoto S, Okino N, Ito M (2012) Versatile transformation system that is applicable to both multiple transgene expression and gene targeting for Thraustochytrids. Appl Environ Microbiol 78:3193–3202

Sakayaroj J, Pang KL, Jones EBG, Vrijmoed LLP, Abdel-Wahab MA (2005) A systematic reassessment of marine ascomycetes *Swampomyces* and *Torpedospora*. Bot Mar 48:395–406

Sakayaroj J, Preedanon S, Supaphon O, Jones EBG, Phongpaichi S (2010) Phylogenetic diversity of endophyte assemblages associated with the tropical seagrass *Enhalus acoroides* in Thailand. Fungal Divers 42:27–45

Sakayaroj J, Pang KL, Jones EBG (2011) Multi-gene phylogeny of the Halosphaeriaceae: its ordinal status, relationships between genera and morphological character evolution. Fungal Divers 46:87–109

Sakayaroj J, Preedanon S, Phongpaichit S, Buatong J, Chaowalit P, Rukachaisirikul V (2012) Diversity of endophytic and marine-derived fungi associated with marine plants and animals. In: Jones EBG, Pang K-L (eds) Marine fungi and fungal-like organisms. De Gruyter, Berlin, pp 291–328

Samiaji J, Bärlocher F (1996) Geratology and decomposition of *Spartina alterniflora* Loisel in a New Brunswick saltmarsh. J Exp Mar Biol Ecol 201:233–252

Samson RA, Hoekstra ES, Frisvad JC, Filtenborg O (2002) Introduction to food- and airborne fungi, 6th edn. Centraalbureau voor Schimmelcultures, Baarn

Santangelo G, Bongiorni L, Pignataro L (2000) Abundance of thraustochytrids and ciliated protozoans in a Mediterranean sandy shore determined by an improved, direct method. Aquat Microb Ecol 23:55–61

Sarma VVS (2012) Diversity and distribution of marine Fungi on *Rhizophora* spp. in mangroves. In: Raghukumar C (ed) Biology of marine fungi. Springer, Berlin, Heidelberg, pp 243–276

Sarma VVS, Vittal BPR (2000) Biodiversity of mangrove fungi on different substrata of *Rhizophora apiculata* and *Avicennia* spp from Godavari and Krishna deltas, east coast of India. In: Hyde KD, Ho WH, Pointing SB (eds) Aquatic mycology across the millennium, Fungal Divers, vol 5, pp 23–41

Sarma VVS, Vittal BPR (2001) Biodiversity of manglicolous fungi on selected plants in the Godavari and Krishna deltas, East coast of India. Fungal Divers 6:115–130

Sathe V, Raghukumar S (1991) Fungi and their biomass in detritus of the seagrass *Thalassia hemprichii* (Ehrenberg) Ascherson. Bot Mar 34:272–277

Sathe-Pathak V, Raghukumar S, Raghukumar C, Sharma S (1993) Thraustochytrid and fungal component of marine detritus. I. Field studies on decomposition of the brown alga *Sargassum cinereum* J Ag. Indian J Mar Sci 22:159–167

Savin MC, Martin JL, LeGresley M, Giewat M, Rooney-Varga J (2004) Plankton diversity in the Bay of Fundy as measured by morphological and molecular methods. Microb Ecol 48:51–65

Schaumann K, Weide G (1990) Enzymatic degradation of alginate by marine fungi. Hydrobiologia 204(205):589–596

Schaumann K, Nulach M, Molitoris HP (1986) Comparative studies on growth and exoenzyme production of different Lulworthia isolates. In: Moss ST (ed) The biology of marine fungi. Cambridge Press, Cambridge, pp 49–60

Schnepf E, Drebes G (1977) Über die Entwicklung des marinen parasitischen Phycomyceten *Lagenisma coscinodisci* (Lagenidiales). Helgoländer Meeresun 29:291–301

Schnepf E, Deichgräber G, Drebes G (1978a) Development and ultrastructure of the marine parasitic oomycete *Lagenisma coscinodisci* Drebes (Lagenidiales) Thallus, zoosporangium, mitosis and meiosis. Arch Microbiol 116:121–132

Schnepf E, Deichgräber G, Drebes G (1978b) Development and ultrastructure of the marine, parasitic oomcete, *Lagenisma coscinodisci* (Lagenidiales) Sexual reproduction. Can J Bot 56:1315–1325

Schoch CL, Sung GH, Volkmann-Kohlmeyer B, Kohlmeyer J, Spatafora JW (2006) Marine fungal lineages in the Hypocreomycetidae. Mycol Res 110:257–263

Schoch CL, Sung GH, Volkmann-Kohlmeyer B, Kohlmeyer J, Spatafora JW (2007) Marine fungal lineages in the Hypocreomycetidae. Mycol Res 111:154–162

Schoch CL, Sung GH, López-Giráldez F (2009) The Ascomycota tree of life: a phylum–wide phylogeny clarifies the origin and evolution of fundamental reproductive and ecological traits. Syst Biol 58:224–239

Scholz B, Guillouc L, Marano AV, Neuhauser S, Sullivan B, Karsten U, Küpper C, Gleason FH (2015) Zoosporic parasites infecting marine diatoms: A black box that needs to be opened. Fungal Ecol 19:59–76

Schubert K, Groenewald JZ, Braun U, Dijksterhuis J, Starink M, Hill CF, Zalar P, de Hoog GS, Crous PW (2007) Biodiversity in the *Cladosporium herbarum* complex (Davidiellaceae, Capnodiales), with standardisation of methods for *Cladosporium taxonomy* and diagnostics. Stud Mycol 58:105–156

Schweikert M (2015) Biology of parasitic heterotrophic nanoflagellates: parasitoids of diatoms. In: Ohtsuka S, Suzaki T, Horiguchi T, Suzuki N, Not F (eds) Marine Protists. Springer, Japan, pp 519–530

Seki H, Fulton J (1969) Infection of marine copepods by *Metschnikowia* sp. Mycopathol Mycol Appl 38:61–70

Sekimoto S, Hatai K, Honda D (2007) Molecular phylogeny of an unidentified *Haliphthoros*-like marine oomycete and *Haliphthoros milfordensis* inferred from nuclear-encoded small- and large-subunit rRNA genes and mitochondrial-encoded cox2 gene. Mycoscience 48:212–221

Sguros PL, Simms J (1964) Role of marine fungi in the biochemistry of the oceans. IV. Growth responses to seawater inorganic macroconstituents. J Bacteriol 88:346–355

Sguros PL, Rodrigues J, Simms J (1973) Role of marine fungi in the biochemistry of the oceans. V. Patterns of constitutive nutritional growth responses. Mycologia 65:161–174

Sharma S, Raghukumar C, Raghukumar S, Sathe-Pathak V, Chandramohan D (1994) Thraustochytrid and fungal component of marine detritus. IV. Laboratory studies on decomposition of the brown alga *Sargassum cinereum* J Ag. J Exp Mar Biol Ecol 175:217–242

Sheppard CRC, Davy SK, Pilling GM (2009) The biology of coral reefs. Oxford University Press, Oxford

Shetye SR, Gouveia AD, Shenoi SSC (1994) Circulation and water masses of the Arabian Sea. Proc Indian Acad Sci Earth Planet Sci 103:107–123

Shields JD (1990) *Rhizophydium littoreum* on the eggs of *Cancer anthonyi*: parasite or saprobe? Biol Bull 179:201–206

Shields JD, Overstreet RM (2007) Diseases, parasites, and other symbionts. Chapter 8. In: Kennedy VS, Cronin LE (eds) The blue crab, *Callinectes sapidus*. Maryland Sea Grant, College Park, MD, pp 299–417

Shinn EA, Smith GW, Prospero JM, Betzer P, Hayes ML, Garrison V, Barber RT (2000) African dust and the demise of Caribbean coral reefs. Geophys Res Lett 27:3029–3032

Short FT, Mathieson AC, Nelson JI (1986) Recurrence of the eelgrass wasting disease at the border of New Hampshire and Maine, USA. Mar Ecol Prog Ser 29:89–92

Short FT, Burdick DM, Wolf J, Jones GE (1993) Eelgrass in estuarine research reserves along the East Coast, USA, Part I: Declines from pollution and disease and Part II: Management of eelgrass meadows. NOAA Coastal Ocean Program Publ

Sigee DC (2005) Freshwater microbiology. Wiley, Chichester

Silber J, Kramer A, Labes A, Tasdemir D (2016) From discovery to production: biotechnology of marine fungi for the production of new antibiotics. Mar Drugs 14:137

Simon M, Alldredge AL, Azam F (1990) Bacterial carbon dynamics on marine snow. Mar Ecol Prog Ser 65:205–211

Simonato F, Campanaro S, Lauro FM, Vezzi A, D'Angelo M, Vitulo N et al (2006) Piezophilic adaptation: a genomic point of view. J Biotechnol 126:11–25

Simpoulos AP (2008) The importance of the omega-6/omega-3 fatty acid ratio in cardiovascular disease and other chronic diseases. Exp Biol Med (Maywood) 233:674–688

Singh AP, Butcher JA (1991) Bacterial degradation of wood cells: a review of degradation patterns. J Wood Sci 12:143–157

Singh P, Raghukumar C (2014) Diversity and physiology of deep-sea yeasts: a review. Kavaka 43:50–63

Singh RS, Saini GK (2008) Pullulan-hyperproducing color variant strain of *Aureobasidium pullulans* FB-1 newly isolated from phylloplane of *Ficus* sp. Bioresour Technol 99:3896–3899

Singh P, Raghukumar C, Verma P, Shouche Y (2010) Phylogenetic diversity of culturable fungi from the deep-sea sediments of the Central Indian Basin and their growth characteristics. Fungal Divers 40:89–102

Singh P, Raghukumar C, Verma P, Shouche Y (2011) Fungal community analysis in the deep-sea sediments of the Central Indian Basin by culture-independent approach. Microb Ecol 61:507–517

Singh P, Raghukumar C, Meena RM, Verma P, Shouche Y (2012a) Fungal diversity in deep-sea sediments revealed by culture-dependent and culture-independent approaches. Fungal Ecol 5:543–553

Singh P, Raghukumar C, Verma AK, Meena RM (2012b) Differentially expressed genes under simulated deep-sea conditions in the psychrotolerant yeast *Cryptococcus* sp NIOCC#PY13. Extremophiles 16:777–785

Singh P, Raghukumar C, Verma P, Shouche Y (2012c) Assessment of fungal diversity in deep-sea sediments by multiple primer approach. World J Microbiol Biotechnol 28:659–667

Singh P, Raghukumar C, Parvatkar RR, Mascarenhas-Pereira MB (2013) Heavy metal tolerance in the psychrotolerant *Cryptococcus* sp isolated from deep-sea sediments of Central Indian Basin. Yeast 30:93–101

Smith DC, Simon M, Alldredge AL, Azam F (1992) Intense hydrolytic enzyme activity on marine aggregates and implications for rapid particle dissolution. Nature 359:139–142

Smith GW, Nagelkerken IA, Ritchie KB (1996) Caribbean sea-fan mortalities. Nature 383:487

Smith GW, Harvel CD, Kim K (1998) Response of sea fans to infection with *Aspergillus* sp (Fungi). Rev Biol Trop 46:205–208

Sogin ML, Morrison HG, Huber JA, Welch DM, Huse SM (2006) Microbial diversity in the deep sea and the underexplored "rarebiosphere". Proc Natl Acad Sci USA 103:12115–12120

Sparrow FK Jr (1969) Zoosporic marine fungi from the Pacific Northwest (USA). Arch Microbiol 66:129–146

Spatafora J, Volkmann-Kohlmeyer B, Kohlmeyer J (1998) Independent terrestrial origins of the Halosphaeriales (marine Ascomycota). Am J Bot 85:1569–1580

Sridhar KR (2009) Fungal diversity of Pichavaram mangroves, Southeast coast of India. Nat Sci 7:67–75

Sridhar KR, Maria GL (2006) Fungal diversity on mangrove woody litter *Rhizophora mucronata* (Rhizophoraceae). Ind J Mar Sci 35:318–325

Sridhar KR, Alias SA, Pang K-L (2012a) Mangrove fungi. In: Jones EBG, Pang K-L (eds) Marine mycology-marine fungi and fungal-like organisms. De Gruyter, Berlin, pp 253–271

Sridhar KR, Karamchand KS, Pascoal C, Cássio F (2012b) Assemblage and diversity of fungi on wood and seaweed litter of seven Norwest Portuguese beaches. In: Raghukumar C (ed) Biology of marine fungi. Springer, Berlin, pp 209–228

Srinivasan MC (2004) Practical mycology for industrial biotechnologists. Tata McGraw-Hill, New Delhi

Stanley SJ (1992) Observations on the seasonal occurrence of marine endophytic and parasitic fungi. Can J Bot 70:2089–2096

Statzell-Tallman A, Belloch C, Fell JW (2008) *Kwoniella mangroviensis* gen nov, sp nov (Tremellales, Basidiomycota), a teleomorphic yeast from mangrove habitats in the Florida Everglades and Bahamas. FEMS Yeast Res 8:103–113

Steele L, Caldwell M, Boettcher AA, Arnold T (2005) Seagrass-pathogen interactions: "pseudo-induction" of turtle grass phenolics near wasting disease lesions. Mar Ecol Prog Ser 303:123–131

Steenkamp ET, Wright J, Baldauf SL (2006) The protistan origins of animals and fungi. Mol Biol Evol 23:93–106

Stentiford GD, Feist FW, Stone DM, Bateman KS, Dunn AM (2013) Microsporidia: diverse, dynamic, and emergent pathogens in aquatic systems. Trends Parasitol 11:567–578

Stephen K, Kurtböke TI (2011) Screening of oomycete fungi for their potential role in reducing the biting midge (Diptera: Ceratopogonidae) larval populations in Hervey Bay, Queensland, Australia. Int J Environ Res Public Health 2011(8):1560–1574

Stief P, Fuchs-Ocklenburg S, Kamp A, Manohar CS, Houbraken J, Boekhout T, de Beer D, Stoeck T (2014) Dissimilatory nitrate reduction by *Aspergillus terreus* isolated from the seasonal oxygen minimum zone in the Arabian Sea. BMC Microbiol 14:35

Stock A, Jürgens K, Bunge J, Stoeck T (2009) Protistan diversity in suboxic and anoxic waters of the Gotland Deep (Baltic Sea) as revealed by 18S rRNA clone libraries. Aquat Microb Ecol 55:267–284

Stoeck T, Epstein SS (2003) Novel eukaryotic lineages inferred from small subunit rRNA analyses of oxygen-depleted marine environments. Appl Environ Microbiol 69:2657–2663

Stoeck T, Taylor GT, Epstein SS (2003) Novel eukaryotes from the permanently anoxic Cariaco Basin (Caribbean Sea). Appl Environ Microbiol 69:5656–5663

Stoeck T, Hayward B, Taylor GT, Varela R, Epstein SS (2006) A multiple PCR-primer approach to access the microeukaryotic diversity in environmental samples. Protist 157:31–43

Strittmatter M, Gachon CMM, Küpper FC (2009) Ecology of lower oomycetes. In: Lamour KH, Kamoun S (eds) Oomycete genetics and genomics: diversity, interactions and research tools. Wiley, Hoboken, NJ, pp 25–46

Strittmatter M, Grenville-Briggs LJ, Breithut L, van West P, Gachon CMM, Küpper FC (2015) Infection of the brown alga *Ectocarpus siliculosus* by the oomycete *Eurychasma dicksonii* induces oxidative stress and halogen metabolism. Plant Cell Environ 39:259–271

Strongman DB, Miller JD, Calhoun L, Findlay JA, Whitney NJ (1987) The biochemical basis for interference competition among some lignicolous marine fungi. Bot Mar 30:21–26

Stuart V, Newell RC, Lucas MI (1982) Conversion of kelp debris and faecal material from the mussel *Aulacomya ater* by marine micro-organisms. Mar Ecol Prog Ser 7:47–57

Suetrong S, Schoch CL, Spatafora JW, Kohlmeyer J, Volkmann-Kohlmeyer B, Sakayaroj J, Phongpaichit S, Tanaka K, Hirayama K, Jones EBG (2009) Molecular systematics of the marine Dothideomycetes. Stud Mycol 64:155–173

Sullivan BK, Sherman TD, Damare VS, Lilje O, Gleason FH (2013) Potential roles of *Labyrinthula* spp in global seagrass population declines. Fungal Ecol 6:328–338

Suryanarayanan TS (2012) Fungal endosymbionts of seaweeds. In: Raghukumar C (ed) Biology of marine fungi. Springer, Berlin, pp 53–68

Suryanarayanan TS, Thirunavukkarasu N, Govindarajulu MB, Sasse F, Janen R, Murali TS (2009) Fungal endophytes and bioprospecting. Fungal Biol Rev 23:9–19

Suryanarayanan TS, Venkatachalam A, Thirunavukkarasu N, Ravishankar JP, Doble M, Geetha V (2010) Internal mycobiota of marine macroalgae from the Tamilnadu coast: distribution, diversity and biotechnological potential. Bot Mar 53:457–468

Sutherland JB, Crawford DL, Speedie MK (1982) Decomposition of 14C-labeled maple and spruce lignin by marine fungi. Mycologia 74:511–513

Sverdrup HU, Johnson MW, Fleming RH (1942) The oceans, their physics, chemistry, and general biology. Prentice-Hall, New York

Swanson D, Block R, Mousa SA (2012) Omega-3 fatty acids EPA and DHA: health benefits throughout life. Adv Nutr 3:1–7

Taboski MAS, Rand TG, Piórko A (2005) Lead and cadmium uptake in the marine fungi *Corollospora lacera* and *Monodictys pelagic*. FEMS Microbiol Ecol 53:445–453

Takai K, Horikoshi K (1999) Genetic diversity of Archaea in deep-sea hydrothermal vent environments. Genetics 152:1285–1297

Takami H (1999) Isolation and characterization of microorganisms from deep-sea mud. In: Horikoshi K, Tsujii K (eds) Extremophiles in deep-sea environments. Springer, Tokyo, pp 3–26

Takasaki K, Shoun H, Yamaguchi M, Takeo K, Nakamura A, Hoshino T (2004) Fungal ammonia fermentation, a novel metabolic mechanism that couples the dissimilatory and assimilatory pathways of both nitrate and ethanol – role of acetyl CoA synthetase in anaerobic ATP synthesis. J Biol Chem 279:12414–12420

Takishita K, Miyake H, Kawato M, Maruyama T (2005) Genetic diversity of microbial eukaryotes in anoxic sediment around fumaroles on a submarine caldera floor based on the small-subunit rDNA phylogeny. Extremophiles 9:185–196

Takishita K, Tsuchiya M, Reimer JD, Maruyama T (2006) Molecular evidence demonstrating the basidiomycetous fungus *Cryptococcus curvatus* the dominant microbial eukaryote in sediment at the Kuroshima Knoll methane see. Extremophiles 10:165–169

Takishita K, Yubuki N, Kakizoe N, Inagaki Y, Maruyama T (2007) Diversity of microbial eukaryotes in sediment at a deep-sea methane cold seep: surveys of ribosomal DNA libraries from raw sediment samples and two enrichment cultures. Extremophiles 11:563–576

Taoka Y, Nagano N, Okita Y, Izumida H, Sugimoto S, Hayashi M (2009) Extracellular enzymes produced by marine eukaryotes, thraustochytrids. Biosci Biotech Bioch 73:180–182

Taylor JD, Cunliffe M (2016) Multi-year assessment of coastal planktonic fungi reveals environmental drivers of diversity and abundance. ISME J 10:2118–2128

Thaler AD, Dover CLV, Vilgalys R (2012) Ascomycete phylotypes recovered from a Gulf of Mexico methane seep are identical to an uncultured deep-sea fungal clade from the Pacific. Fungal Ecol 5:270–273

Thorsen MS (1999) Abundance and biomass of the gut-living microorganisms (bacteria, protozoa and fungi) in the irregular sea urchin *Echinocardium cordatum* (Spatangoida: Echinodermata). Mar Biol 133:353–360

Thurber RV, Willner-Hall D, Rodriguez-Mueller B, Desnues C, Edwards RA, Angly F et al (2009) Metagenomic analysis of stressed coral holobionts. Environ Microbiol 11:2148–2163

Tillmann U, Hesse KJ, Tillmann A (1999) Large scale parasitic infection of diatoms in the Northfrisian Wadden Sea. J Sea Res 42:255–261

Tocher DR (2015) Omega-3 long-chain polyunsaturated fatty acids and aquaculture in perspective. Aquaculture 449:94–107

Toledo-Hernandez C, Zuluaga-Montero A, Bones-Gonzalez A, Rodriguez JA, Sabat AM (2008) Fungi in healthy and diseased sea fan (*Gorgonia ventalina*): is *Aspergillus sydowii* always the pathogen? Coral Reefs 27:707–714

Tong SL, Miao HZ (1999) A new species of marine yeast *Kluyveromyces penaeid* isolated from the heart of penaeid shrimp *Penaeus chinensis*. J Mar Biol Assoc UK 79:559–561

Torta L, Lo Piccolo S, Piazza G, Burruano S, Colombo P (2015) *Lulwoana* sp, a dark septate endophyte in roots of *Posidonia oceanica* (L) Delile seagrass. Plant Biol 17:505–511

Torzilli AP, Andrykovitch G (1986) Degradation of *Spartina* lignocellulose by individual and mixed cultures of salt marsh fungi. Can J Bot 64:2211–2215

Torzilli AP, Sikaroodi M, Chalkley D, Gillevet PM (2006) A comparison of fungal communities from four salt marsh plants using automated ribosomal intergenic spacer analysis (ARISA). Mycologia 98:690–698

Traer K (1979) The consumption of *Posidonia oceanica* Delile by echinoids at the isle of Ischia. In: Jangou M, Balkema AA (eds) Proceedings of the European colloquium on echinoderms, pp 241–244

Tsirigoti A, Küpper F, Gachon C, Katsaros C (2014) Cyto skeleton organisation during the infection of three brown algal species, *Ectocarpus siliculosus*, *Ectocarpus crouaniorum* and *Pylaiella littoralis*, by the intracellular marine oomycete *Eurychasma dicksonii*. Plant Biol 16:272–281

Tsui CKM, Vrijmoed LLP (2012) A re-visit to the evolution and ecophysiology of the Labyrinthulomycetes. In: Cruzado DA (ed) Marine ecosystems. InTech, Rijeka, pp 161–176

Tsui CKM, Marshall W, Yokoyama R, Honda D, Lippmeier JC, Craven KD, Peterson PD, Berbee ML (2009) Labyrinthulomycetes phylogeny and its implication for the evolutionary loss of chloroplasts and gain of ectoplasmic gliding. Mol Phylogenet Evol 50:129–140

Tubaki K, Asano I (1965) Additional species of fungi isolated from the Antarctic materials JARE 1956–1962. Sci Rep Ser E 27:1–14

Turner RD (1973) Wood-boring bivalves, opportunistic species in the deep sea. Science 180:1377–1379

Tutin TG (1938) The autecology of *Zostera marina* in relation to the wasting disease. New Phytol 37:50–71

Ulken A (1968) Über zwei marine niedere Pilze vom Meeresboden der Nordsee. Veröff Inst Meeresforsch Bremerh Suppl 3:71–74

Ulken A (1983a) Distribution of Phycomycetes in mangroge swamps with brackish waters and wates of high salinithy, Chapter 8. In: Teas HJ (ed) Biology and ecology of mangroves tasks for vegetations science, vol 8. Springer, The Netherlands, pp 111–116

Ulken A (1983b) Phycomyceten im Watt des Jadebusens. Veröff Inst Meeresforsch Bremerh 19:177–183

Ulken A (1986) Estimation of thraustochytrid propagules in two mangrove swamps. Bot Mar 29:85–89

Valiela I, Wilson J, Buchsbaum R, Rietsma C, Bryant D, Foreman K, Teal J (1984) Importance of chemical composition of salt marsh litter on decay rates and feeding by detritivores. Bull Mar Sci 35:261–269

van der Wielen PW, Bolhuis H, Borin S, Daffonchio D, Corselli C, Giuliano L, D'Auria G, de Lange GJ, Huebner A, Varnavas SP, Thomson J, Tamburini C, Marty D, McGenity TJ, Timmis KN, BioDeep Scientific Party (2005) The enigma of prokaryotic life in deep hypersaline anoxic basins. Science 307:121–123

Van Dover CL, Ward ME, Scott JL, Underdown J, Anderson B, Gustafson C, Whalen M, Carnegie RB (2007) A fungal epizootic in mussels at a deep-sea hydrothermal vent. Mar Ecol 28:54e62

van Hannen EJ, Mooij W, van Agterveld MP, Gons HJ, Laanbroek HJ (1999) Detritus-dependent development of the microbial community in an experimental system: qualitative analysis by denaturing gradient gel electrophoresis. Appl Environ Microbiol 65:2478–2484

Van Ryckegem G, Verbeken A (2005a) Fungal diversity and community structure on *Phragmites australis* (Poaceae) along a salinity gradient in the Scheldt estuary (Belgium). Nova Hedwigia 80:173–197

Van Ryckegem G, Verbeken A (2005b) Fungal ecology and succession on *Phragmites australis* in a brackish tidal marsh I Leaf sheaths. Fungal Divers 19:157–187

Van Ryckegem G, Gessner MO, Verbeken A (2007) Fungi on leaf blades of *Phragmites australis* in a brackish tidal marsh: diversity, succession, and leaf decomposition. Microb Ecol 53:600–611

Van Uden N, Branco RC (1963) Distribution and population densities of yeast species in Pacific water, air, animals, and kelp off Southern California. Limnol Oceanogr 8:323–329

Vaupotič T, Plemenitaš A (2007) Differential gene expression and HogI interaction with osmoresponsive genes in the extremely halotolerant black yeast *Hortaea werneckii*. BMC Genomics 8:280

Vaupotič T, Gunde-Cimerman N, Plemenitas A (2007) Novel 3'-phosphoadenosine-5'-phosphatases from extremely halotolerant *Hortaea werneckii* reveal insight into molecular determinants of salt tolerance of black yeasts. Fungal Genet Biol 44:1109–1122

Velmurugan N, Lee YS (2012) Enzymes from marine fungi: current research and future prospects. In: Jones EBG, Pang KL (eds) Marine fungi and fungal-like organisms. Walter de Gruyter, Berlin, pp 441–474

Venkatachalam A, Thirunavukkarasu N, Suryanarayanan T (2015) Distribution and diversity of endophytes in seagrasses. Fungal Ecol 13:60–65

Ventosa A, Arahal DR (2002) Physico-chemical characteristics of hypersaline environments and their biodiversity. In: Encyclopedia of life support system. Paris: EOLSS Publishers. http://www.eolss.net/sample-chapters/c03/e6-73-04-01.pdf

Verma AK, Raghukumar C, Verma P, Shouche YS, Naik CG (2010) Four marine-derived fungi for bioremediation of raw textile mill effluents. Biodegradation 21:217–233

Vijaykrishna D, Jeewon R, Hyde KD (2006) Molecular taxonomy, origins and evolution of freshwater ascomycetes. Fungal Divers 23:351–390

Viswakiran Y, Thakur NL, Raghukumar S, Yennawar PL, Anil AC (2001) Spatial and temporal distribution of fungi and wood-borers in the coastal tropical waters of Goa, India. Bot Mar 44:47–56

Vohnik M, Borovec O, Kolařík M (2015a) Communities of cultivable root mycobionts of the seagrass *Posidonia oceanica* in the Northwest Mediterranean Sea are dominated by a hitherto undescribed pleosporalean dark septate endophyte. Microb Ecol 71:442–451

Vohnik M, Borovec O, Župan I, Vondrášek D, Petrtýl M, Sudová R (2015b) Anatomically and morphologically unique dark septate endophytic association in the roots of the Mediterranean endemic seagrass *Posidonia oceanic*. Mycorrhiza 25:663–672

Vrijmoed LLP (2000) Isolation and culture of higher filamentous fungi. In: Hyde KD, Pointing SB (eds) Marine mycology – a practical approach, Fungal Divers Res Ser 1. Fungal Diversity Press, Hong Kong, pp 1–20

Vrijmoed LLP, Hodgkiss IJ, Thrower LB (1986) Occurrence of fungi on submerged pine and teak blocks in Hong Kong coastal waters. Hydrobiologia 135:109–122

Wagner-Merner DT, Duncan WR, Lawrence JM (1980) Preliminary comparison of Traustochytriaceae in the guts of a regular and irregular echinoid. Bot Mar 23:95–97

Wainwright W, Sherbrock-Cox V (1981) Factors influencing alginate degradation by the marine fungi: *Dendryphiella salina* and *D arenaria*. Bot Mar 24:489–492

Walker AK, Campbell J (2010) Marine fungal diversity: a comparison of natural and created salt marshes of the north-central Gulf of Mexico. Mycologia 102:513–521

Walker DI, McComb AJ (1988) Seasonal variation in the production, biomass and nutrient status of Arnphibolis antartica (Labill) Sonders ex Aschers and Posidonia australis hook f in Shark Bay, Western Australia. Aquat Bot 31:259–275

Wallander H, Ekbladb A, Godbold DL, Johnson D, Bahra A, Baldriane P, Björkf RG, Kieliszewska-Rokickag B, Kjøllerh R, Kraigheri H, Plassardj C, Rudawska M (2013) Evaluation of methods to estimate production, biomass and turnover of ectomycorrhizal mycelium in forests soils – a review. Soil Biol Biochem 57:1034–1047

Walsh J (1983) Death in the sea: enigmatic phytoplankton losses. Prog Oceanogr 12:1–86

Wang G, Johnson ZI (2009) Impact of parasitic fungi on the diversity and and functional ecology of marine phytoplankton. In: Columbus F (ed) Marine phytoplankton. Nova Science, Hauppauge, NY, pp 215–222

Wang L, Chi Z, Wang X, Liu Z, Li J (2007a) Diversity of lipase-producing yeasts from marine environments and oil hydrolysis by their crude enzymes. Ann Microbiol 57:495–501

Wang X, Chi Z, Yue L, Li J, Li M, Wu L (2007b) A marine killer yeast against the pathogenic yeast strain in crab (*Portunus trituberculatus*) and an optimization of the toxin production. Microbiol Res 162:77–85

Wang G, Li Q, Zhu P (2008) Phylogenetic diversity of culturable fungi associated with the Hawaiian Sponges *Suberites zeteki* and *Gelliodes fibrosa*. Antonie Van Leeuwenhoek 93:163–174

Wang G, Wang X, Liu X, Li Q (2012) Diversity and biogeochemical function of planktonic fungi in the ocean. In: Raghukumar C (ed) Biology of marine fungi. Springer, Berlin, Heidelberg, pp 71–88

Wang FP, Lu SL, Orcutt B, Xie W, Chen Y, Xiao X, Edwards K (2013) Discovering the roles of subsurface microorganisms: progress and future of deep biosphere investigation. Chin Sci Bull 58:456–467

Wang X, Singh P, Gao Z, Zhang X, Johnson ZI (2014) Distribution and diversity of planktonic fungi in the West Pacific Warm Pool. PLoS One 9(7):e101523

Wang Y, Barth D, Tamminen A, Wiebe MG (2016) Growth of marine fungi on polymeric substrates. BMC Biotechnol 16:3

Ward OP, Singh A (2005) Omega-3/6 fatty acids: alternative sources of production. Process Biochem 40:3627–3652

Ward JR, Kim K, Harvell CD (2007) Temperature drives coral disease resistance and pathogen virulence. Mar Ecol Prog Ser 329:115–121

Warren BA (1994) Context of the suboxic layer in the Arabian Sea. Proc Indian Acad Sci Earth Planet Sci 103:301–314

Wasser SP, Grishkan I, Kis-Papo T, Buchalo AS, Volz PA, Gunde-Cimerman N, Zalar P, Nevo E (2003) Species diversity of the Dead Sea fungi. In: Nevo E, Oren A, Wasser SP (eds) Fungal life in the Dead Sea. ARG Gantner Verlag, Ruggell

Weber D (2009) Endophytic fungi, occurence and metabolites. In: Anke T, Weber D (eds) The Mycota XV physiology and genetics. Springer, Berlin, pp 153–195

Webster J, Weber RKS (2007) Introduction to fungi. Cambridge University Press, Cambridge

Wegley L, Edwards R, Rodriguez-Brito B, Liu H, Rohwer F (2007) Metagenomic analysis of the microbial community associated with the coral *Porites astreoides*. Environ Microbiol 9:2707–2719

Weir-Brush JR, Garrison VH, Smith GW, Shinn EA (2004) The relationship between Gorgonian coral (Cnidaria: Gorgonacea) diseases and African dust storms. Aerobiologia 20:119e126

West AJ, Lin C-W, Lin T-C, Hilton RG, Liu S-H, Chang C-T, Lin K-C, Galy A, Sparkes RB, Hoviusf N (2011) Mobilization and transport of coarse woody debris to the oceans triggered by an extreme tropical storm. Limnol Oceanogr 56:77–85

Winwood RJ (2013) Algal oil as a source of omega-3 fatty acids. In: Jacobsen C, Nielsen NS, Horn AF, Sørensen A-DM (eds) Food enrichment with omega-3 fatty acids. Woodhead Publishing, Oxford, pp 389–404

Wong MKM, Vrijmoed LLP, Au DWT (2005) Abundance of thraustochytrids on fallen decaying leaves of *Kandelia candel* and mangrove sediments in Futian National Nature Reserve, China. Bot Mar 48:374–378

Wu Y-R, He T-T, Lun J-S, Maskaoui K, Huang T-W, Hu Z (2009) Removal of benzo a pyrene by a fungus *Aspergillus* sp BAP14. World J Microbiol Biotechnol 25:1395–1401. doi:10.1007/s11274-0090026-2

Wu Q, Zhang M, Wu K, Liu B, Cai J, Pan R (2011) Purification and characteristics of fucoidanase obtained from *Dendryphiella arenaria* TM94. J Appl Phycol 23:197–203

Xu W, Pang K-L, Luo Z-H (2014) High fungal diversity and abundance recovered in the deep-sea sediments of the pacific ocean. Microb Ecol 68:688–698

Yarden O (2014) Fungal association with sessile marine invertebrates. Front Microbiol 5:1–6

Yarden O, Ainsworth TD, Roff J, Leggat W, Fine M, Hoegh-Guldberg O (2007) Increased prevalence of ubiquitous Ascomycetes in an acroporid coral (*Acropora formosa*) exhibiting symptoms of brown band syndrome and skeletal eroding band diseases. Appl Environ Microbiol 73:2755–2757

Yokochi T, Honda D, Highashihara T, Nakahara T (1998) Optimization of docosahexaenoic acid production by *Schizochytrium limacinum* SR21. Appl Microbiol Biotechnol 49:72–76

Yokoyama R, Honda D (2007) Taxonomic rearrangement of the genus *Schizochytrium* sensu lato based on morphology, chemotaxonomic characteristics, and 18S rRNA gene phylogeny (Thraustochytriaceae, Labyrinthulomycetes): emendation for *Schizochytrium* and erection of *Aurantiochytrium* and *Oblongichytrium* gen nov. Mycoscience 48:199–211

Yokoyama R, Salleh B, Honda D (2007) Taxonomic rearrangement of the genus *Ulkenia* sensu lato based on morphology, chemotaxonomical characteristics, and 18S rRNA gene phylogeny (Thraustochytriaceae, Labyrinthulomycetes): emendation for *Ulkenia* and erection of *Botryochytrium*, *Parietichytrium*, and *Sicyoidochytrium* gen nov. Mycoscience 48:329–341

Zajc J, Zalar P, Plemenitas A, Gunde-Cimerman N (2012) The Mycobiota of the Salterns. In: Raghukumar C (ed) Biology of marine fungi. Springer, Berlin, pp 133–158

Zalar P, Gunde-Cimerman N (2014) Cold-adapted yeasts in Arctic Habitats. In: Buzzini P, Margesin R (eds) Cold-adapted yeasts biodiversity, adaptation strategies and biotechnological significance. Springer, Berlin, Heidelberg

Zalar P, de Hoog GS, Schroers H-J, Frank JM, Gunde-Cimerman N (2005a) Taxonomy and phylogeny of the xerophilic genus *Wallemia* (Wallemiomycetes and Wallemiales, cl et ord nov). Antonie Van Leeuwenhoek 87:311–328

Zalar P, Kocuvan MA, Plemenitaš A, Gunde-Cimerman N (2005b) Halophilic black yeasts colonize wood immersed in hypersaline water. Bot Mar 48:323–326

Zhang C, Kim SK (2010) Research and application of marine microbial enzymes: status and prospects. Mar Drugs 8:1920–1934

Zhang X-y, Tang G-l, Xu X-y, Nong X-h, Qi S-H (2014) Insights into deep-sea sediment fungal communities from the East Indian Ocean using targeted environmental sequencing combined with traditional cultivation. PLoS One 9(10):e109118

Zhou Z, Takaya N, Nakamura A, Yamaguchi M, Takeo K, Shoun H (2002) Ammonia fermentation, a novel anoxic metabolism of nitrate by fungi. J Biol Chem 277:1892–1896

Zinjarde SS, Pant A (2002) Emulsifier from a tropical marine yeast, *Yarrowia lipolytica* NCIM 3589. J Basic Microbiol 42:67–73

Zuccaro A, Schoch CL, Spatafora JW, Kohlmeyer J, Draeger S, Mitchell JI (2008) Detection and identification of fungi intimately associated with the brown seaweed *Fucus serratus*. Appl Environ Microbiol 74:931–941

Zwirglmaier K (2005) Fluorescence in situ hybridisation (FISH) – the next generation. FEMS Microbiol Lett 246:151–158

MIX
Papier aus verantwortungsvollen Quellen
Paper from responsible sources
FSC® C105338

If you have any concerns about our products,
you can contact us on
**ProductSafety@springernature.com**

In case Publisher is established outside the EU,
the EU authorized representative is:
**Springer Nature Customer Service Center GmbH**
**Europaplatz 3, 69115 Heidelberg, Germany**

Printed by Libri Plureos GmbH
in Hamburg, Germany